Wiley Survival Guide in Global Telecommunications

SIGNALING PRINCIPLES, NETWORK PROTOCOLS, AND WIRELESS SYSTEMS

WILEY SURVIVAL GUIDES IN ENGINEERING AND SCIENCE

Emmanuel Desurvire, Editor

Wiley Survival Guide in Global Telecommunications

SIGNALING PRINCIPLES, NETWORK PROTOCOLS, AND WIRELESS SYSTEMS

Emmanuel Desurvire

WILEY-INTERSCIENCE

A John Wiley & Sons, Inc., Publication

Library of Congress Cataloging-in-Publication Data:

Desurvire, Emmanuel, 1955-
   Wiley survival guide in global telecommunications: signaling principles,
   network protocols, and wireless systems / Emmanuel Desurvire.
       p. cm.
   Includes bibliographical references and index.
   ISBN 0-471-44608-4 (cloth)
   1. Signal processing. 2. Computer networks. I. Title.

TK5102.9.D48 2004
621.382'2--dc22

                                                        2004041197

Printed in the United States of America

10 9 8 7 6 5 4 3 2 1

Contents

CHAPTER 2
Telephony and Data Networking, 91

CHAPTER 3
An Overview of Core-Network Transmission Protocols, 133

Foreword

This book is for the curious... Most of us open the refrigerator, take out a bottle of milk and never wonder how it stays cold. Or we may notice the sun rise or set without ever needing a wider explanation of the daily celestial phenomenon that governs our lives. If this description fits you, you probably need read no further; you are excused. But if you do wonder about the technological and scientific marvels we encounter every day, this *Survival Guide in Global Telecommunications* may provide considerable personal satisfaction and, depending on your professional or academic status, can offer practical benefits as well.

The book (in two independent volumes) is addressed to a diverse spectrum of readers, who will find it accessible at different levels. The format is designed for the nontechnical reader who is curious about the telephone and the Internet. It is also designed for the nontechnical reader who encounters technology in his professional life, for example: at home, setting up an office computer; in government, legislating or regulating telecom technology; in law, litigating telecom-related patents or contracts; in finance, investing in high-tech companies. In addition, university students will find the book a handy resource. The book is also addressed to the technically savvy reader who wants to quickly get up to speed in new areas of telecommunications that arise due to rapid career evolution or the prospect of future job movement.

A large impediment to entering an unfamiliar technical field is the jargon that has been developed by the practioners over many years as a form of shorthand. The author, Emmanuel Desurvire, is careful to define each new term in language familiar to a beginner. He also uses these terms repeatedly in context, such that their meanings become intuitively familiar. Another impediment to learning about a particular subject in an ordinary text is the need to wade through the introductory chapters in order to comprehend the few pages of interest. Each chapter in the *Survival Guide* is constructed to be self-contained. Thus, the curious reader can jump right in and read on until she is satisfied or snowed.

I have enjoyed 50 years as a research engineer and physicist, mainly studying optical fiber telecommunications at AT&T Bell Labs (at one time the premier telephone research lab in the world). Much of my enjoyment came from understanding how things work. I also got a great deal of pleasure from making novel contributions of my own. I have no idea where these drives come from, but like the enjoyment of food or friendship, they are real. Although I retired from Bell Labs in 1996, I still found this book personally fascinating for several reasons.

On leafing through the Contents, I found topics relevant to my current interests as a consultant and visiting professor. Some topics, I had known from decades earlier but had forgotten their underpinnings; others, I had never really understood or found a satisfactory explanation. In each case, the discussion in the text was clear and correct, and was presented in an effective tutorial fashion, starting with basics and adding complexity.

My first job after retiring from Bell Labs was as a Congressional Fellow on the professional staff of the U.S. Congress. One assignment was in the Congressional Research Service (CRS) of the Library of Congress whose mission is to provide Congress with background information on the myriad aspects of its legislative and oversight duties. In 1996, the Internet became a commercial-, rather than a government-managed, service. Questions of Internet regulation, taxation, security, governance, encryption and wire-tapping were just popping up. I was asked to prepare an Issue Brief describing the Internet for the staffs of representatives, senators, and committees. It is important to know that of the 535 members of Congress, only a handful have any technical training and their overburdened staffs are largely 20- or 30-somethings trained in liberal arts or political science (which is not a science). Thus, the CRS Issue Briefs are designed to give them a quick tutorial on anticipated technical subjects. After a month or so, I presented my 20-page draft to my boss. She said the limit was 6 pages and I couldn't use technical words like *digital*. It was a struggle, but she did manage to boil my explanation down to 6 pages, although there was no sensible way to avoid *digital*. Today, of course, Congress does know the word "*digital*" at some level. However, I dare say that it would benefit from the nice discussions and examples presented in the *Survival Guide*.

Another of my post-retirement jobs has been as expert witness in patent cases involving telecom issues. Most patent lawyers have some technical training but cannot know every field they are likely to encounter, which is why they hire an expert. The expert may consult with lawyers to help prepare their case, and may also be called upon to testify in deposition or in court to provide technical opinions. Communicating technical ideas in simple language to attorneys can be difficult, but it is a piece-of-cake compared with presenting a technical argument to a jury of high school graduates in the brief time allowed by the judge. Clearly, the attorneys would benefit from reading the explanations offered in *Survival Guide*, and might even want to cite it in their briefs.

Up until about 10 or 15 years ago, the telecom business was managed, designed, and operated by experienced engineers and specialists. The telecom bubble has created numerous job opportunities for an army of self-styled consultants, CEOs, CTOs, CFOs, marketing advisors, strategic planners, and other emerging professionals, whose technical backgrounds are often very weak. As demonstrated in the examples above, these people will find the *Survival Guide* an invaluable tool in getting their jobs done.

The two volumes of Emmanuel Desurvire's *Survival Guide* are a remarkable achievement in my opinion. First, I have wrestled with presenting technical ideas to the uninitiated in class, Congress, and court and know that it requires careful planning, ingenuity, and patience. Desurvire has created a novel instructional instrument for the purpose. Second, I have known Emmanuel since 1986, when he was an exceptionally talented researcher in my department at Bell Labs. In 1990, he became a professor at Columbia University and, in 1993, he joined Alcatel, a global telecom supplier, in France where he was a researcher and, later, a predevelopment project manager. With this background, he has realized an unprecedented feat: he has covered the entire telecom field through these two books, in order to explain to both young and mature minds how things work, he has succeeded in presenting complex ideas in an entertaining style with a minimum use of mathematics.

Thanks to its effective format and pleasing teaching style, I believe that the first two pioneering volumes of the *Survival Guide Series in Global Telecommunications* will find a place on a diversity of reference shelves around the world. Indeed, I expect that it will serve as the template for sequels on other topics.

IVAN P. KAMINOW

Holmdel, New Jersey
June 2004

Preface

These days, the words telecommunications and telecom are often identified with the concept of new technologies, which bears various subliminal meanings. These concern the enthusiastic hopes and illusions of our budding *Information Age*, the promise of a rapidly deployed global-communications culture, of instant person-to-person connectivity, of better and faster interaction between two parties anywhere in the world. Overall, the new and inescapable technology, which reaches everyone at home or at work, is the **Internet**, the panacea of all private or professional communications needs. *What* is being communicated is normally more important than *how* it is communicated. But, paradoxically, communication means and its bandwidth performance (the effect of being instantaneous) remain the user's top concern. What is important is to get a maximum of information from which to select. These new telecom technologies attempt to catch up to this growing perception and need.

The rapid evolution of telecom networks and related broadband access services now seems to differentiate humankind into five basic categories, in growing order of population:

- The rare telecom generalists, who conceptually grasp the full picture, from close familiarity with some of the technologies to accurate market analysis;
- The more diversified experts who deploy, integrate and manage telecom systems, the global and local, incumbent or competitive operators and the service providers;
- The greater number of individuals, who are technical contributors, scientists and engineers, who concentrate their top-level expertise on microscopic aspects of network sublayers, software applications and hardware circuit/board design;
- The majority of end users, and consumers, professionals or private, who are essentially unaware of telecom technologies' features and the intricacies of their complex, pyramidal integration, but are skilled in using their commodities;
- The populations in developed and underdeveloped countries, who rarely or never use any of these technologies, because of the lack of any telecom infrastructure, or their belonging to an older generation, or their underclass situation.

The penultimate, privileged category conceive of "telecom" as being a mere *commodity*, just like transportation. They do not need to have a background in railway engineering or air-traffic control, or to acquire conductor/pilot licenses in order to take trains or planes. Similarly, the telecom generalists of the first category, do not need either an engineering background or an international scientific reputation to employ their skills (though some rare individuals may enjoy both!). Generalists may not even know anything technical at all, but they master the global concepts, the latest developments in programs and trends, to the point of sounding like top technology experts. At this level, considerations of technology integration, evolution and limits, market projections and investment opportunities are all taken into account, which defines an expertise field of its own. To such a category also belongs the rarer academic generalist, who can provide a six-hour tutorial on any aspect of telecom to any audience level, sometimes with talent. But generalists also make mistakes, since their views may be obscure, confused or biased, and their sources may be the same press in which other generalists express themselves, forming a closed informationless loop. The intelligent ones often visit the labs, talk to the engineers, visit the troops. They don't look at top management as a promoting machine where "yes, sir or madam" remains the master keyword. They like to share their knowledge and views with broad audiences, having understood that the more information you freely give, especially to the young, the more you may acquire.

The third and intermediate population, that is, the telecom scientists and engineers, are people whose over-specialization and heavy-duty agenda keep them away from any global picture whatsoever. As top experts and sometimes prime contributors to new technologies, they enjoy the creativity, the inventiveness, the challenges and success of their projects and products. They may even experience the awe of the engineering beauty of their realizations, which keeps their vocation alive and makes them work hard, including at weekends. From any reasonable viewpoint, such a population is key to any future development and innovation in telecom. Yet we are talking of an endangered species. With the current evolution, there are fewer and fewer opportunities to make a career in telecom based on science or engineering. The result is an accelerated over-specialization which is merely project- or product-oriented. As technologies grow in complexity, it is important that instead of over-specializing, experts do cultivate a general background in the whole field. By default, the risk is of rapidly becoming obsolete as a key specialist, should the associated products be phased out, owing to new market orientations or radical evolutions.

Yet, in order to meet the market and be "on time" in product development cycles, experts have very little leisure to expand their background. Because of this, most of them do not (or are physically unable to) pay attention to technologies that are too remote from their area of responsibility. Their immediate boss will say that they are not paid for self-education and growth, but for project effectiveness. The same situation applies to innovation: innovation yes why not, but only within the project framework. Upper management experiences the same constraints, from a broader expertise standpoint but with fewer career risks. Mobility encourages the most inspired to climb the ladder anyway, leaving their expert teams behind. In big corporations with many intermediate decision levels and a transverse structure, the result is a loss of potential, if not a lack of technical vision and reactivity. Moreover, issues of experts' "employability," "retraining" and technical "career management" become increasingly acute. The palliative solution is that employees in the technical field, at any operational level, should be obliged constantly to broaden their expertise, beyond their field of immediate responsibility. The corporation must believe that it is worth investing in this human potential. Even if people may leave to start their own businesses or join more creative teams (as in the normal cause of events), younger generations will be attracted to the potential, so that the net flow and benefits are globally positive. In any case, telecom engineering specialists are at high risk if they become over-specialized, and trust a big corporation, or a small business, to provide them lasting job security or possibly make them wealthy.

With the development of telecom as a commodity, the above concerns and needs are real. It would be useless to attempt to solve the issue by the force of a single "retraining" handbook, however big or small, written by all-knowing and all-mighty generalists. What a book could provide, however, is a rapid introduction. This would build mental bridges towards new domains, integrate or reactivate pieces of prior knowledge, bring familiarity with new concepts (not simply silly "acronyms"!), stimulate curiosity and thirst for exploring new directions. Our "survival guide" approach may represent what is needed to escape the risk of over-specialization and competency stagnation, to reach out towards new technical horizons. It is also addressed to scientists and engineers, working outside telecom, or young people just graduating, who may consider joining the telecom adventure. Where to start in this thick jungle and where it leads one, are key problems to all. We will come back to this later.

End users and customers have it best of all. Operators are competing to lower their prices and increase service quality. The telecom commodity has a price, but no more than that. There is a long way to go to fiber-to-the-home, broadband services on demand, and multimedia fixed/mobile communications, although progress is now irresistible. Once service providers have achieved the right account balance between capital investments, operating costs and service revenues, a new world of telecom will start on a fresh basis, to the great benefit of the aforementioned categories.

We are left with the last category, the people who, for differing reasons of geography and social status, cannot have access (yet) to the new telecom commodity. For them, the picture may also change. Submarine fiber-optic cables have been extensively deployed over the last decade, connecting the most remote locations to the Internet, which is complemented at local or continental scale by GEO satellite coverage. Wireless infrastructures may also compensate for the failed development, degradation, or obsolescence of old telephone systems. With appropriate help, these countries will be able to catch up rapidly with the Internet, including in schools and universities, and not just in business. In both advanced and less advanced developing countries, telecom technologies create the so-called *digital divide*. Should this neologism help society to propose new solutions, it would then be useful. Underclasses may not need e-mail and Internet browsing so much as the reassurance that their children will enjoy equal training and aware in the new telecom technologies for equal education, vocation, and job/career potential. Teachers and professors may also use the Internet to facilitate their communications, develop on-line education programs, and learn new effective ways to educate and train their classes.

And what about this book? It is **not** a pocket guide, **not** a guided tour, **not** a crash course, **not** an illustrated glossary of acronyms, and **not** a kind of Bible reference. The ambition is not to provide a layer of varnish on otherwise superficial understanding. Rather, it is to explain how things work and open up conceptual horizons. If the reader goes through the whole of this book, s/he is on track towards a better career realization, and why not, moving in the direction of the top two aforementioned levels. The idea is to disperse the feeling of being overwhelmed by telecom complexity, and maybe more important, **not** being afraid of it. Even better would be to find it exciting and challenging. Telecom is a beautiful science. It does not have that reputation, but it really is, and this is what this book attempts to convey. Telecom is full of clever and fun stuff, provided one pays due attention and takes due interest.

A move towards a higher-level of insight or into a deeper level of understanding requires some effort, concentration and acquisition. This book does not provide **ad-hoc** recipes as a substitute for true expertise and training. Rather, it guides the reader directly into the depth (and again, fun!) of the issues, without the unnecessary burden of cross-references, historical descriptions, and heavy-duty mathematics.

Concerning mathematical formalism, we made it a self-imposed rule that all engineering and physics should be explained exclusively with elementary operations and functions $(+, -, :, \times, \%,$ sine, log, exp, etc.). These functions are available from any scientific pocket calculator for instant verification. Top-level generalists only recognize $(+, -, :, \times, \%)$, which makes ultimate sense for business but these are low-level, if not worthless, knowledge references to technology specialists. Our challenge has been to explain waves and signal processing with reasonably simple mathematics, such as sine waves, which even high-school students are able to understand. We have dared to strip away from the telecom field any derivatives and integrals, which proves that one could teach and learn more without such an advanced background.

We have included a wealth of easy and practical exercises the purpose of which is to illustrate the concepts, to correctly assess the magnitude of the effects involved (from 10^{-12} in optics to 10^{100} in cryptography) and to become familiarized with standard units. We certainly believe (and we are not the first) that no engineering can coexist with fuzzy concepts, mistaken magnitudes and mishandled standard units. We also believe that engineering experience is not fully integrated unless one is able to address practical applications without gross mistakes. The path to becoming a new telecom expert in a new telecom field is narrow, but we believe that asking oneself the right questions, and developing a true interest for the real "stuff," with curiosity and sustained effort, is probably the safest way to go.

In most books, normal practice has it that the author describes the contents in the introduction: what the reader will find, and where. Here, we shall be innovative. This book contains different fields of knowledge that are identified by the chapter titles. Any chapter can be addressed separately, with little or no cross-reference. Should the reader be interested first in *signaling* and *coding*, *local-area networks*, *core-network protocols* (such as TCP/IP) or *wireless cellular fixed*, *mobile* or *satellite networks*, there is practically no need to go through any preceding chapters. At the end of each chapter, we have provided a summary of the main keywords and acronyms. We call these lists **My Vocabulary**. These summaries are not subindexes. They just represent a list of concepts that are associated with the field covered in the corresponding chapter. If the reader goes through the list at first, it looks like a headache. But, after reading the chapter, the list should be clarified nearly "crystal clear." If a few items still do not make sense, this could mean that the concept was not acquired or remains difficult. Another application of these **My Vocabulary** lists is to use the items as search keywords in the Internet. The inexperienced reader might be surprised by how much knowledge and clarification can be gained by such self-guiding exploration. There are many academic and professional sites from which tutorials, white papers and other hotline news can be freely consulted and downloaded. Our Bibliography list, far from being exhaustive, constitutes an indicative start which includes real Web-site "jewels," but we won't say which ones, the purpose here being not to distribute points or compile a "hot list." Besides, we trust that the reader knows how to use Web links and rapidly generate his/her personalized URL database.

Could one ever become an expert in *all* fields of telecom? Obviously, the answer is no, unless we redefine what expertise means. As a matter of fact, the telecom concept has become sufficiently elusive that we may dare to innovate. This is because telecom's expertise is no longer a matter of scientific/technologic specialization, let alone being a heroic contributor of devices, systems, standards or service/software applications. Rather, it is a matter of integrating all technology aspects into one coherent viewpoint, being aware not so much of the picky details as of the great essentials. Since after the telecom deregulation, these essentials concern more economic than scientific issues, the engineer has been overtaken by the "marketeer," to be taken over in turn by the "client" and the "investor."

Another factor contributing to the demise of engineering expertise is the increasing role of software and the widespread view according to which every technology should follow "Moore's law," to become a mere and cheaper commodity. In the companion volume of this book, we show at least how Moore's law could be revised for wavelength division multiplexing (WDM). Based upon this deceptive faith of self-fulfilling Moore's laws in every direction and field, which "innovation seminars" have a tendency to overdo, new engineers should not tarry on the basic rules of telecom, but instead view things globally, with an emphasis on specialized market directions and new opportunities thereof. In big corporations, the sum of all these decaying factors has led the pure engineering/scientist role to a very low level of esteem. Such a fate would be acceptable if telecom networks did not rest upon 30 years of pure engineering science. The public now regards software and applications as "technology," confusing the *service* with the *physical or hardware channel*. Telecom technology can hardly be reduced to software and services, but the physical layer, without which there would be no telecom, can hardly claim to represent the whole. The network-layer model, where the physical layer has been conventionally put at the bottom, is one more repelling factor for engineering vocations! But such a perspective has changed with the developments of Internet on WDM and, possibly, quantum communications. In telecom, physics and engineering may always keep the upper hand as a faithful servant, regardless of the business implications or plans, a feature that has been verified at least with the radio, the transistor, the laser, the optical fiber, and semiconductor chips.

Can engineers absorb so much complexity as to master all "conceptual layers" of telecom networks? The practical and radical answer is no. This is why this book is published under a series called "Survival Guides." Let's face it: during the busy (and sometimes exhausting) course of our professional and family lives there is so little time we can dedicate for our own education, that we can't hope to master everything, for example music, if it is not our bread-earning job. But we surely can understand better how things work, with limited time investment, for our own benefit and potential career orientations. This may be one definition of the scientific spirit: inquire, then ask the right questions and find the answers, and possibly dig further if not satisfied.

My sincere hope is that this book will help engineers and scientists to "survive" by catching up to, and rising above the fear of inadequacy, and the mistrust of new technologies and standards. All the above through minimal conceptual effort. The reward is to get a thrilling sense of understanding some of the important things that every aspect of telecom touches upon, being able to make the right decisions, to train or teach, and be better prepared for the telecom future.

EMMANUEL DESURVIRE

May 2004

Acronyms

AAA Authorization and accounting [center]
AALn AAL cell type (ATM)
ABR Available bit rate (ATM)
AC Alternative current
ACF Access control field (SMDS)
ADC American digital cellular (mobile)
ADM Add-drop multiplexing
ADPCM Adaptative differential pulse code modulation
ADSL Asymmetric digital subscriber line
AEL Accessible emission limit (laser safety)
AES Advanced encryption standard
AFS Andrew file system (internet)
AGC Automatic gain control
AH Authentication header (IP)
AIM America-Online (AOL) instant messenger
AM Amplitude modulation
AMF Apogee motor firing (satellite)
AMI Alternate mark inversion
AMPS Advanced mobile phone systems
AMS ATM mobility server
AMSS Aeronautical MSS (satellite)
APNIC Asia-Pacific Network Information Centre (IP)
APR Automatic power reduction (laser safety)
ARIN American registry for Internet numbers
ARP Address resolution protocol (TCP/IP)
AS Authentication server
ASAM ATM subscriber access multiplexer

ASBC Adaptative subband coding
ASCII American standard code for information exchange
ASK Amplitude shift keying
ASN.1 Abstract syntax notation 1 (TCP/IP)
ASP Application service provider
ATM Asynchronous transfer mode
ATMARP ATM address-resolution protocol
AU Administrative unit (SDH)
AuC Authorization center (GSM)
AUG Administrative unit group (SDH)
B3G Beyond 3G
BBS Bulletin board system
BCH Bose-Chaudhuri-Hocquenghem (code)
BER Bit error rate
B-ISDN Broadband integrated services digital network
B-FWA Broadband FWA
BLEC Building local-exchange carrier
BMS Broadband messaging service
BOL Beginning of life (system)
BRA Basic rate access
BRI Basic rate interface
BSC Base station switching center (GSM)
BSS Basestation subsystem (GSM)
BSS Broadcast satellite service
BTS Base transmitter station (GSM)
BWA Broadband wireless access
B&W Black and white (picture)
C×D Capacity-distance figure of merit
C-HTML Compact HTML
CAP Competitive access provider

CAP Carrierless amplitude-phase (modulation)

CATV Common-antenna (or community-access) television

CBC Chain block coding

CBDS Connectionless broadband data service

CBR Constant bit rate

CCBS Call connection to busy subscriber

CCK Complementry code keying (WLAN)

CCS Hundred calls second

CD Compact disk

CD-ROM Compact disk, read-only memory

CDMA Code-division multiple access

cdma2000 3G CDMA system family

CDV Cell delay variation (ATM)

CDVT Cell delay variation tolerance (ATM)

CEQ Customer equipment

CES Circuit emulation service (ATM)

CGI Common gateway interface [script] (internet)

CIR Committed information rate

CLEC Competitive local-exchange carrier

CLI Caller line identification

CLP Cell loss priority (ATM)

CLR Cell loss ratio (ATM)

CMI Coded mark inversion

CN Core network (UMTS)

CNI Convergence sublayer indicator (ATM)

CO Central office

CP Customer premises

CPE Customer premises equipment

CPU Central processing unit

CR Clock recovery

CRC Cyclic redundancy check (code)

CRNC Controlling RNC (UMTS)

CRS Cell relay service (ATM)

CS Convergence sublayer (ATM)

CSMA/CA Carrier-sense multiple access (protocol) with collision avoidance

CSMA/CD Carrier-sense multiple access (protocol) with collision detection

CSPDN Circuit-switched public data networks

CTD (max, mean) Cell transfer delay, maximum or mean (ATM)

CTE Channel translating equipment

CUG Closed users group (satellite)

CVSD Continuously variable slope delta

cw Continuous wave

D4 Superframe in T1 systems

D-AMPS Digital AMPS (mobile)

DAC Double attachment concentrator (FDDI)

DACS Digital access cross-connect

DAS Double attachment station (FDDI)

DAVIC Digital Audio-Video Council

dB Decibel

dBHz Decibel-Hertz

dBi Decibel with isotropic-antenna power reference

dBK Decibel-Kelvin

dBm Decibel-milliwatt

DBR Distributed Bragg-reflector (laser)

dBW decibel-watt

DCE Data circuit-termination equipment

DCE Distributed computing environment (internet)

DCS-1800 Digital communications system at 1,800MHz

DD Direct detection

DDM Direction-division multiplexing

DDoS Distributed DoS

DES Data encryption standard

DGPS Differential GPS

DHCP Dynamic host control protocol (WLAN)

DHSS Direct-sequence spread-spectrum

DLCI Data link connection identifier
DLEC Data local-exchange carrier
DMT Discrete multi-tone (coding)
DMUX Demultiplexing/demultiplexer
DNS Domain name system (IP)
DoD Department of Defense (USA)
DoS Denial of service
DPCM Differential pulse-code modulation
DPSK Differential phase shift keying
DQDB Distributed-queue dual bus (protocol)
DRNC Drifting RNC (UMTS)
DS1..DS4 PDH levels corresponding to T1..T4
DSI Digital speech interpolation
DSL Digital subscriber line
DSSS Direct-sequence spread-spectrum
DTE Data terminal equipment (DTE)
DTMF Dual-tone multi frequency
DVB-IP DVB for IP
DVB-RCS DVB return-channel over satellite
DVB-S Digital video broadcast by satellite
DVB-TV DVB for TV
DVD Digital video disk
DWDM Dense wavelength-division multiplexing
DWMT Discrete-wavelet multi-tone (modulation)
DXC Digital cross-connect
E1..E4 PDH levels in Europe
E-GPRS Enhanced GPRS
E-HSCSD Enhanced HSCSD
EBCDIC Extended binary coded decimal interchange code
ECC Error-correction coding / error-correcting code
ECP Excessive cross-posting
EDGE Enhanced data rate for global/ GSM evolution
EESS Earth-exploration satellite service

EGNOS European Geostationary Navigation Overlay Service
EHF Extremely low frequency
EIR Equipment identity register (GSM)
EIR Excess information rate
ELF Extra low frequency
EM Electromagnetic [wave, field]
EMI Electro-magnetic interference
EMP Excessive multiposting
EMSS Enhanced MSS
E/O Electrical-to-optical conversion
EOL End of life (system)
ER Extinction ratio
ESA European Space Agency
ESF Extended superframe
ESP Encapsulating security payload (IP)
ETACS Extended TACS (mobile)
ETSI European Telecommunications Standards Institute
EU European Union
F-NMT France NMT (mobile)
FA-OH Frame-alignment overhead (DW)
FAQ Frequently asked questions
FAX Facsimile (machine)
FDD Frequency-division duplexing
FDD-DMT Frequency-division duplexed DMT
FDDI Fiber distributed data interface
FDM Frequency-division multiplexing
FDMA Frequency-division multiple access
FEC Forward error correction
FEXT Far-end cross-talk
FFT Fast Fourier transform
FHSS Frequency hopping spread-spectrum
FIFO First-in, first out
FIR Fast infrared (wireless IrDA)
FM Frequency modulation
FOMA Freedom of mobile multimedia access
FPLMTS Future public land mobile telecommunications system

FRBS Frame-relay bearer service (ATM)
FSK Frequency shift keying
FSO Free-space optics
FSR Free spectral range
FSS Fixed satellite service
FTP File transfer protocol (TCP/IP)
FWA Fixed wireless access
FYI For your information
GEO Geosynchronous Earth orbit
GFC Generic flow control (ATM)
GGSN Gateway GSN (UMTS)
g.hsdsl (Single/double-pair) high-speed DSL
GigE Gigabit Ethernet
GLONASS Global navigation satellite system
GMSC Gateway MSC (UMTS)
GMSK Gaussian minimum shift keying
GMT Greenwich mean time
GOS Grade of service
GPRS General radio packet service (mobile)
GPS Global positioning system
g.shdsl Single-pair high-speed DSL
GSM Global system for mobile telecommunications
GSO Geostationary satellite orbit
GTE Group translating equipment
HALE High-altitude long-endurance [airship/craft]
HALEP HALE platform [airship/craft]
HALO High-altitude long-endurance [aircraft]
HAP High-altitude platform [technology/system]
HAPS High-altitude platform system
HBE Hub baseband equipment (satellite)
HDB3 High-density bipolar code of order 3
HDLC High-level data-link control protocol (Internet/SDH)

HDTV High-definition TV
H/E Head-end (HFC)
HEC Header error control (ATM)
HEO Highly elliptical Earth orbit (or high-Earth orbit)
HF High frequency
HFC Hybrid fiber coaxial system
HLR Home location register (GSM)
HomePNA Home Phone Network Alliance
HomeRF Home RF (working group)
HP High precision (navigation signal) [GPS]
HPBW Half-power beam width
HSCSD High-speed circuit-switched data service (mobile)
HTML Hypertext markup language (internet)
HTTP Hypertext file transfer protocol (internet)
HTTPS HTTP over SSL
IAD Integrated access device
IANA Internet assigned numbers authority
ICANN Internet corporation for assigned names and numbers
ICI Intercarrier interference
ICMP Internet control message protocol
ICO Intermediate circular orbit (telephone system)
ICQ "I seek you" Internet chat program
ICV Integrity check value (IPsec)
ID Identification document
IDEA International data encryption algorithm
IDSL ISDN-DSL
IEC International Engineering Consortium
IEEE Institute of Electrical and Electronics Engineers
IETF Internet Engineering Task Force
IF Intermediate frequency
IKE Internet key exchange
ILEC Incumbent local-exchange carrier

IM Intensity modulation

IM-DD Intensity modulation and direct detection

IMT-2000 International mobile telecommunications 2000

INMARSAT International maritime satellite organization

INTELSAT International telecommunications satellite [organization]

IP Internet protocol

IPsec IP security protocol

IPv4, IPv6 IP addressing versions 4 or 6

IR Infrared

IRC Internet relay chat

IrDA Infrared Data Association

IrLAN IrDA protocol for LAN connection

IS-95x Standard CDMA system (cdmaOne) version $x = a$ or b

ISAKMP Internet Security Association key management protocol

ISD Information spectral density

ISDN Integrated services digital network

ISI Intersymbol interference

ISM Industrial, scientific and medical [wireless band]

ISM-2.4 ISM band at 2.4GHz

ISM-5.8 ISM band at 5.8GHz

ISO Internations Standards Organization

ISP Internet service provider

ISS Intersatellite service

ISS International Space Station

ISX Internet service exchange

ITFS Instructional fixed television services

ITRS International Technology Roadmap for Semiconductors

ITU International Telecommunications Union

ITU-R ITU for radio systems

ITU-T ITU for telecommunications systems

IXC Interexchange carrier

J1..J4 PDH levels in Japan

JDC Japanese digital cellular (mobile)

JPEG Joint photographic experts group

J-TACS Japanese TACS (mobile)

LAN Local area network

LANE LAN emulation (ATM)

LAP-B Link access procedure (balanced)

LCD Liquid-crystal display**LCF** Laser control field (APON)

LCN Logical channel number

LEC Local exchange carrier

LED Light-emitting diode

LEO Low Earth orbit

LEPA-HDSL Local-exchange primary access for HDSL (ISDN)

LF Low frequency

LHS Left-hand side

LIS Logical IP subnetwork (IP over ATM)

LLC Logical link control (sublayer)

LMDS Local multipoint distribution system (wireless)

LMSS Land MSS (satellite)

LO Local oscillator

LOS Line of signt (radio)

LPC Linear predictive coding

LW Long wave (radio)

LXC Local exchange carrier

LY Light-year

MAC Medium access control (sublayer)

MAN Metropolitan area network

MBS Maximum burst size (ATM)

MC Multiple carrier (3G mobile)

MCR Minimum cell rate (MCR)

MCS (L1,L2) Mobile communication system (generations)

MDS Multipoint distribution service

ME Mobile equipment (UMTS)

MEO Middle Earth orbit

METSS Meteorological satellite service

MF Medium frequency

MIB Management information base (SNMP)

MIC Message integrity check (PEM)

MILP Mixed-integer linear program (VTD)

MIME Multimedia/multipurpose Internet mail extension

MIR Medium infrared (wireless IrDA)

MMDS Multipoint multichannel distribution system (wireless)

MMS Multimedia messaging service

MMSS Maritime MSS (satellite)

MOS Mean opinion score

MPE Maximum permissible exposure (laser safety)

MPEG Moving pictures experts group

MPLS Multiprotocol label switching

MPλS Multiprotocol lambda switching (MPLS)

MPOA Multiprotocol over ATM

M-PAM M-ary PAM (format)

M-PSK M-ary PSK (format)

MQAM M-ary QAM (format)

MSC Mobile switching center (GSM)

MSS Mobile satellite services/systems

MTSO Mobile telephone switching office

MTU Maximum transmission unit

MUF Maximum usable frequency

MUX Multiplexing/multiplexer

MW Medium wave (radio)

MxU MTU or MDU

NADC North American digital cellular (mobile)

NAMTS Nippon automatic mobile telephone system

NASA National aeronautics and space Administration (USA)

NAT Network address translation (WLAN)

NAVSTAR/GPS Navigation system with time and ranging/GPS

NCC Network control center (satellite)

NCN Nearest common node [handover] (ATM)

NE Network element

NF Noise figure

NFS Network file server (internet)

NFS Number field sieve

NHRP Next hop routing protocol (IP over ATM)

NIST National Institute of Standards and Technologies

NMS Network managing station

NMT Nordic mobile telephony

NNE Network node equipment

NNSS Navy navigation satellite system

NOC Network operations center (satellite, WBA, MDS)

NRT Non real time (ATM)

NRZ Nonreturn to zero

NRZI Nonreturn to zero inverted

NSS Network switching subsystem (GSM)

NT1, NT2 Network terminating equipment (ISDN)

NVT Network virtual terminal (Telnet)

OBEX Object exchange (wireless)

OC Optical circuit (SONET)

OC Optical channel (SDH/SONET)

OCC Optical connection controller (ASON)

O/E Opto-electronic (modulation, conversion, regeneration)

OFDM Orthogonal FDM

OH Overhead

OLEC Optical local-exchange carrier

OMC Operations and maintenance center (GSM)

OOK On-off keying

OSI Open systems interconnection (model)

OSPF Open shortest path first (TCP/IP)

OTS Optical transmission section (SDH/SONET)

PABX Private automatic branch exchange

PAD Packet assembler/disassembler

PAL Phase alternating line (TV standard)

PAM Pulse amplitude modulation
PAN Personal area network
PBX Private-branch exchange
PC Personal computer
PCM Pulse code modulation
PCN Personal communications
networks (mobile)
PCR Peak cell rate (ATM)
PCS Personal communications services
(mobile)
PCT Private communication
technology
PDA Personal digital assistant (mobile)
PDC Personal digital cellular (mobile)
PDF Probability density function
PDU Protocol data unit (ATM)
PEM Privacy-enhanced mail
PHY (FDDI) Physical layer (protocol)
PIN Personal identification number
PLL Phase locked loop
PM Phase modulation
PMD Polarization-mode dispersion
PMD (FDDI) Physical media
dependent (protocol)
POP Point of presence
POTS Plain old telephony service
PPP Point-to-point protocol (Internet/
SDH)
PPS Precise positioning service
PRA Primary rate access
PRBS Pseudo-random bit sequence
PRI Primary rate interface (ISDN)
PRN Pseudo-random noise
PSK Phase shift keying
PSPDN Packet-switched public data
networks
PSTN Public switched telephone
network
PTI Payload type identifier (ATM)
PTT Post, telegraph and telephone
PVC Permanent virtual circuit
QAM Quadrature amplitude
modulation
QoS Quality of service
QPSK Quadriphase PSK
R&D Research and development

RAN Residential access network
RARP Reverse address-resolution
protocol (IP)
RAS Remote access server (internet)
RC4 Rivest cipher 4 (WLAN, WAP,
SSL)
RDSS Radio-determination satellite
service
REXEC Remote execution command
protocol (internet)
RF Radio frequency
RFC Request for comments
RFI RF interference
RHS Right-hand side
RIP Routing information protocol
(TCP/IP)
RIPE Réseaux IP Européens
RIPEM Riordan PEM
RIR Regional Internet registries
rms Random mean-square (value)
RMTS Radio mobile telephone system
RNC Radio network controller
(UMTS)
RNR Receiver not ready
RNS Radio network subsystem
(UMTS)
RNSS Radio-navigation satellite service
ROM Read-only memory (IC)
RR Receiver ready
RS Reed-Solomon (code)
RSH Remote shell protocol (internet)
RT Real time (ATM)
RTT Radio transmission technology
(3G mobile)
RZ Return to zero
SA Security association Ipsec
SAR Segmentation and reassembly
(ATM)
SAS Single attachment station (DDI)
S-CDMA Synchronous CDMA (HFC)
SCG Server gated cryptography
SCR Sustained cell rate (ATM)
SDH Synchronous digital hierarchy
SDM Space-division multiplexing
SDMA Space-division multiple
access

SDMT Synchronized DMT

SEAL Simple and easy adaptation layer (ATM)

SESAME Secure European system for applications in multivendor environment

SGSN Serving GPRS support node (UMTS)

SHF Super high frequency

SIM System/subscriber identification module (GSM)

SIR Serial infrared (wireless IrDA)

SIT Satellite-interactive terminal

SMDS Switched multigigabit digital service

SME Small and medium enterprise

SMS Short message service (GSM)

SMT Station management (protocol)

SMTP Simple mail transfer protocol (TCP/IP)

SN Sequence number (ATM)

SNCC Subnetwork control center (satellite)

SNCP Subnetwork connection protection (=UPSR)

SNMP Simple network management protocol (TCP/IP)

SNP Sequence number protection (ATM)

SNR Signal to noise ratio

SONET Synchronous optical network

SP Standard precision (navigation signal) [GPS]

SPI Security parameter index (IPsec)

SPE Synchronous payload envelope (SONET)

SPF Shortest-path first algorithm (IP)

SPI Security parameter index (IP)

SPS Standard positioning service

SRNC Serving RNC (UMTS)

SSP Storage service provider

STE Signaling terminal equipment (X.25)

STE Supergroup translating equipment (FDM)

STM Synchronous transport module (SDH)

STS Space transportation system

STS Synchronous transport system (SONET)

STT Secure transactions technology

SV Space vehicle

SVC Switched virtual circuit

SW Short wave (radio)

SWAP Standard wireless access protocol

T1..T4 PDH levels in North America

TACS Total access communications systems

TASI Time assignment speech interpolation

TBP Time-bandwidth product

TCH/Fx Traffic-channel, full rate at x kbit/s

TCP Transmission control protocol

TCPSat TCP satellite

TDD Time-division-duplexing

TDD-DMT Time-division-duplexed DMT

TDM Time-division multiplexing

TDMA Time-division multiple access

TE Traffic engineering

TE Transverse electric

TFTP Trivial file transfer protocol (Internet)

TGV *Très grande vitesse* [very high speed] (train)

THP Tominson-Harashima precoding (CAP)

TIS/PEM Trusted information systems PEM

TKIP Temporary-key integrity protocol (WLAN)

TLE Two-line elements format (satellite)

TLS Transport layer security

TLV Type-length-value option format (IP)

TOA Time of arrival (GPS)
ToS Theft of service
TU Tributary unit (SDH)
TUG Tributary unit group (SDH)
TV Television
UAE United Arab Emirates
UAV Unmanned aerial vehicle
UBR Undefined bit rate (ATM)
UCT Universal coordinated time
UDLR Unidirectional link routing (satellite)
UDP User datagram protocol (TCP/IP)
UE User equipment (UMTS)
UHF Ultra high frequency
ULSR Unidirectional line-switched ring
UMTS Universal mobile telecommunications system
UNI User network interface (ATM)
U-NII Unlicensed national information infrastructure (wireless)
UPSR Unidirectional path-switched ring
URL Uniform resource locator
USB Universal serial bus [port]
USDC U.S. digital cellular (mobile)
USHR Unidirectional self-healing ring
USIM UMTS subscriber identity module
UTP Unshielded twisted pair
UTRAN UMTS terrestrial radio access network
UV Ultraviolet
VBR Variable bit rate
VC Virtual channel, circuit or call
VC Virtual container (SDH)
VCI Virtual channel identifier (SMDS and ATM)
VFIR Very fast infrared (wireless IrDA)
VHF Very high frequency
VLF Very low frequency
VLR Visitor location register (GSM)
VoATM Voice over ATM
VoD Video on demand
VP Virtual path

VPI Virtual path identifier (ATM)
VPN Virtual private network
VSAT Very small aperture terminal (satellite)
VT Virtual tributary (SONET)
VTD Virtual topology design
VTD-MILP Mixed-integer linear program for VTD
VTOA Voice and telephony over ATM
WAN Wide-area network
WAP Wireless Internet access
WATM Wireless ATM
W-CDMA Wideband CDMA (UMTS)
WDM Wavelength division multiplexing
WDP Wireless datagram protocol
WDSL Wireless DSL
WECA Wireless Ethernet Compatibility Alliance
WEP Wired equivalent privacy (WLAN)
Wi-Fi Wireless Fidelity (trademark for 802.11b)
Wi-Fi5 Wireless Fidelity (trademark for 802.11a)
WLAN Wireless LAN
WLL Wireless local loop
WML, WML2 Wireless markup language (and version 2)
WON Wireless optical networks
WWW World Wide Web
XHTML Extended HTML (wireless Internet)
XOR Exclusive OR
ZM Zone manager [node] (ATM)
1G...4G First to fourth (mobile-telephone) generation
2B1Q Two-binary one-quaternary (modulation/coding format)
3GPP Third-generation partnership project
3R Signal regeneration by signal Repowering, Retiming and Reshaping
4-PAM Quaternary PAM (modulation/coding format)

Introduction
The Network Cloud

During the year 2000, I went through an unusual experience at the company's cafeteria. People were busy walking around, picking up food and drink from the booths. As I was doing the same, my attention was suddenly caught by a strange scene. A young Asian man was walking alone, erratically, eyes rolling and unfocused, waving one hand in the air with the other in his back pocket, engaged in an animated conversation with himself. In such a situation, one immediately feels some unease, as if witnessing a madman who has lost control in a public place. But he did not really look like a madman neither did he seem drunk, and the scene and place and looks were not right. What could HE ever be doing?! Getting a bit closer, it took me a few moments to realize that a near-invisible, minute wire was connecting his mouth, ear and something in his pocket. He was engaged in a hands-free, wireless telephone conversation! The weird thing is that he was having this conversation in the midst of this busy crowd all too hungry to pay any attention, too concentrated himself to feel self-conscious. Apparently, this was an important call, maybe with a "big" customer. Lesson of the day: the telecom network has become so all-pervasive and user-flexible that we hardly even realize it, even as trained engineers. You can be called upon by your customers, boss, colleagues, friend, spouse, anywhere and anytime, especially if you choose to carry with you this network terminal, as light as a pen or a credit card. The signal quality is such that you don't need to isolate yourself from the noise, and you can be reached deep inside any building, whatever you are doing, from anywhere in the world.

This is where we start. Let's imagine that the person at the other end of the line is driving a car in some country a few time zones away, or seated on a flight to Paris, or at his/her office desk. Once established, the connection between the two persons is seamless, instantaneous and as immaterial as a cloud of nothingness, as networks are often pictured and as illustrated in Figure I.1. Nothingness means that practically any communication means could be activated from within this transparent or immaterial cloud, as illustrated in Figure I.2. At the entry edge of the cloud, intelligent "boxes" pick up the incoming signals and convert them into something meaningless for human beings, but wholly meaningful and intelligent for the network: *bits*, short for "binary digits." The signal bits are aggregated into various packets, frames,

Wiley Survival Guide in Global Telecommunications: Signaling Principles, Network Protocols, and Wireless Systems, by E. Desurvire
ISBN 0-471-44608-4 © 2004 John Wiley & Sons, Inc.

FIGURE I.1 A telephone connection through the "network cloud."

streams or combinations thereof, packaged or encapsulated with label information (headers, control bits, trailers), compressed at higher speeds (bit-rate conversion), mixed with other incoming signals (multiplexing), until they reach the core of the cloud, referred to as the "backbone" for the physical infrastructure, or the "core" for the network hierarchy level. Such a backbone transports the resulting

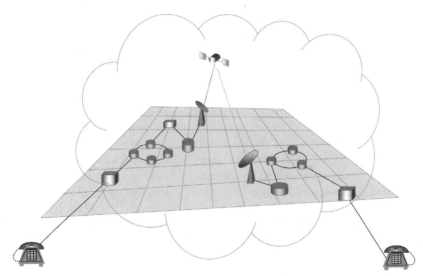

FIGURE I.2 Detail of the "network cloud," showing access/gateway stations (cubes), local sub-networks (rings), wireless edge network and radio-satellite connection, forming a highly complex yet virtually seamless and quasi-instant communication medium.

information (of which a substantial fraction is for network use only) anywhere in the world, via terrestrial or sub-marine cables or satellites, depending upon the physical distance to cover through the immaterial cloud, to reach the other edge. What is said about this phone call also applies to intercomputer or data communications, whether or not initiated, activated and controlled by actual human beings.

This description points to the tremendous complexity of any global communication process and the corresponding network operation, regardless of the type of signal to be exchanged (namely: voice, video or data). The first characteristic of the network is that its existence and reality rests upon a multiplicity of old and new *hardware and software technologies*. The technologies include telephone wires, microwave links, coaxial cables, optical fibers, satellites, digital switches, mainframe computers, hybrid electronics and integrated circuits and all their interconnections.

The second characteristic is *network intelligence*. The network always *knows* where to send your signal message. It makes sure that it will reach its destination through the most secure and fastest route, regardless of the underlying technology and transmission protocol available on the different path segments. This is why so many bit-processing stages and conversions occur through the cloud, which enable not only the precise routing function for the message, but its management with millions of simultaneous other network users. The only way to achieve such an apparent miracle is to structure the network cloud into several hierarchical layers, as shown in Figure I.3. Schematically, these network layers are the *access*, the "*metro*" (for metropolitan), and the *core*. Within the metro layer, one may also distinguish the *metro edge* (interface with the access) and the *metro core*

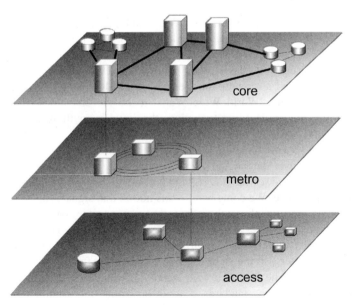

FIGURE I.3 Network segmentation in three functional layers (bottom to top): access, metropolitan and core.

(central functions, traffic aggregation or deaggregation, and interface with the core). The meaning of the different boxes shown in the figure is not important at this introduction stage.

Each layer can be operated nearly independently, with decentralized management and control, and by several owner agents (operators) and service providers (vendors). Each layer has its own ingrained technology favorites, according to the *legacy* (the system's deployment history and heritage), the evolving needs and the technology progress. Clearly, the core requires the most powerful technologies, that is, those providing the highest *bandwidth*, or bit-per-second transmission capability. *Lightwave technologies*, also referred to as *Optics*, or *Photonics* at component, system and backbone network-layer levels, meet such a need. At the bottom end, the access layer, the type of *user* and *user bandwidth* needs determine which technology is best suitable, from microwave to radio to electrical or even optical. Users can be either *fixed* or *mobile*. Mobile users can also *roam* the network, for instance using the same cellular telephone or laptop computer in different places and even different countries.

Schematically, there exist two global categories of network user, namely *residential* (home, personal, private, communities, associations, institutions, government..) or *business* (office, corporation, enterprise, retailing, agencies..). But the difference between the two may not be so clear cut. The possibility of *telecommuting* (more accurately termed *teleworking*) turns home into office, and residential block into small business center. To each user category corresponds a broad, ever-progressing variety of communications *services*, of which the ancestral *telephone*, whether fixed or mobile, is the most common and routinely used. But voice services are not limited to mere "phone calls": there are also voice mail, toll-free numbers, automatic call back and caller identification, among several other possibilities.

In the most developed countries, *cable-TV* and *Internet* have become indispensable services to tens or hundreds of millions of private users, while in the least developed, a single telephone line could be shared by hundreds of inhabitants. The Internet has come into the network picture as a revolution, both as a telecom-technology driver and a serendipitous service offer. Most people erroneously conceive the Internet, the "World Wide Web," to be the above-described network cloud. Yet, the concept is partly true, since the cloud is anything you want to put in it, as we have seen. By itself, the Internet also creates its own network, hence the spider-like web denomination. But in reality, the Internet is a new way of packaging transmission over the old and traditional network. It may borrow the routes of our *plain old telephone service* (POTS) and find a path through any network channel it may find appropriate. We come here to this notion that signals can be transmitted not only though various physical media (atmosphere, outer space, electrical wire, optical fiber), but also under different intelligent arrangements. The rules governing such arrangements are referred to as *protocols*, as are the way governments receive and treat their visitors, in the right order of importance, function and precedence. Protocols should not be confused with *standards*, which are the rules that define their specifications. The standards may differ from one continent to another (e.g., United States, Europe, Asia), while the resulting protocols may or may not be mutually compatible at different traffic-hierarchy levels. A basic example is

SONET (for North America) and SDH (for Europe, Asia and other). These two international standards happen to be compatible at some traffic concentration levels. But even then, their framing protocols remain wholly different.

The analogy between communications and transportation networks is often helpful for making one-to-one comparisons. Indeed, cars, trucks, trains, airplanes or boats can be viewed as representing as many different transportation "protocols" for people and goods, each following their own traffic rules and priorities (the standards). Protocols may or may not be compatible. Cars or trucks can be put into trains or boats to go faster, farther or cheaper, immediately recovering their autonomy at the other end of the route. Air travelers may take the subway to reach the airport and a bus shuttle or taxi or car rental to reach their final destination. Each time they switch protocol and adopt the local standard, which is one of the fun experiences of travelling (ever tried the Tokyo subway for the first time during rush hour?). Likewise, voice can be transported through the *Internet protocol* (IP), as referred to by the **VoIP** acronym. In our network cloud, the same signal payload my swap protocols several times. The network elements receiving and sending signals must also be "multi-protocol", meaning that they recognize several protocol types as they come in (of course they have been designed to handle this at this network point). One important function of network elements is protocol conversion, usually at the transition point between two network layers (e.g., core to edge, edge to metro, etc., and the reverse). Another function is *traffic aggregation* (upstream) or *deaggregation* (downstream), which may involve both protocol conversion and bandwidth conversion to higher/lower bit-rates within a given protocol. Finally, protocols may evolve within a given standard. For instance, the Internet is moving from IPv4 to IPv6. Contrary to general belief, it is not just a new addressing system, but a powerful way to use the standard, which for instance may consist in encapsulating IP packets into other IP packets after encrypting them, a secure Internet over the Internet. Standards may also evolve, but not necessarily with *backward compatibility*. For instance, the cryptography standard DES will be changed into a radically different advanced version, AES. Yet this will not prevent one from still using DES, the point being that computers recognize which standard is used. As a matter of fact, telecom could be entirely described through a suite of standards and protocols, with their complex history and intricacies. Hence a need for some abstraction and focus. At the opposite end, telecom could be viewed as entirely physical, a mesh of wires, cables, fibers, radio links and terminals handling different flows of data. Such a perspective is also entirely correct. Would not it be nice to have a deep appreciation of both perspectives? This is the kind of reconciliation that this book is attempting.

Communication networks are thus characterized by multilayered, multifunctional, and multistandard complexity. It is quite difficult to apprehend such a global picture at once, because of the inevitable confusion one often makes between the different applications, services, protocols and technologies. The *Internet* is the best example of this intertwined conception. When we "log on" the Internet, we use a computer application which connects to a service (a server with its web pages and search engines), which in response generates IP packets to feed our PC. But all these different stages remain invisible, albeit spurious losses of connection and exasperating downloading times occasionally remind us of the *physical reality* of the network. A failed connection could also be explained by the child who played with the telephone plug in the basement! Those experiences of wire-and-plug fixing also remind us of the benefits of basic engineering, and that the Internet is not just abstract software.

To the nonspecialist, and also the professional, another factor of confusion is introduced by *marketing*. Indeed, telecom marketing has this tendency to name technologies after a product brand, and not the reverse. As a result, the same technology can bear multiple brand names, making it absolutely unclear what it is about and how it actually works. Market considerations could re-name the same technologies, for example what you see here is *not* a car. The truth is, that telecom is not rocket science with dependable engineering reference and designation. The combined scientific and engineering aspects of telecom fully deserve the *hi-tech* label, as compared to other innovative but more trivial markets. Yet, "telecom technologies" do not come down to mere science and engineering. They are deeply rooted in evolving services and customer applications. This is to the point that version X.Y of this cellular phone is promoted as a "new technology," because it is compliant to a new standard and has extra command buttons or display features. This is where (we believe) most people with scientific and engineering training, especially those with the highest education, often find it hard to rationalize and see clearly into the telecom mess. For such people, things make sense only if there is a fundamental starting point, a logic track, a rational chain of deductions and conclusions based upon common principles and definitions, and possibly historical factors. The development of telecom technologies hardly follows any scientific logic (we would like the opposite) because it is plugged in to irrational market fluctuations and instant needs. But telecom has its own fundamentals and scientific rationale, and this is what this book attempts to summarize and convey. Telecom also has its own history based upon market, legacy, product success, service evolution, and technology innovations. There is probably no other field in modern society in which anyone has something sensible to say, from the mere user to the advanced specialist, while at the same time being unaware of the global picture.

Writing a book that would fully describe telecom, from technologies to standards and service applications could be viewed as a *risky* if not a pretentious endeavor. Why? Because telecom, unlike applied and exact sciences, is market-sensitive, which involves inputs from many independent contributors, participants, competitors, investors and overall, market strategies. For this reason alone, telecom is *not* a science. Rather, it is an evolving body of knowledge, with subjective and incomplete perceptions. The tragedy of the so-called "Internet bubble" showed the limitations of market analysis and predictions, which would be a joke if not for the loss of jobs and the waste of capital. Any telecom handbook, no matter how thick and how many contributors, would never be able to present the field in a decisively final and satisfactory way to all these participants. As a matter of fact, it would have to be rewritten every six months, or even more frequently, each time a new standard or service appears on the market.

Being aware that in telecom no statement can ever be final, and most important of all, that market is "just" one aspect of the field, we may then try to rationalize and consolidate its body of knowledge. The task apparently looks immense and hopeless. But we can reduce it to a minimum, making an inventory of all technology fields one after another, and extract what seems to represent the core concepts. A first difficulty, but easy to overcome, is *subjectivity*. Can technologies be dissociated from their brands or mother companies (has everything been ever invented in *that* specific place)? Are not technologies evolved from the works of several independent teams or inventors? Could one avoid going through a lengthy account of historical background and cross-references? The answer to all these questions is an emphatic yes, as long as the goal is to explain how thinks work, and not to pretend that the solutions came overnight to people and teams "skilled in the art." A second level of difficulty is the language barrier, or more precisely, the technology *jargon*. This jargon takes the form of more or less meaningful alphanumerical scripts or *acronyms*, such as 802.11b, RC-4, RS(255,231), or EDFA. The first is a leading wireless-LAN standard, the second a popular encryption algorithm, the third a frequently used error-correction code and the last a ubiquitous optical fiber amplifier. Could one make sense of and memorize all of these? The answer is a partial yes, should the *concept* always precede the acronym. An expert should not be reduced to a living glossary of acronyms, but he or she should at least be able to tell what acronyms mean in practical terms, and point to their immediate equivalents or derivatives. One may forget what GSM or UMTS mean but not that they are cellular-telephone standards, and be able to tell the difference without having to borrow presentation slides from a colleague. Finally, the supreme impediment to understanding telecom is *mathematical formalism*. Some experts seem to have the talent to render things obscure and deterring, or even worse, to convey the feeling that there is no other way to grasp a telecom concept than to suffer intense maths derivations, with a toolbox of integrals, non-linear equations and perturbation methods! Although respectable and essential, these branches of knowledge do not need to be introduced to telecom (after all, $E = mc^2$ may be sufficient for an introduction to relativity). On the other hand, formalism is required when it comes to measuring performance and understanding system limits. This is what market and investors need the most from engineers. As often in physics, the answer should be simple and easy to grasp (regardless of its possible academic complexity). Imagine you must explain to your CEO a new discovery, without advanced notice and only two slides. Good engineering and gaining management support also require indispensable communications skills!

For all of the above reasons, the author prepared this book according to the following principles. No use shall be made of:

Formulae involving differential/integral calculus;

Mathematical derivations other than using $\pm$, $\times/:$, % and elementary functions that can be pressed on a pocket calculator;

Quotations of $/Euro market figures, unless meaningful as a stand-alone or historical reference;

Comparisons of competing technologies;

Predictions concerning market trends;

> *Quotations of companies or products, unless generic or historic (e.g., Bell Labs transistor);*

The author neither does claim to:

> *Convey any "insider" viewpoint, implying any form of higher-source knowledge or company bias;*
>
> *Be 100% accurate and complete in describing current or future standards;*
>
> *Suggest market trends for any technology;*
>
> *Provide reference grounds for business decision making.*

This book, which is entirely based upon publicly available information, intends only to be a helping tour guide, either as an introduction to new fields of expertise or a stimulation towards new vocations or professional evolution, or just for checking out one's knowledge. Hence the series brand name, "Survival Guide" destined for a new practice we could call telecom bushwalking.

Before entering a new telecom field jungle, a first check up might be to test one's *vocabulary*. We prefer this term to "jargon," because the spirit is to communicate, not to annoy or conceal. At the end of each chapter, we have introduced a **My Vocabulary** list. The reader will find below such a list to show the principle right from this basic introduction. We may suggest this approach as a new self-teaching or class-teaching method, being aware that this may represent yet another case of reinventing the wheel. The individual reader may cross out some words from the list, but only those for which the meaning and significance are absolutely clear. He or she may even distinguish the "reasonably" clear ones. Finally, the ones that he or she would tend to avoid should be circled. These are the most important, because they point to *knowledge gaps*, or even worse, *cognitive* knowledge gaps. For class use, the teacher may ask the students to volunteer a few words and acronyms on a piece of paper. The collection could then be pinned on separate boards by different subgroups. Each subgroup would then have to figure out a good conceptual definition and present the results to the class. If need be, a 15-mn Internet search, right on the premises, may be allowed. A class vote for the team that made the best contribution, with some token reward, might conclude the session. Try this in your company as well. The management will appreciate the absence of consulting fees and travel expenses for a very high return in employee training and team motivation.

Finally, all chapters in this book (apart from this Introduction) end with *exercises*. These exercises, which are not "problems," are recommended as a means for the reader to prove that a given concept has been understood. There is no shame in looking directly at the solution if one is definitely stuck or does not even understand the question. Comprehensive and extensively detailed "Solutions to the Exercises" can be found at the end of the book and are included as another tool for the teacher to use. The class participants, equipped with pocket calculators (or working without, as computing wizards) have only 10 mn to find the solution. The class looks at the results and goes together through the proper demonstration. It might be important to look at the reasons why some went wrong. Orders of magnitude? Algebra? Wrong numbers? Wrong reasoning? This could be yet another way to prepare good engineers. Mouse clicks are no substitute to this indispensable form of intelligence and skill.

We hope that this book, which covers *signaling principles*, *network protocols* and *wireless systems*, will be useful to the novice student and the skilled expert as well. (Note: The other fields of *broadband access*, *optical communications components and systems*, and *cryptography*, are covered in a second, indpendent volume in the series.) Telecom is a nice field to be in or to work for. Like music, it has a background and rules of its own, and the more one is familiar with them, the more one is in a position to appreciate, to contribute and to innovate.

a MY VOCABULARY

Can you briefly explain each of these words or acronyms in the telecom network context?

Access	Edge	Microwave	Residential
Acronym	Header	Multiplexing	Routing
Aggregation	Internet	Network (layer)	Signal
Backbone	IP	Packet	Standard
Bandwidth	Layer	Payload	Transport
Bits	Legacy	Photonics	Voice
Business	Lightwave	POTS	VoIP
Core	Metro	Protocol	

CHAPTER 1

Signal Modulation, Coding, Detection and Processing

In this chapter, we take a close look at how information signals are generated, coded and detected, under either analog or digital form. We shall learn about the conversion from one to the other, about the unavoidable noise, and the resulting bit errors. Yet some error-correction schemes can be implemented which leads to the ultimate concept of channel information capacity.

In telecommunications science, the principles of *signal modulation, coding, detection* and *processing* represent an entire domain of their own. It is a challenge to attempt to summarize from this domain what is most important to telecommunications engineers, or to people wanting to get a first exposure to these principles; there is so much in it and so much to explain. The task is made even more difficult if we choose not to rely upon extensive mathematical derivations. A minimum mathematics level is required, but we have restricted it to the mere handling of sine, cosine, logarithm and exponential functions, at times when needed and appropriate. While this chapter, of limited size, does not provide the academic background that it takes entire books to cover, we are confident of providing a good exposure to the aforementioned principles, with seamless understanding of concepts.

We begin (Section 1.1) with a description of *waves and analog signals*. What are waves, and how can one put some information onto them, whether made of electrical, radio or light signals? Focusing on telephony applications, the description leads to *voice channel multiplexing*, the possibility of conveying hundreds of analog phone calls in a single physical transmission line.

But lines can also be used to transmit bits of information. Hence the importance of *binary* or *digital coding* options (Section 1.2). How are analog waveforms

Wiley Survival Guide in Global Telecommunications: Signaling Principles, Network Protocols, and Wireless Systems, by E. Desurvire
ISBN 0-471-44608-4 © 2004 John Wiley & Sons, Inc.

changed into binary numbers or bit words? What are the different possibilities of binary modulation, from amplitude, phase and frequency options? And can we code more information by *multilevel signaling* using the above formats? These questions are adressed in this section.

Consistently with the previous section, we address the issue of *voice sampling*, namely the conversion of analog voice signals into digital signals (Section 1.3).

The concept of *channel noise* (Section 1.4) is not elusive when formulated with relatively simple arguments of mean, variance and Gaussian-distributed probabilities. For binary communications, this leads to the *eye diagram* and the measurement of *bit-error rate*.

The *transmission* and *detection* of binary signals (Section 1.5) share the same basic communications system layout. However, it is important to distinguish direct and coherent detection schemes, which do not have the same complexity and transmission-fidelity performance. We adress then the issue of system power budget, and the need to restore some aspects of signal integrity, whether through *in-line amplification* (repowering) or full *3R regeneration* (repowering, retiming and reshaping). We also introduce the *noise figure* characteristics, which apply to any type of transmission element (degrading or amplifying) and determine minimal system requirements, in particular with periodically amplified lines.

A way to overcome the limitations introduced by channel noise, is to implement *error-correction coding* (Section 1.6). We first consider *linear block codes*, then *cyclic codes*, then review the different code types.

The *information capacity* of communication channels (Section 1.7) leads to more abstraction, but within the aforementioned rules of keeping mathematics absolutely minimal. The fundamental results from Shannon's information theory, which concern *channel capacity* and error-rate correction capability in the presence of channel noise, are presented.

■ 1.1 WAVES AND ANALOG SIGNALS

From early childhood, everyone is familiar with waves. One of the purest waveforms is the one generated by the fall of a pebble into a quiet pond. We see a bunch of concentric circles forming from the impact point and growing bigger with time. In fact, what happens is that the water surface is moving up and down and the ripple propagates at a certain speed, about a meter per second. The perfect circles that are formed show that the wave's speed or *velocity* is the same in every direction. If we could see the pond in the vertical plane (such as through an aquarium window) this is what we would see (Figure 1.1). The set of circles correspond to two groups of identical "wave-packets" moving away from each other. The packets have finite dimensions, corresponding to the fact that (1) the wave started from some point in time (the impact of the pebble on the pond surface) and (2) the resulting oscillation is progressively acquired then damped by the medium. After a while, the circles have completely vanished, showing that the wave's initial energy has been absorbed by the water medium through which it propagated (see Exercises at the end of chapter).

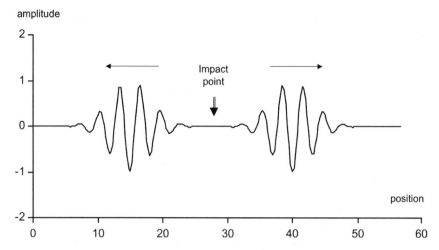

FIGURE 1.1 Ripple caused by the fall of a pebble onto a pond surface, as viewed in the vertical plane. Two wave-packets move away from each other at some characteristic velocity.

This example is a rapid but intuitive introduction to this phenomenon called "wave." We are also familiar with sound, but the corresponding oscillations are too fast to be sensed or visualized. We perceive sounds as having "pitch" or music *tones*, rather than being vibrations (except for construction workers!). Likewise, light is an *electromagnetic wave* whose oscillations are sensed by the eyes as colors, which represent a very limited range called the "visible." We will come back to this notion of tones versus time oscillations when considering the *frequency domain*.

1.1.1 Sinusoids and Waveforms

Let's now recall the basic mathematical concepts associated with waves. The elementary function $f(t)$ defining an oscillation at time t is the *sinusoid*. It is written:

$$f(t) = A \sin(\omega t + \varphi) \tag{1.1}$$

In such a definition, A is dimensionless and is called the wave *amplitude*; ω is called the *circular frequency* (or *pulsation*), and its unit is radians per second (rad/s). The radian is a dimensionless angle measure (π rad $= 180°$, $\pi = 3.14159\ldots$). The pulsation is related to the oscillation *frequency* ν (in *cycles per second*, or *Hertz*, or Hz, or s^{-1}), the *period* $T = 1/\nu$ (in seconds, or s), the wave *velocity* c (in meter per second, or m/s), and the *wavelength* $\lambda = c/\nu$ (in meter, or m), through the following identities:

$$\omega = 2\pi\nu = \frac{2\pi}{T} = 2\pi\frac{c}{\lambda} \tag{1.2}$$

The phase φ is a number defining how the wave is shifted with respect to the arbitrary time reference $t = 0$. The *cosine* function is the same as the sine function with a $\pi/2$ advance phase shift, i.e., $\cos(\omega t - \pi/2) = \sin(\omega t)$. Figure 1.2 shows plots of both sine and cosine functions with $A = 1$, $\omega = 1$, $\nu = 1/2\pi$ Hz, $T = 2\pi$s, and $\varphi = 0$. It is seen that the sine/cosine wave oscillates with a ± 1 amplitude excursion. The maxima or minima of the wave meet periodically every $T = 2\pi$ seconds. By definition, the *power* carried by the wave is given by A^2, within a proportionality constant connecting to physical units. With waves, power and *intensity* have similar meanings. More precisely, intensity (also called power density) is defined as the incident power per unit surface, as expressed in Watts per square meter or centimeter (W/m^2 or W/cm^2).

It is useful at this point to recall the different ways to express units in powers or fractions of ten. We are familiar with kilowatts/kilometers and miliwatts/millimeters, for instance, the Greek prefixes *kilo* or *milli* meaning a thousand or a thousandth, respectively. But do we know so well all the powers of ten ($10^{\pm N}$) used in telecommunications engineering? The definitions summarized in Table 1.1 are definitely to be known by heart! In the rest of this book, such definitions will be used extensively.

Power can also be defined in *decibels*. By definition, the decibel equivalent of power P is given by the formula

$$P_{\mathrm{dB}} = 10 \log_{10}\left(\frac{P_{\mathrm{W}}}{P_{\mathrm{W}}^{\mathrm{ref}}}\right) \tag{1.3}$$

where $P_{\mathrm{W}}^{\mathrm{ref}}$ is a reference power level and $\log_{10}$ is the decimal logarithm. Therefore, decibels do not give absolute values of power levels, but their ratio with respect to some reference. To contract this power-ratio scale, the decimal logarithm (times 10) is used. For instance if $P_{\mathrm{W}}^{\mathrm{ref}} = 1W$, the powers of $10\,\mathrm{mW} = 10^{-2}\,\mathrm{W}$ and

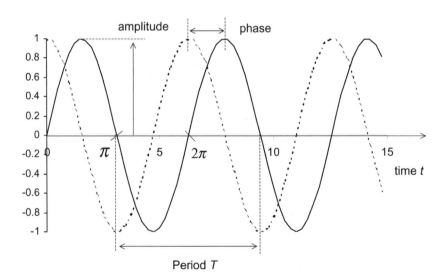

FIGURE 1.2 Plots of sine (full line) and cosine (dashed line) functions, assuming $A = \omega = 1$, $T = 2\pi$, $\varphi = 0$ showing amplitude, period and relative phase of the two oscillations.

TABLE 1.1 **Powers of ten for meters, seconds, Hertz and Watts, as most frequently used in telecommunications engineering**

		picosecond	10^{-12} s				
nanometer	10^{-9} m	nanosecond	10^{-9} s			nanowatt	10^{-9} W
micrometer	10^{-6} m	microsecond	10^{-6} s			microwatt	10^{-6} W
millimeter	10^{-3} m	millisecond	10^{-3} s			milliwatt	10^{-3} W
meter	10^{0} m	second	10^{0} s	Hertz	10^{0} s^{-1}	Watt	10^{0} W
kilometer	10^{3} m			kilohertz	10^{3} s^{-1}	kilowatt	10^{3} W
				megahertz	10^{6} s^{-1}		
				gigahertz	10^{9} s^{-1}		
				terahertz	10^{12} s^{-1}		

$1 \text{ kW} = 10^3 \text{ W}$ correspond to $P_{\text{dB}} = 10 \log_{10}(10^{-2}/1) = -20 \text{ dB}$ and $P_{\text{dB}} = 10 \log_{10}(10^3/1) = +30 \text{ dB}$, respectively. In telecommunications, the reference power level is $P_{\text{W}}^{\text{ref}} = 1 \text{ mW}$. The corresponding decibel unit is called the *decibel-milliwatt*, or dBm. Thus, we have

$$P_{\text{dBm}} = 10 \log_{10}\left(\frac{P_{\text{mW}}}{1 \text{ mW}}\right) \tag{1.4}$$

As examples, $1 \, \mu\text{W}$ and 100 mW are -30 dBm and $+20 \text{ dBm}$, respectively. It is useful to memorize some of the dBm values. It turns out that $10 \log_{10}(2) = 3.010 \approx 3$, which is convenient to use. So multiplying/dividing powers by factors of two correspond to $+3$-dBm or -3 dBm shifts. Thus 200 mW is $+23$ dBm, 400 mW is $+26$ dBm, etc. (see exercise at end of chapter).

Waves defined by equation (1.1) are called *monochromatic* (from the Greek *mono* or single and *chroma* or color). This means that the oscillation is based upon a single, perfectly-defined "color" or frequency. One of the characteristics of *laser light* is precisely that it is very close to monochromatic. In the general case, waves (or *waveforms*) can be conceived as the superposition of elementary oscillations defined by monochromatic waves:

$$f(t) = A_1 \sin(\omega_1 t + \varphi_1) + A_2 \sin(\omega_2 t + \varphi_2) \ldots \tag{1.5}$$

Figure 1.3 shows an example of the superposition of three waves having different amplitudes, frequencies and phases. It is seen that the resulting waveform exhibits a complex pattern of peaks and dips, each having its own periodicity. This pattern is the result of constructive (same-sign wave) or destructive (opposite-sign wave) additions, or interferences, between the different components of the wave. By definition, we say that this wave has three discrete *frequency tones*, ω_1, ω_2, and ω_3, with respective powers of A_1^2, A_2^2 and A_3^2. The word tone also applies if the waves are not purely monochromatic. This happens if the phase and frequency are defined with some relative uncertainty (e.g., 5%). Such an uncertainty is caused by phase and frequency noise.

The previous example provides an intuitive definition of the concept of waveform *spectrum*, which is the wave representation in the *frequency domain*,

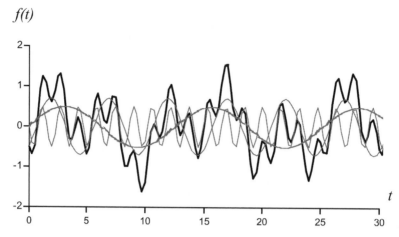

FIGURE 1.3 Waveform defined by the superposition of three sinusoids [equation (1.3)], with parameters $A_1 = 0.5, A_2 = 0.7, A_3 = 0.5, \omega_1 = 0.5, \omega_2 = 1.25, \omega_3 = 4,$ and $\varphi_1 = 0, \varphi_2 = -\pi/4, \varphi_3 = \pi$ (thick line). Each of the three sinusoid components is also shown as a thin line.

as opposed to the *time domain*. The familiar rainbow is nothing but the spectrum of sunlight, the colors corresponding to its frequency components or tones. Unlike in the previous example, the color spectrum is continuous because sunlight (unlike laser light) contains an infinity of frequency components. The fact that there are seven groups of colors just indicates that our eye has a similar perception of different frequency subgroups. Light emitted by atoms, such as in a fluorescent tube, is made up of discrete frequency components, corresponding to specific transitions between atomic energy levels. Sodium-vapor lights installed along highways have a single yellow color, but in fact the light is made up of two closely spaced tones ($\Delta\nu = 520\,\text{GHz}$ or $\Delta\lambda = 0.6\,\text{nm}$) that the eye cannot resolve.

Time- and frequency-domain representations are central to telecommunications engineering. The mathematical conversion from time-domain to frequency-domain is called the *Fourier transform*. Since this book is not meant to use advanced mathematical formalism, we will leave Fourier transforms here, but just recall that they can be performed rapidly with small-size computers or PCs, depending upon the number of time/spectral resolution points. The corresponding algorithm is referred to as *fast Fourier transform* (FFT), and is at the basis of digital *radio-frequency* or *microwave spectrum analyzers* which, with optical-frequency analyzers and oscilloscopes (time-domain) form the basic measuring apparatus of telecom laboratories.

1.1.2 Analog Waveform Modulation

Now that we have a clearer representation of waves, we must turn to the concept of *waveform modulation*. Modulating a wave is the action of modifying its characteristics with some kind of pre-established pattern. The wave is then called a *carrier*, since it "carries" some kind of intended message, or information. There are three ways to do this, since waves have amplitude, frequency and phase characteristics (we shall leave here *polarization*, which defines two orthogonal oscillation directions in space, as another degree of freedom). Figure 1.4 shows examples for each case, namely *amplitude modulation* (AM), *frequency modulation* (FM) and *phase modulation* (PM). Intensity modulation (IM) consists of modulating the carrier's power, in contrast with amplitude. Since usual (square-law) detectors measure the wave intensity, A^2, rather than the wave amplitude, A, AM and IM are closely related. However, there are other types of wave-measuring devices, referred to as *coherent detectors* (see Section 1.5), which measure both amplitude and phase, in which case AM/IM, FM/IM or PM/IM conversion are defined by more complex rules, as described in the section. If the modulation is sinusoidal (or a linear superposition of sinusoidal waveforms), it is referred to as *analog modulation*. This is the case for AM in Figure 1.4. If the modulation consists in jumps of amplitude, frequency or phase, it is referred to as *digital modulation*, as shown in FM and PM in the figure. If the jumps are of two kinds, the modulation is said to be *binary*. Binary modulation is described in the next section.

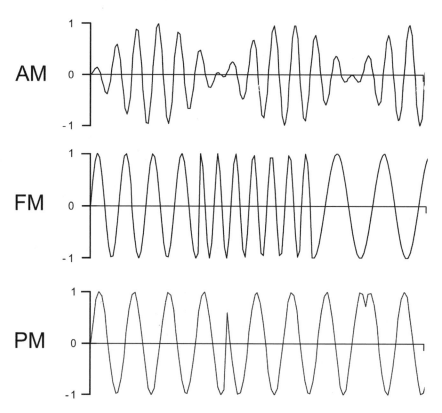

FIGURE 1.4 Three basic types of carrier modulation: amplitude (AM), frequency (FM), and phase (PM). In this example, AM modulation is sinusoidal, while FM or PM modulations correspond to discrete jumps having different frequency or phase values.

A fundamental property is that amplitude modulation generates new frequency tones, called *side-bands*. This can be shown simply by considering sinusoidal modulation at frequency ω_M of a carrier at frequency ω_0 (note that it is common practice to use the term frequency to designate pulsation, remembering the 2π proportionality). The waveform can be expressed as:

$$f(t) = A_M \sin(\omega_M t + \varphi_M) \times A_0 \sin(\omega_0 t + \varphi_0) \tag{1.6}$$

To simplify, assume that $A_0 = 1$ and $\varphi_M = \varphi_0 = 0$, giving

$$f(t) = A_M \sin(\omega_M t) \sin(\omega_0 t) \tag{1.7}$$

Checking out our basic trigonometry formulae, we find the rule

$$\sin(a) \sin(b) = \frac{\cos(a - b) - \cos(a + b)}{2} \tag{1.8}$$

thus from equation (1.6) and using the property $\cos(-x) = \cos(x)$:

$$f(t) = \frac{A_M}{2} \cos[(\omega_0 - \omega_M)t] - \frac{A_M}{2} \cos[(\omega_0 + \omega_M)t] \tag{1.9}$$

The result we have obtained shows that the modulated waveform is made of two tones at frequencies $\omega_0 \pm \omega_M$ (isn't this remarkable?). Thus the carrier and the modulation frequencies have virtually disappeared from the wave. The result is a superposition of two waves shifted from each other by $\Delta\omega = \omega_0 + \omega_M - (\omega_0 + \omega_M) = 2\omega_M$. We call them *modulation side-bands*. Note that the two side-bands have the same amplitude, but not the same phase. Indeed, using the property $-\cos x = \cos(x + \pi)$, equation (1.9) is equivalent to

$$f(t) = \frac{A_M}{2} \cos[(\omega_0 - \omega_M)t] + \frac{A_M}{2} \cos[(\omega_0 + \omega_M)t + \pi] \tag{1.10}$$

showing that the side-bands are "out of phase" from each other. Side-bands can in fact exhibit any arbitrary phase relation, depending upon the relative phases of the carrier and the modulation signals.

In order to further refine the side-band concept, assume now that the modulation is not 100%. This can be done by leaving a fraction of the carrier without modulation, or formally:

$$f(t) = A_M[1 + m \sin(\omega_M t)] \sin(\omega_0 t) \tag{1.11}$$

In the above definition, m is called the *modulation index*. The same calculation as previously yields the following expansion:

$$f(t) = A_M \cos\left(\omega_0 t - \frac{\pi}{2}\right)$$

$$+ m \frac{A_M}{2} \{\cos[(\omega_0 - \omega_M)t] + \cos[(\omega_0 + \omega_M)t + \pi]\} \tag{1.12}$$

In this result, it is seen that the waveform is made of three tones: the unmodulated carrier at frequency ω_0, with amplitude A_M, and the two side-bands at frequencies

$\omega_0 \pm \omega_M$, with equal amplitudes $mA_M/2$. We note that both side-bands carry the same modulation signal, while the carrier does not have any information.

Assume next that our modulation waveform is not a pure sinusoid, but (as in the general case) a superposition of sinusoids of amplitudes $A_1, A_2, \ldots, A_n$ and frequencies $\omega_1, \omega_2, \ldots, \omega_n$. The previous derivation also applies. It is easily checked that the resulting spectrum consists in a series of side-bands located at $\omega_0 \pm \omega_1$, $\omega_0 \pm \omega_2, \ldots, \omega_0 \pm \omega_n$ and having amplitudes $mA_1/2, mA_2/2, \ldots mA_n/2$. The two sets of side-bands located below or above the carrier frequency form two bigger side-bands, referred to as upper and lower, and which are mirrors of each other, as shown in Figure 1.5. We note from the figure that the side-bands occupy a certain frequency space, which is called the *modulation bandwidth*. The concept of modulation bandwidth (or bandwidth for short) is central to all telecommunications. It defines the frequency range which one actually needs to code, transmit and receive information. If the coder/transmitter, the transmission medium or the receiver/detector do not have enough bandwidth, some tones will be lost in the process, corresponding to signal degradation and some information loss. It is therefore a prerequisite that all system elements (transmitter, line, receiver) be properly matched with the required signal bandwidth.

1.1.3 Frequency-Division Multiplexing with Voice Channels

In the previous subsection, we saw that analog modulation transforms a carrier at frequency ω_0 into a set of two side-bands located near $\omega_0 \pm \omega_M$, where ω_M is

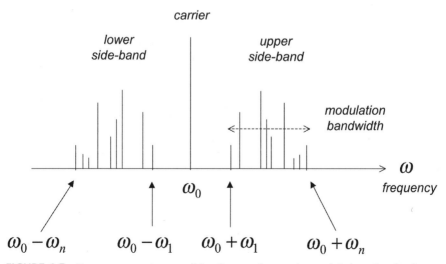

FIGURE 1.5 Frequency spectrum resulting from analog carrier modulation, showing lower and upper side-bands mirroring each other. The modulation bandwidth is indicated by the dashed arrrow.

some average or center modulation frequency. Consider *voice*. It is an analog signal whose frequency spectrum ranges from 20 Hz to 20 kHz. The shape of the spectrum depends upon the person and what he or she could be saying or singing. In a typical conversation, however, the voice spectrum spreads between 300 Hz and 4KHz (outside this band, voice would sound be very grave or very high!). We can then build a telephone link by modulating an electrical current with this voice signal. Such a modulation is basically the principle of a microphone. The signal needs to be electronically amplified. The output is then connected to a copper wire (because it is a good conductor). At the other end, the signal is electronically amplified again, to compensate for the power loss caused by the wire's resistance. The output is connected to a loudspeaker. The person near the speaker "hears" what the person says to the microphone. If we implement the same connection in the other direction, twisting a second copper wire around the first one, we have completed our telephone system.

This telephone installation requires as many twisted-wire cables as there are persons to call each other. This is precisely what happened in the early days of the telephone system. The streets were progressively invaded by a mesh of aerial wires going in all directions. In the telephone central office, people connected the wires together for each possible party-to-party call. But this approach faces limits as the number of telephone subscribers grows. Clearly, we need individual twisted-wires for each user, but at some point their combination into a single cable becomes physically impractical. If at this point, we had the possibility of conveying several voice signals into a single wire, this would create a substantial economy of cabling and resources! The way to achieve this is to *multiplex the voice signals in the frequency domain*. By definition, multiplexing is the action of combining several signals in the same physical channel. If we were to multiplex voices in the time domain, the result would be unintelligible, like a table conversation where everybody speaks at the same time. Yet, have you noticed that it can still work more or less if you look at the person facing you? But this is because you have identified (or previously knew) the person's voice pattern, and the ear recognizes this pattern through the rest of the noise. Also, the brain identifies the words and can make sense of that conversation out of the words flowing around. Finally, seeing the person's face, lips and expressions can certainly help. Yet, imagine what this would give through a telephone link. Instead, let's consider multiplexing in the frequency domain.

We upgrade our primary telephone system by using the apparatus shown in Figure 1.6. In this new arrangement, several voices are being captured. Each microphone's output is used to modulate a carrier at a given frequency ω_{0n} ($n = 1 \ldots k$). The resulting outputs are combined (added) through a multiplexer. Before combining, however, they are passed through a special filter. Figure 1.7 shows what has thus been accomplished. The initial voice spectrum of a given person is seen to be centered near a mean frequency ω_M (namely 2 kHz). The voice spectra are not exactly the same for each person, but the mean frequency and overall bandwidth are similar. After carrier modulation, we recognize the familiar double-side-band spectrum, centered near ω_{0n}. The filter we used is of the *band-pass* type. This means that it blocks the tones which fall outside a certain

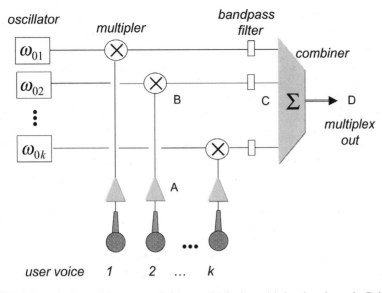

FIGURE 1.6 Principle of frequency-division multiplexing with k voice channels. Points A, B and C are referred to in Figure 1.7.

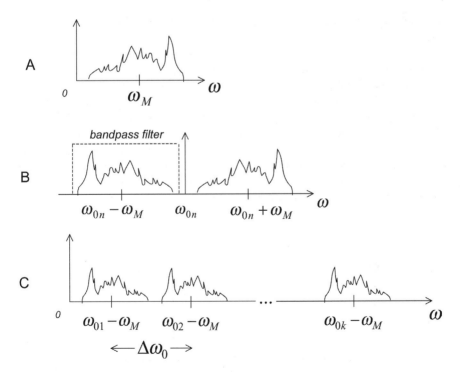

FIGURE 1.7 Spectra corresponding to points A, B and C in Figure 1.6: (A) baseband voice, (B) modulated carrier with upper side-band rejection through bandpass filtering, and (C) resulting multiplex of k frequency-division multiplexed voice channels.

region. Here, the filter only passes the lower side-band. Such a filtering operation is called *upper side-band rejection*. Then we combine all the resulting signals and obtain the spectrum shown at the bottom of the figure. The different voice channels appear at regular spacings, if we have chosen our carriers to be equally spaced $(\omega_{0,n+1} - \omega_{0,n} = \Delta\omega_0$ for all n). The carrier spacing must be chosen slightly greater than the voice baseband (4 kHz) in order to avoid overlap between modulated voice channels. The resulting signal, which contains the information relevant to all voice channels, can then be transmitted over some distance through a single wire. This entire operation is called *frequency-division multiplexing*, or FDM, as applied to analog voice channels. At the other end of the line, the different voice channels can be retrieved by a simple bandpass-filtering operation. In real analog systems, carrier frequencies are not attributed on a one-to-one subscriber basis. Rather, they are randomly attributed, as available, each time a call is placed. Thus, the overall system bandwidth does not need to be large. Indeed, ten-thousand simultaneous analog voice channels corresponds to a bandwidth of only $10,000 \times 4$ kHz, or 40 MHz, which can be easily handled by specially shielded twisted-pair cables or coaxial cables. For more information on telephone systems and analog FDM, see Chapter 2.

■ 1.2 DIGITAL SIGNALS AND CODING

The principle of *digital signaling* is intimately associated with signal *coding*. Basically, digital signals are not waves, as in the analog case seen in the previous subsection, but numbers. Numbers can be represented and transmitted under the form of multilevel signals, the voltage of which changes at a certain clock rate. The voltage can be made proportional to the number's magnitude (Figure 1.8). The drawback of such an approach is that as many discrete voltage levels as the maximum number to be transmitted are required, for instance 26 at least if the numbers are to represent letters in the Latin alphabet. Further, channel noise and distortion may cause random changes in the signal voltage, which introduces errors if the levels are closely spaced.

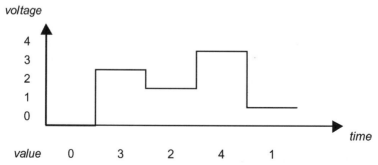

FIGURE 1.8 Multilevel, discrete signaling, used to transmit decimal symbols (0, 3, 2 . . .).

Another approach is to use a two-level signaling, based on *bits* (short for *binary digit*) whose voltages are either zero or nonzero. The nonzero bit is referred to as "mark" and the zero bit as "space". The advantage of the approach is that we only have two levels to identify. Intuitively, this type of signaling can handle more distortion and noise, as long as we can make the mark voltage sufficiently high. But how can we transmit numbers with only two coding levels? The answer is that any number can be represented by a string of zeros and ones, which is referred to as base 2 or *binary* representation. Let us have a closer look at the way one represents numbers.

1.2.1 Binary Number Representation

Take the number 1789, which is the year of the French Revolution. One understands from this ordained sequence of numbers "one thousand plus seven hundred plus eighty (eight times ten) plus nine," or identically:

$$1789 = 1 \times 10^3 + 7 \times 10^2 + 8 \times 10^1 + 9 \times 10^0 \tag{1.13}$$

This number representation utilizes powers of 10 as the base, hence its name *decimal system*. For this, we need 10 symbols (0,1, 2, 3, 4, 5, 6, 7, 8, 9). The numbers 10, 11, 12 and up are not symbols, but two-symbol (or N-symbol) strings where the one to the right is the number of times we have, or count 10^0 and the one to the left at position N is the number of times we have, or count 10^{N-1}. A basic reason for choosing the decimal system is that we can count numbers with our ten fingers. Mickey Mouse would have preferred a base of 8! Let us look now at the binary representation. The first four integer powers of two are the following:

$$2^0 = 1, 2^1 = 2, 2^2 = 4, 2^3 = 8 \tag{1.14}$$

Thus, in the binary representation the number 13 is written as

$$13 = 1 \times 2^3 + 1 \times 2^2 + 0 \times 2^1 + 1 \times 2^0 = \underline{1101}_2 \tag{1.15}$$

In the equation (1.15), the term $\underline{xxxx}_2$, as underlined, means a string of $1/0$ symbols in the binary representation, or identically: $\underline{1101}_2 = \underline{13}_{10}$. If it is clear that we are in the decimal or in the binary number representation system, the underline and subscripts can be omitted, which we routinely do in both systems. As is easily checked, the maximum number that can be represented with a string of N bits is $2^N - 1$, for example, 15 and 255 for strings of $N = 4$ and $N = 8$ bits, respectively. By definition, an 8-bit string is called a *byte*, or equivalently, an *octet*. Strings of bits are referred to as *binary words* or *words* (e.g., "a 2-byte word").

It is convenient to reduce the length of binary words by introducing the *hexadecimal* format. This format is just an equivalent way to write any group of four bits. The correspondence between hexadecimal and binary formats is the following:

$$
\begin{array}{llll}
0 = 0000 & 4 = 0100 & 8 = 1000 & C = 1100 \\
1 = 0001 & 5 = 0101 & 9 = 1001 & D = 1101 \\
2 = 0010 & 6 = 0110 & A = 1010 & E = 1110 \\
3 = 0011 & 7 = 0111 & B = 1011 & F = 1111
\end{array}
$$

We observe that numbers of value 0–9 are represented by the usual decimal digits, while numbers 10–15 are represented by the letters A–F. An 8-bit word (one byte) can thus be represented in hexadecimal under the form XY, where X represent the 4 bits of higher weights and Y the 4 bits of lower weigths. For instance $A1 = 10100001$, $7B = 01111011$ and $EF = 11101111$. We note that the hexadecimal format is not a substitute to the fundamental binary system, but just an equivalent and convenient representation.

Since we now have a representation scheme allowing us to represent any number with two elementary symbols, we can transmit the signal information through two-level voltages (mark $= 1$, space $= 0$). In engineering terms, the "1" and "0" voltage actually mean $V_{max} \neq 0$ and $V_{min} = 0$, respectively, or any set of two different voltages. We shall then use our previous example, but with binary numbers. Assume that the maximum number to be represented is $15 = 2^4 - 1$, so signals must be transmitted by strings of 4 bits. The result is shown in Figure 1.9. By convention, the bit of lowest weight (the first one from the right, corresponding to 2^0) is always transmitted first. For clarity, therefore, the Figure shows the time evolving from right to left, unlike in Figure 1.8. If the two time scales in Figures 1.8 and 1.9 are identical, for example, one decimal number or one 4-bit string transmitted per second, we can say that the symbol rates are identical. But in the second case, the *bit rate* (number of bits per second) is four times higher than the symbol rate. This is the price we had to pay for reducing the 5-level signaling of Figure 1.8 to the binary signaling of Figure 1.9. Since 4 bits/symbol are sufficient to transmit the same information as a 15-level signal, while showing less risk of error and being much more simple to implement, it is clear that binary signaling is the winning approach.

Since decimal numbers up to $2^N - 1$ can be represented by N bits, we can invent a coding table for representing alphanumeric characters. For instance, we could code the numbers 0–9 and the letters A–Z into the binary words 000000 to 001001 for the ten numbers, and 001010 to 100011 for the 26 letters. This would require of course 6-bit coding since $31 = 2^5 - 1 < 36 < 2^6 - 1 = 63$. The oldest example of alphanumeric coding is *Morse code*, used in old telegraphs and in certain military communications. The code is based on "dit" $= \bullet$ and "da" $= -$ (or "dot" and "dash," respectively) for the two bit values. The different possibilities of dits and das make it possible to generate at least 39 code words for the 26 letters $+ 10$ numbers $+ 3$ punctuation marks (./,/?). One surely knows the S-O-S signal ("save our souls") as $\bullet\bullet\bullet \; - - - \; \bullet\bullet\bullet$. Each word is separated by short pauses, which are meant for unambiguous identification, also remembering that

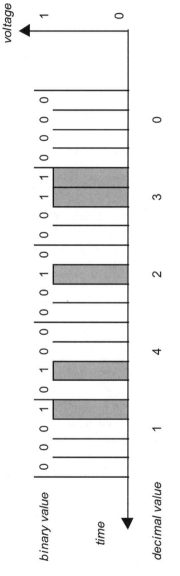

FIGURE 1.9 Same example as in Figure 1.8, but using binary signaling in 4-bit strings or symbols. Marks representing "1" binary value are shown in gray (maximum signal voltage), and spaces representing "0" binary value are shown in white (minimum or zero voltage). Bits are transmitted from right to left, starting with the lowest-weight bits in each string/symbol.

the messages are conveyed and read by humans. As a diversion, observe that Morse code is fancier than it looks. Indeed:

contrary to usual binary coding, the Morse code symbols have different word lengths. For instance, we have $E = \bullet$ and $T = -$ (the shortest letter words) and $J = \bullet - - -$, $Q = - - \bullet -$, $X = - \bullet\bullet -$, $Y = - \bullet - -$, $Z = - - \bullet\bullet$ (the longest letter words). This choice is because in English and in most western languages, the letters E and T are the most frequent ones, while J, Q, X, Y and Z are the least frequent. Because the shortest code words correspond to the most frequent letters, this saves time in message transmission, in contrast with the approach consisting of coding with 6-bit words. Another coding trick is that letter words occupy up to 4 bits, while number words are exactly 5-bit long (e.g., $0 = - - - - -$, $1 = \bullet - - - -$ and $5 = \bullet\bullet\bullet\bullet\bullet$, $6 = -\bullet\bullet\bullet\bullet$). This makes it easier to distinguish between letters and numbers and avoids the risk of confusion for numbers. Thus, Morse code is easy to memorize and is very efficient for rapid communications between human entities.

If we were to code upper-case and lower-case letter symbols into Morse-like words, the approach would not be advantageous anymore, especially if it is to be used by fast machines. Binary coding of alphanumeric characters (and other special keyboard or processing commands) has been standardized for telex communications and PC networks use. Telexes use the *Baudot Code* (as named after Emile Baudot, a telegraph pioneer). This code, also called IA2, is based on 32 different 5-bit words. There are two character-shift commands which make it possible to code either capital letters or figures (numbers, punctuation, etc.), and virtually multiplies by two the number of symbol-coding possibilities from the same 5-bit set of $2^5 - 1 = 31$ words. The *Baud rate* is defined as the number of times per second a signal is being modulated, with *Baud* as the unit. For binary modulation, the baud rate, modulation rate and bit rate are all equal ($1 \text{ Baud} = 1\text{bit/s}$). In a following section, we will see that there are other coding types based on multilevel signaling, for which the bit rate is higher than the Baud rate. As the telex typically operates at 50 Bauds, the transmission corresponds to $50/5 = 10$ characters per second, which is close to the speed of human speech.

The two main alphanumeric codings used in data communications are ASCII (American standard code for information interchange), which uses 7-bit words and EBCDIC (extended binary coded decimal interchange code), which is an extension of ASCII using 1-byte words. Table 1.2 shows an extract of the ASCII table for common keyboard characters. For instance, the letter A is coded as 1000001.

1.2.2 Binary Coding into Waveforms

We have seen that any type of number or symbol can be transmitted by coding them into strings of marks and spaces (bits), which are easy to identify at the other end of the line, as long as they are well defined from each other (e.g., voltage difference). In the previous example, we have used a two-voltage scheme to unambiguously identify the marks (1) and the spaces (0). But there are many other possible ways to do this even more imaginatively, which are made explicit in Figure 1.10, and are

TABLE 1.2 ACSII code table (extract)

	← bit word			7	0	0	0	0	1	1	1	1
				6	0	0	1	1	0	0	1	1
4	3	2	1	5	0	1	0	1	0	1	0	1
0	0	0	0				sp.	0	@	P	\	p
0	0	0	1				!	1	A	Q	a	q
0	0	1	0				"	2	B	R	b	r
0	0	1	1				#	3	C	S	c	s
0	1	0	0				$	4	D	T	d	t
0	1	0	1				%	5	E	U	e	u
0	1	1	0				&	6	F	V	f	v
0	1	1	1				'	7	G	W	g	w
1	0	0	0				L	8	H	X	h	x
1	0	0	1				)	9	I	Y	i	y
1	0	1	0				*	:	J	Z	j	z
1	0	1	1				+	;	K	[	k	{
1	1	0	0				'	<	L	\	l	|
1	1	0	1				−	=	M	]	m	}
1	1	1	0				.	>	N	^	n	~
1	1	1	1				/	?	O	−	o	

described as follows:

- *Nonreturn to zero* (NRZ): high signal is 1, low signal is 0. A string of consecutive 1s or 0s is reflected by no signal level change.
- *Nonreturn to zero inverted* (NRZI): change of signal polarity (low to high or high to low) means 0, no polarity change means 1.
- *Return to zero* (RZ): same as in NRZ, but marks occupy half of the bit slot. A string of consecutive 1s is a comb of regularly-spaced, identical pulses, unlike in NRZ. By definition, the ratio of duration where the signal is high and low is the *duty cycle*, which here is 50–50 or 50%.
- *Coded-mark inversion* (CMI): 0s are represented by an up-to-down level change through the bit slot, while 1s are represented by either level following the previous bit. In the figure, the fist bit (0) must have been preceded by a high-level 1, while the second bit (1) is consistently at low level.
- *Manchester coding*: transition from high to low and low to high within the bit slot correspond to 0s and 1s, respectively.
- *Differential Manchester coding*: same principle as in Manchester coding, but 0s are coded by a level transition from the bit start-point and 1s from the bit mid-point from the level reached in the previous bit. In the figure, the first bit (0) must have been preceded by a 1 ending at high level, like the 1 in the fifth bit.
- *Miller coding*: 1s are represented by a transition at mid-point from either level, while 0s correspond to no transition at all.

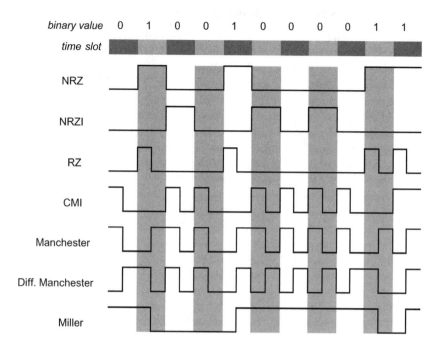

FIGURE 1.10   Principal binary codes corresponding to the number 01001000011, from top to bottom: nonreturn to zero (NRZ), nonreturn to zero inverted (NRZI), return-to-zero (RZ), coded mark inversion (CMI), Manchester, differential Manchester, and Miller. Alternate time slots are shaded in order to highlight the amplitude changes within the bit slots.

If the low signal (0 bit) is identically zero in terms of bit power, this is called unipolar modulation. For NRZ or RZ signals, *unipolar* modulation is also called ON/OFF (N)RZ keying, also referred to as OOK. If the change from marks to spaces or the reverse is a change of amplitude sign, the modulation is said to be *bipolar*.

The above are referred to as either signal *modulation formats* or *coding* types or both. The word modulation means that such codes are used to change the amplitude of a wave *carrier*, like flashing on and off a torch light, at regular time intervals (the bit period). The action of coding has two possible meanings. Familiarly, it is often identified with carrier modulation (e.g., the laser signal is "coded" with NRZ data.). But the strict meaning of coding is to convert the input data to be transmitted (e.g., $1101_2 = 13_{10}$) into a string of bits (marks and spaces) having any of the above shapes and level-changing rules. Clearly, NRZ, RZ and Manchester codes have a one-to-one correspondence with the input signals (marks and spaces individually identifiable as 1s or 0s), while the other code types use the information of consecutive bits. The advantage of this more sophisticated way of coding is to avoid the generation of long bit strings with identical level. In NRZ and RZ, long strings of 0s or 1s are virtually equivalent to an "idle" state or absence of signal, which may cause problems of clock recovery or temporary loss of synchronization at the receiving end. In optical transmission, strings of 1s correspond to rectangular light pulses with constant intensity, which is known to be unstable under nonlinear effects in the fiber. This is why RZ (or RZ-like) pulses with at maximum 50% duty cycle are generally used in long-haul transmission. In contrast, the other codes shown in Figure 1.10 (except Miller) do not permit the voltage to be constant over more than one and a half bit slots, which alleviates the problem of long one-level strings. In the description, we have considered electrical pulses coded (or modulated) through voltage changes. In radio and optical communications, the same coding principles can be applied to the electromagnetic waves, but signal bit changes take the form of *amplitude modulation* (AM), *intensity modulation* (IM), *frequency modulation* (FM) or *phase modulation* (PM), leading to a wide variety of coding options.

Other types of binary code can make use of *multilevel signaling*, which leads to more complex M-ary modulation formats (see forthcoming subsection). Considering here only a simple case, the *alternate mark inversion* (AMI) code uses zero voltage for 0 bits and $\pm V$ for 1 bits, which represents three possible amplitude levels. Two successive 1s, regardless of the number of 0s in between have opposite voltages. The advantage of this coding is that the net average current through the line is zero. By using an alternate version of AMI, referred to *high-density bipolar* (HDBn) code, it is possible to prevent long strings of 0s which generate the aforementioned *clock-recovery/synchronization* problems. The idea is to code the 0 coming in the n place in the 0 sequence under the form of a 1, but in violation of the rule. The violation consists in setting this 1 at the same level as the 1 that precedes the string of 0s. Accordingly, the receiver at the end of the line recognizes the violation, and therefore interprets the bit as meaning a 0.

We have described several ways to encode binary numbers into binary signals. This can be done for instance by modulating the amplitude of a carrier, turning it ON for code marks and OFF for code spaces. Upon reaching a square-law detector (see Section 1.5), the amplitude-modulated (AM) signal wave is converted into an intensity-modulated (IM) signal current. As we have seen in Section 1.1, however, a wave can also be modulated in frequency and phase, referred to as frequency (FM) and phase (PM) modulations, respectively. For binary signal coding, one refers to AM as *amplitude-shift keying* (ASK), to FM as *frequency-shift keying* (FSK) and PM as *phase-shift keying* (PSK), which are three basic *modulation formats*.

Figure 1.11 shows examples of bit sequences in the ASK, FSK and PSK formats. It is seen that the bit value is determined by two possible states in the wave amplitude, frequency or phase. In the PSK example, the phase shift is π, meaning that the amplitude of the wave is inverted. This type of PSK is thus equivalent to AM with two levels ± 1. We also see that FSK waveforms are made of two different tones. If the FSK signal is passed through a narrow-band filter which selects only one of the two tones, the resulting output is an ASK signal (zero bits corresponding to the blocked tone). These examples show that there are lots of options to encode 1/0 signal bits onto a waveform. How the ASK, FSK or PSK wave signals can be detected and reconverted into 1/0 electrical bits is an issue addressed in Section 1.5.

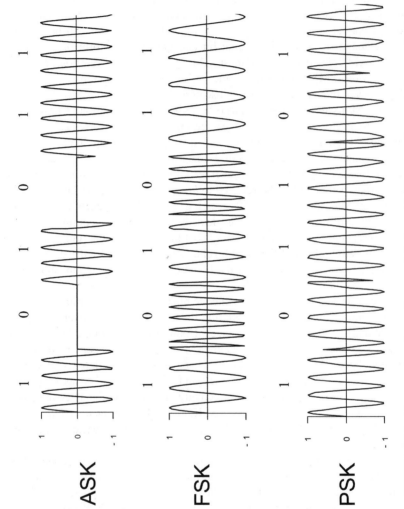

FIGURE 1.11 Three types of waveform modulation with binary coding: unipolar amplitude (ASK), frequency (FSK) and phase (PSK) shift keying.

1.2.3 Multilevel Coding and M-ary Modulation

The previous modulation formats concerned binary modulation, as achieved by changing the wave's amplitude, frequency or phase with two possible values. In fact, one can generate more fancy signals by simultaneously modulating the waveform amplitude and phase, for instance. To make things even more complex, it is possible to use more than two levels for amplitude and phase, as referred to *multi-level coding* or *M-ary modulation* (M > 2).

We start with the simplest case of multi-level coding, considering first AM then PM. In the previous subsection, we have described ASK. Such a format can also be called *binary pulse-amplitude modulation* (PAM). We can then form M-ary PAM signals by modulating the wave with $M = 2^k$ possible amplitudes. A word of k bits corresponding to a given amplitude in this amplitude set is called a *symbol*. Thus, the symbol has a duration of k times the bit period. An example of quaternary PAM (4-PAM) is provided in Figure 1.12. The figure shows both coding-level values (the *baseband* signal) before modulation and the resulting AM waveform (the *bandpass* signal). The diagram shows the symbolic correspondence between the four levels and four possible 2-bits-word values (called *dibits* or, also, *quats*). Such a one-to-one mapping is arbitrary. The one shown is called the *Gray code*. The mapping rule is that every adjacent symbols differs from its neighbors by a single bit of lowest possible weight, which is easily verified in the figure diagram. An alternative name of this 4-PAM/Gray code is 2B1Q, for *two-bits, one quat(-ernary)*.

The same principle as PAM can be applied to phase modulation. The M-ary signals are coded with $M = 2^k$ possible phases. The binary version of this modulation is PSK.

We shift next to multilevel AM and PM modulation, as combined simultaneously. Assume the same number of AM and PM levels, for instance 2, 4,

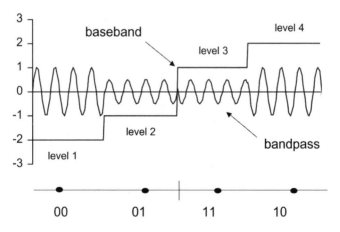

FIGURE 1.12 Quaternary signaling (4-PAM or 2B1Q) showing both baseband signal from lowest to highest values and resulting bandpass signal waveform. The diagram below provides the bit-word value or symbol correspondence with the four amplitude levels (Gray code).

8 ... corresponding to $M = 4, 16, 64$ ... different symbols. The resulting waveform can be expressed under the general form:

$$f(t) = A_m \sin(\omega t + \phi_m) \tag{1.16}$$

Using the rule $\sin(a + b) = \sin a \cos b + \sin b \cos a$ to decompose the above definition, we see that simultaneous AP/PM with discrete coefficients A_m, ϕ_m is equivalent to amplitude modulate two carrier waves which are in quadrature (sine and cosine), and with discrete coefficients a_m, b_m:

$$\begin{aligned} f(t) &= A_m \cos \phi_m \sin(\omega t) + A_m \sin \phi_m \cos(\omega t) \\ &\equiv a_m \sin(\omega t) + b_m \cos(\omega t) \end{aligned} \tag{1.17}$$

with $b_m/a_m \equiv \tan \phi_m$. Because the waveform is the sum of two AM quadratures, such a modulation technique is called *quadrature PAM*, or *quadrature amplitude modulation*, or QAM.

The two-dimensional space diagram representing the different QAM level combinations and associated Gray-code symbols is called a *constellation*. Figure 1.13 show the constellation of 16-ary QAM, here a_m, b_m have four degrees of freedom each. The constellation is called rectangular, since the increments in a_m, b_m are equal. Another possibility would be to have a_m, b_m increments chosen such that the outer points in this constellation would follow a circle pattern, but the rectangular approach is more practical since it is equivalent to combining two PAM signals. Figure 1.14 shows an example of a quaternary QAM waveform with dibit symbols 00, 01, 10 and 11 in orderly sequence.

Since in QAM there is one symbol, or Baud, per modulation period with $M = 2^k$ levels, the bit rate is k times (or $k = \log_2 M$) the symbol rate. For 16-ary or 64-ary QAM, the bit rate is thus four times or six times the symbol rate, respectively. Thus

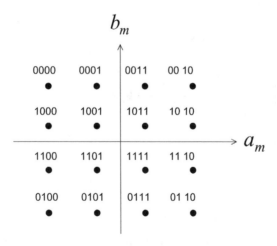

FIGURE 1.13 Rectangular constellation diagram for 16-ary QAM and Gray-code symbol equivalence.

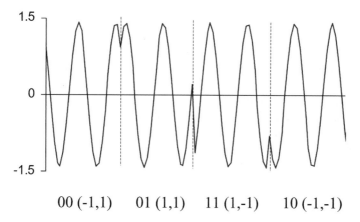

00 (-1,1)   01 (1,1)   11 (1,-1)   10 (-1,-1)

FIGURE 1.14 Quaternary-QAM bandpass waveform and Gray-code symbol equivalence. The constellation coordinates are shown in parentheses.

QAM is another means to efficiently carry digital data at relatively lower AM rates. It is therefore extensively used in *modems*, which are devices that connect PCs to telephone lines. The modem function is to achieve digital-to-analog conversion or the reverse (hence the name MOdulator-DEModulator). The QAM output can thus be regarded as being an "analog" signal with multiple, discrete phase and amplitude jumps. QAM demodulation into binary data requires coherent detection (see Section 1.5).

■ 1.3 ANALOG-TO-DIGITAL VOICE CONVERSION

The transition from analog voice signals (FDM waveforms) to digital voice data (bytes) requires some kind of conversion algorithm. As also described in Section 1.7, the *Nyquist theorem* states that for proper analog-to-digital (A/D) conversion, analog signals must be sampled at twice the highest frequency component. For voice channels, the highest frequency component is near 4 kHz, while the lowest is near 300 Hz. The human hearing range actually extends from 20 Hz to 20 kHz, but normal voice pitch and conversation are clearly resolved within the 300 Hz–4 kHz range. Therefore, the Nyquist *sampling rate* must be set to 2 × 4 kHz = 8 kHz, corresponding to 8,000 samplings per second. Since by convention each digital sampling is coded into a single *byte* (8 bits), the resulting data rate is 8,000 × 8 bits/ s = 64 kbit/s. This explains why 64 kbit/s is the elementary bit rate for voice channels, from which multiplexing hierarchies can be built (see Chapters 2 and 3).

1.3.1 Pulse-Code Modulation

The technique of analog-to-digital conversion of voice channels is called *pulse-code modulation*, or PCM. The underlying concept of PCM is to sample the voice's waveform amplitude at regular intervals (*sampling period* T_{samp}) and convert the sample values into bytes. Because the resulting digital signal only represents samples of the analog input, the process is called *quantization*. Figure 1.15 shows an example of 8-level quantization and its application to a waveform. One can see from the sampling that some parts of the waveform structure are forever lost. The reconstructed waveform will not be a faithful counterpart of the original, therefore it is said that the process introduces *quantization noise*.

Formally, one can express the result of quantization $x(t)$ of a waveform $f(t)$ through the formula:

$$x(t_n) = f(t_n) + e(t_n) \tag{1.18}$$

where $t_n = nT_{samp}$ is the sampling time, and $e(t_n)$ is the error associated with the quantization. If the quantization increment is Δ, the error is in the interval $(-\Delta/2, +\Delta/2)$ and its probability distribution $p(e)$ is uniform, or $p(e) = 1/\Delta$. It can be checked that the variance of this distribution, which defines the noise power (see section 1.4), is $\sigma^2 = \Delta^2/12$. We now relate this result to the number of bits used in the quantization. If N bits are used, then the increment is $\Delta = 1/2^N$. The noise power is thus $\sigma^2 = 1/(12 \times 2^{2N})$, or in decibels:

$$P_{QN}(\text{dB}) = 10 \log_{10}[1/(12 \times 2^{2N})] = -6.02N - 10.79 \tag{1.19}$$

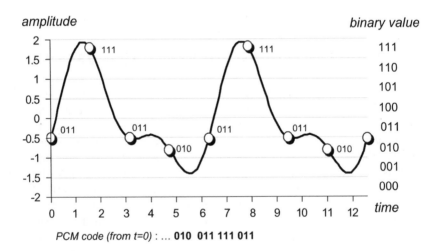

FIGURE 1.15 Principle of pulse-code modulation (PCM), or analog-to-digital conversion. The waveform is sampled at twice its maximum frequency ω_{max} (*). Eight quantization levels are defined with 0.5 amplitude increments. The corresponding 3-bit binary values (000 to 111) are shown at right. The open circles correspond to the sampled points. The PCM code, starting from $t = 0$, is shown at the bottom. [(*) here $\omega_{max} = 2$ or $T_{samp} = 2\pi/(2\omega_{max}) = 1.57$.]

It is seen from this result that the quantization noise decreases with the number of bits (or as the increment size, $\Delta = 1/2^N$, is made smaller). For instance, 4-bit and 8-bit codings would give quantization noises of $P_{QN} = -6.02 \times 8 - 10.79 = -59$ dB, and $P_{QN} = 6.02 \times 4 - 10.79 = -35$ dB, showing that the noise is $59 - 35 = 24$ dB (250 times) lower with 8 bits quantization. Clearly, the noise can be brought to arbitrarily low levels with a large number of bits, leading to near-absolute conversion fidelity. However, we are limited by the maximum bit rate that our system can handle or what bandwidth we are "willing" to attribute to a single voice channel. Another improvement technique would consist of increasing the sampling rate while keeping the same quantization. But here again, we increase the number of signal bits generated in a given time interval, which is equivalent to assigning to the channel a higher bit rate or bandwidth. We need therefore to find another method to improve the sampling accuracy, but without increasing the sampling rate or the number of quantization levels.

It has been observed that in voice waveforms, amplitude changes occur more frequently at low amplitude levels. One can use this property to adapt the quantization increments according to the amplitude level, using smaller increments in the low-amplitude region. Because using a nonuniform or nonlinear quantization scale is not very practical, it is possible instead to nonuniformly *compress* the waveform to be sampled and apply a uniform quantization. Here are two (out of the many) possible compression laws, which are called *μ-law* and *A-law*, respectively. The first is continuous over the amplitude domain:

$$\left| f_{comp}(t) \right|_{\mu-law} = \frac{\log(1 + \mu|f(t)|)}{\log(1 + \mu)} \tag{1.20}$$

while the second distinguishes two amplitude regions:

$$\left| f_{comp}(t) \right|_{A-law} = \begin{cases} \dfrac{A|f(t)|}{1 + \log A} & \text{if } |f(t)| \leq \dfrac{1}{A} \\[2ex] \dfrac{1 + \log(A|f(t)|)}{1 + \log A} & \text{if } \dfrac{1}{A} < |f(t)| \leq 1 \end{cases} \tag{1.21}$$

In the above definitions, the waveform $f(t)$ is normalized to unity and log is the natural logarithm ($\log x = 2.302 \log_{10} x$). Figure 1.16 shows the result of the two transformations on a sinusoidal waveform for different values of the compression parameters μ and A. It is seen that compression dramatically increases the slope of the original waveform near the zero-crossing points, while the high amplitude regions near the extremes are flattened. This results in a sampling grid which has narrower and broader increments in the low-amplitude and high-amplitude regions, respectively.

The μ-law is the standard adopted in North America, with $\mu = 255$. In Europe, the standard is the A-law with A = 87.56. It can be shown that this μ-law provides a 24 dB (250 times) reduction of quantization noise. With 8-bit nonlinear quantization, the noise is thus reduced from the previous -59 dB level to $-59-24$ dB = -83 dB. But let's look at this improvement differently. Recall, from equation (1.19) that the quantization noise has a component of $6.02N$, where N is

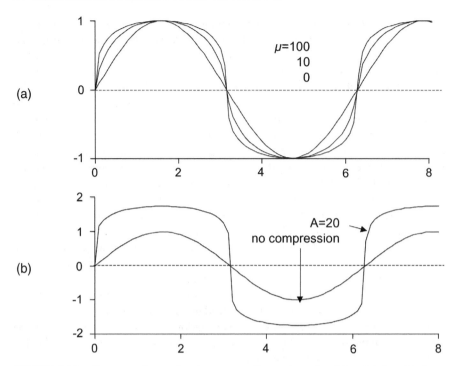

FIGURE 1.16 Compressed waveforms corresponding to the μ-law (a) and A-law (b) algorithms, with $\mu = 10-100$ and A = 20. The original waveform is a sinusoid.

the number of coding bits. Thus, it is theoretically possible to achieve the same signal quality (equivalent noise) with N bits using uniform quantization as with N' bits using compressed quantization, according to the identity:

$$P_{QN}(dB) = -6.02N - 10.79 = -6.02N' - 10.79 - 24 \qquad (1.22)$$

Letting $N = 8$, we find from the above $N' = 4$, meaning that compression represents a saving of two in bandwidth. More sophisticated algorithms (but not necessarily exactly matching the above noise reference) can lead to a five-fold bandwidth reduction, as briefly described below. Put simply, with the appropriate coding algorithms one can encode voice channels into $(1/2) \times 64 = 32$ kbit/s down to $(1/4) \times 64 = 16$ kbit/s with similar end-to-end quality.

As the signal is reconstructed into an analog waveform, the reverse operation, or non-linear expansion, is required. The action of compressing/expanding analog waveforms is known by this intriguing word: *compansion*.

1.3.2 Differential and Adaptative PCM

With analog voice signals, the PCM samples actually exhibit a high correlation. This means that adjacent samples are often very similar, because at the Nyquist sampling rate the signal does not vary rapidly. Thus, PCM signals have a certain amount of redundancy, which could be removed in order to encode the channel at a lower bit rate. A way to achieve this is to sample, not the waveform amplitude, but its amplitude difference with the previous sample or set of samples. Such a technique is called *differential* PCM, or DPCM. A minimal version of DPCM consists of coding the difference between one sample and the next with only two bits. The resulting value is 0 if the difference is negative or null, and 1 if it is positive, as illustrated in Figure 1.17. This DPCM approach is referred to as *delta modulation*. It is seen from the figure that the resulting staircase approximation of the waveform involves a certain amount of quantization noise, which increases as the waveform slope (rate of amplitude change) increases. Minimization of this noise can be achieved by optimizing the sampling rate and the sampling step size. But a more efficient approach would consist of adapting the sampling interval according to the slope variation, i.e., shorter intervals at higher slopes. This approach is referred to *adaptive* DPCM, or ADPCM. It can be shown that through ADPCM, voice can be coded with acceptable quality at code rates as low as 2 kbit/s. In order to obtain this performance, however, complex coding algorithms for removing redundancy and minimizing quantization noise are required. In practice, a ten-fold increase in the number of processing operations (add/multiply) is associated with halving the code rate. Thus, a four-fold reduction of voice channel rate (64 kbit/s to 16 kbit/s) results in a one hundred-fold increase in signal-processing complexity. It is worth briefly explaining how this can be achieved. The approach is referred to as *adaptive subband coding* (ASBC). The voice signal is analyzed in M independent frequency sub-bands at a low sampling rate (e.g. 2 kHz). Interestingly, noise is not perceived by the human hear if it is − 15 dB below the signal in that subband. This property allows for significant quantization error, as results from lower sampling rates. The number of bits to be assigned for subband coding is made to vary dynamically with the voice's spectral characteristics. Thus, two bits can be assigned to the voice subbands with fewer spectral contents, and 3 to 5 bits for the predominant

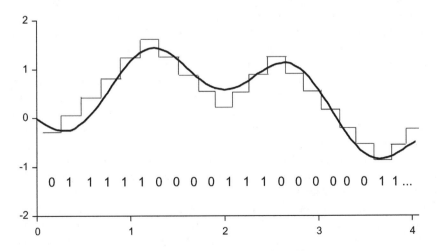

FIGURE 1.17 Principle of delta modulation, with resulting binary sequence.

subbands. Assuming for instance four subbands with bit assignment of 1,1,2,4, the channel can be coded with 8 bits. At 2 kHz sampling rate per band, we obtain a voice channel at $8 \times 2 = 16$ kbit/s.

Another adaptative technique is *continuously variable slope delta* (CVSD). It is similar to ADPCM, but the encoded information is not the (adapted) incremental change of amplitude but (adapted) incremental change of slope. CVSD can be implemented at bit rates as low as 9.6 kbit/s, under which the reconstructed voice signals are not intelligible. Its most frequent use is at 32 kbit/s, allowing a two-fold and high-quality voice-channel compression.

1.3.3 Other Conversion Techniques

There are many other possible techniques for voice-to-data conversion. Two are *linear predictive coding* (LPC) and *digital speech interpolation* (DSI).

The first technique, LPC, is based upon the principle of predicting voice patterns. Voice is made up of some personal variation of typical pitch/intensity patterns. These patterns can be synthesized electronically, which reduces the amount of information to encode. Voice conversion consists of extracting the patterns and encoding the remaining waveform "residue". The residues can be encoded into voiced or un-voiced symbols, which correspond to sounds requiring or not requiring vocal cords, respectively. Conversion back through the analog signal is made by the reverse process, reconstructing the different elements into a voice-like waveform. With LPC, voice channels can be coded down to 2.4 kbit/s. Since the resulting voice signals may not sound as faithful but rather impersonal, the applications of low bit-rate LPC concern electronic voice mail, military communications and games.

The second technique, DSI, replicates in the digital format what *time-assign-ment speech interpolation* (TASI), utilized with analog voice signals. The principle is based on the fact that the average talker is in fact silent for more than 40% of the time this means that ordinary conversations are 20% silent on average. This extra bandwidth can then be utilized by other voice circuits. The principle is to time-share a smaller number of voice circuits, momentarily disconnecting inactive talkers and reconnecting the active ones. The terminals keep track of the talker/listener circuit re-assignments and switch the time slots in due sequence, thus creating a virtually uninterrupted connection. If too many talkers need to be recon-nected, the overflow is clipped or frozen out for a short duration until circuits become available. This is referred to as *competitive clipping*. The statistics showed that with a ratio of two to one in the number of users per number of voice channels competitive clipping is hardly perceptible. A second clipping effect, *connection clipping*, is the momentary loss of circuit connection due to the failure of one terminal to assign a channel at the required speed. The first TASI implementation came with the 1959 transatlantic telephone cable system, TAT-1, in which 72 users shared 32 voice channels. When applied to digital signals (DSI), the bandwidth savings are even more substantial. Indeed, 120 DSI-coded con-versations can be carried over 24 circuits at 64 kbit/s (T1 carrier, see Chapter 2, Section 2.3). This represents a *five-fold* voice-rate compression, which is equivalent to 12.8 kbit/s use per circuit.

End-to-end voice quality after A/D and D/A conversions can be subjectively rated through a *mean opinion score* (MOS), which scales from 1 to 5. The perfect MOS of 5 means that one cannot detect any quality degradation. A MOS of 4 is achieved with 64 kbit/s PCM and 32 kbit/s DPCM, and that reasonably high quality level is in fact nearly achieved with 16 kbit/s ASBC.

■ 1.4 CHANNEL NOISE

The concept of *noise* is very familiar to anyone. We experience it as those annoying sounds coming from the streets, or heard on a distant radio station, or seen as snow-flakes on a poor-quality TV reception. Yet, as a mathematical concept noise is not as easily managed as it may seem from physical experience. In waveforms, noise represents a certain amount of uncertainty in amplitude, frequency or phase. Such an uncertainty is reflected by random departures from the nominal waveform characteristics. We can only say that there is some *probability* that the waveform will have a given amplitude, frequency or phase when we measure it.

Consider amplitude and phase noise on a sinusoidal waveform. One can define this noisy waveform $f(t)$ as follows:

$$f(t) = [A + x_A] \sin(\omega t + \Phi + x_\Phi) \tag{1.23}$$

with A, Φ being fixed amplitude and phase parameters, and x_A, x_Φ being the corresponding random changes. Figure 1.18 shows three examples of sine waveforms with amplitude noise, phase noise, or both. We can recognize the sinusoid in spite of the presence of noise, as long as the random changes are not too large with respect to the nominal waveform amplitude and phase. Note that phase noise does not affect the maximum wave amplitude (here ± 1), but rather the waveform synchronization or its position in time. It is seen that phase noise also causes amplitude noise, but this noise is zero at the ± 1 maxima and is maximum at the zero crossings.

1.4.1 Signal Mean and Variance

One can formally characterize noisy signals by two important parameters, namely the *mean* and the *variance*. which we shall now define. Figure 1.19 (top) shows an example of noisy signal, which is a set of sample points. We can observe several things. First, there are as many points above as below the horizontal axis, showing that their distribution is *symmetrical* with respect to the $x = 0$ amplitude. Since there are as many points on each side, the average or *mean value* of x, which we write $\langle x \rangle$, is identically zero. Finally, we observe that the points are clustered near the horizontal axis, being mostly confined inside the region $x = -2$ to $x = +2$. But there are more points near the axis than away from it, which is represented by the symmetric, bell-shaped curve on the right. The bell curve $p(x)$ is equivalent to a histogram of the distribution of points. Therefore, it peaks about the origin $x = 0$ and vanishes for large x. The points shown at bottom of the figure correspond to the previous points with amplitudes squared. It is seen that

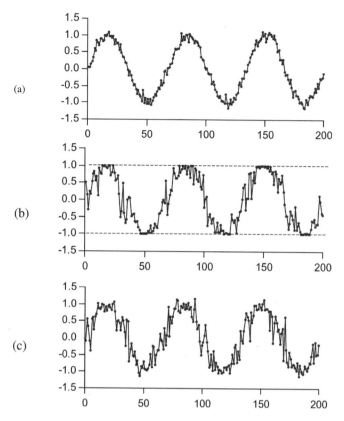

FIGURE 1.18 Sinusoidal waveform with (a) amplitude noise, (b) phase noise, and (c) both amplitude and phase noise.

the mean value of x^2, which we call $\langle x^2 \rangle$, is equal to 0.01. We observe that most of the points are clustered in the squared-amplitude region $\langle x^2 \rangle - 0.01$ to $\langle x^2 \rangle + 0.01$, which provides a definition of the cluster's size.

We can now introduce the definitions of noise *variance* σ^2 and *standard deviation* σ, according to the following:

$$\sigma^2 = \langle x^2 \rangle - \langle x \rangle^2 \tag{1.24}$$

$$\sigma = \sqrt{\sigma^2} = \sqrt{\langle x^2 \rangle - \langle x \rangle^2} \tag{1.25}$$

In the previous example, since the mean is zero, the variance is $\sigma^2 = \langle x^2 \rangle = 0.01$, corresponding to a standard deviation of $\sigma = 0.1$. The number σ thus characterizes the width of the distribution. It is seen from the figure that most of the points are actually located in the $-\sigma$, $+\sigma$ amplitude interval. The bell curve thus provides an intuitive definition of the *probability* of occurrence associated with amplitude x. We can therefore call it a *probability distribution function*, or PDF.

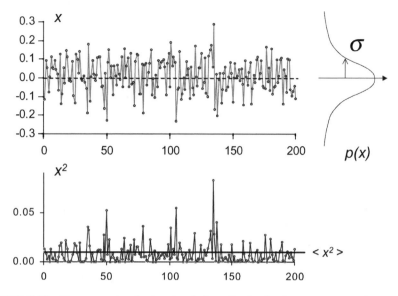

FIGURE 1.19 Top: example of noisy signal of amplitude x having zero mean and a standard deviation of $\sigma = 0.1$. The bell-shaped curve shown at right, $p(x)$ represents the probability distribution for the amplitude, x. Bottom: plot of the squared-amplitudes, x^2, corresponding to the previous set, showing a mean value of $\langle x^2 \rangle = \sigma^2 = 0.01$.

1.4.2 The Gaussian or Normal Probability Distribution

As a matter of fact, a large variety of random processes can be characterized by bell-shaped PDFs, and specifically by a PDF called the *Gaussian distribution*, after the mathematician Karl Friedrich Gauss. There are several ways to formally express a Gaussian PDF, but the most standard one used in probability calculus is the *normal distribution*. It is defined as follows:

$$p(x) = \frac{1}{\sigma\sqrt{2\pi}}\exp\left(-\frac{(x - \langle x \rangle)^2}{2\sigma^2}\right) \tag{1.26}$$

Since the function $\exp(-u^2)$ takes the form of a bell shape centered about the origin $u = 0$, the normal distribution is centered about $x = \langle x \rangle$ with a maximum of $1/(\sigma\sqrt{2\pi})$. For $x = \pm\sigma$, the PDF is equal to $\exp(-1/2)/(\sigma\sqrt{2\pi})$, which is about 0.6 times lower than the peak value. See the Exercises at end of chapter for other characteristics.

1.4.3 Eye Diagram of Binary Signals

Let us consider now a digital signal waveform, namely a string of 1/0 bits, which has accumulated a certain amount of background noise. Figure 1.20 shows an

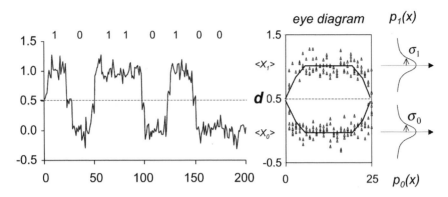

FIGURE 1.20 Left: string of binary pulses with equal intensity noise in the 1 and 0 bits ($\sigma = 0.1$). Right: corresponding eye diagram, decision level d, 1/0 symbol mean values $\langle x_0 \rangle$, $\langle x_1 \rangle$, and symbol probability distributions $p_0(x)$, $p_1(x)$ with standard deviations σ_0, σ_1.

example of such a bit sequence with maximum intensity of unity and a noise background of standard deviation $\sigma = 0.1$, chosen as equal for the 1 and 0 bits. Because the intensity noise is not too important, we can still easily distinguish the 1 and the 0 bits. For each bit observed, the operation that we visually perform is a decision whether the bit intensity is above (bit = 1) or below (bit = 0) a certain threshold d, which here is located somewhere in the middle. If the intensity noise were more important, it could be much more difficult to decide on the actual bit values, and this decision process could inherently be mistaken. A way to visualize this problem of bit-value interpretation is to superpose the bit waveforms onto one another. This results in a plot having a width equal to the bit period, which is called the *eye diagram*, as shown in Figure 1.20. As the name indicates, the resulting plot takes the shape of an eye with minimum and maximum corresponding to the mean intensity values, $\langle x_0 \rangle$, $\langle x_1 \rangle$. The left and right sides of the eye correspond to the rising and trailing edges of the bits, respectively. The clear region in the center, or aperture, is defined as the eye's opening. Figure 1.21 shows for illustration an eye wide-open (RZ data), and the corresponding PDF as monitored and completed with a digital sampling oscilloscope.

The eye can be more or less open, depending upon the importance of the noise in the upper/lower levels and rising/trailing edges. The PDFs associated with each symbol, $p_0(x)$, $p_1(x)$, determine the degree of *eye opening*. Contrary to what its name seems to imply, the eye opening is not a function of the distance $\Delta = \langle x_1 \rangle - \langle x_0 \rangle$, but of this distance relative to the total noise $\sigma_0 + \sigma_1$. Clearly, a relatively large total noise, i.e., $\Delta/(\sigma_0 + \sigma_1) < 1$ would give an eye having practically no blank zone between the 0 and 1 levels, unlike the ones shown in Figures 1.20 and 1.21. In this case, this corresponds to *eye closure*, the least wanted feature in a binary communication system.

The previous considerations represent an advanced introduction to the much simpler concept of *signal-to-noise ratio*, or SNR. Indeed, for a waveform of mean

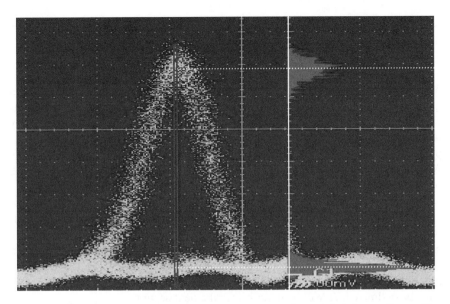

FIGURE 1.21 Return-to-zero eye diagram, as monitored from a digital sampling oscilloscope, showing relatively wide eye opening and associated probability distributions (right) for the 0 (bottom) and 1 (top) symbol bits.

power $\langle x \rangle$ and noise σ, the *linear* SNR is simply defined as:

$$SNR = \frac{\langle x \rangle^2}{\sigma^2} \qquad (1.27)$$

The SNR is commonly defined in the *decibel* scale. Hence,

$$SNR = 10 \log_{10} \frac{\langle x \rangle^2}{\sigma^2} \equiv 20 \log_{10} \frac{\langle x \rangle}{\sigma} \qquad (1.28)$$

It is seen from the decibel definition that for the linear SNR one can indifferently use the definition $SNR_1 = \langle x \rangle / \sigma$ or $SNR_2 = \langle x \rangle^2 / \sigma^2$. However, it must be recalled that the decibel SNR involves noise power (σ^2) and therefore the factor of "2" (according to $2 \times 10 \log_{10} (SNR_1)$), must not be forgotten when converting into decibels the definition $SNR_1 = \langle x \rangle / \sigma$. As we have seen from the example of binary signals, the analysis does not reduce to a mere SNR evaluation. This is because we have two different SNRs to consider, i.e., one for each binary symbol. Intuitively, we can already infer that the right SNR-like definition to use in this case is $SNR^* = \Delta / (\sigma_0 + \sigma_1)$, which represents a measure of eye opening.

We have discussed the issue of amplitude/intensity noise. What about phase and frequency noise? A first consideration is that frequency noise can be identified to random deviations in the signal phase. Indeed, consider the waveform

$$f(t) = A \sin[(\Omega + \omega)t + \Phi] \qquad (1.29)$$

where Ω, Φ are the nominal frequency and phase, and ω a random frequency deviation or noise. We can rewrite this definition as

$$f(t) = A \sin[\Omega t + \varphi(t)] \tag{1.30}$$

where $\varphi(t) = \Phi + \omega t$ is a time-dependent phase noise (since ω is random). Thus frequency noise converts into phase noise, and the reverse is also true. If a waveform is characterized by a given phase noise $\varphi(t)$, it can be associated with the frequency noise $\omega = d\varphi/dt$, where d/dt means the time derivative. As Figure 1.18b illustrates, phase (or frequency) noise results in random time shifts in the signal envelope. Assume now a digital waveform similar to that shown in Figure 1.18, but with only phase noise. The result is a string of "perfect" square-shaped pulses, but with random positions about the nominal bit-frame centers. We would then get an eye diagram with perfectly defined top and bottom levels, but noisy rising and trailing edges. The decision as to identifying 0 or 1 bits would not depend on amplitude being below or above a power level, but on the bit center being below or above a certain arrival time. This is referred to as the phase decision level. When both phase and intensity noise are present, both phase and intensity decision levels must be optimized, as will be described in the next section.

■ 1.5 BINARY TRANSMISSION AND DETECTION

The previous section has shown that amplitude (or intensity) noise degrades signal quality and, ultimately, leads to errors in the interpretation of bit values. We shall focus now on the actual process whereby 1/0 bit values are *encoded* into a waveform, *transmitted* through a line, then *received* at the line's end and reconverted into the original data. For simplicity we shall only consider *binary transmission*, for which signals only have two possible values (mark and space).

1.5.1 Transmission System Elements

The generic layout of a binary transmission system is shown in Figure 1.22. The whole system comprises a *transmitter*, a *line* (or *trunk*) and a *receiver*. We shall closely examine the features of each of these elements.

Transmitter. Electrical bits to be transmitted can represent voice, data or video signals with different coding algorithms. The bits are spaced in time by the period T_{bit}, corresponding to the *bit rate* $B = 1/T_{bit}$. The output consists in bit sequences (N bits) of various lengths (NT_{bit}). This signal, referred to as *baseband*, is used to modulate a *carrier*, which can be electrical, radio-frequency (RF), microwave or light-wave. The modulated carrier is referred to as the *bandpass* signal. A key feature of the *modulator* is its ability to make the carrier's envelope a faithful replica of the input bit stream. The modulator's response must actually be somewhat greater than $f = B$, in order not to introduce distortion in the resulting signal. Indeed, B is only the maximum rate at which bit values are changing (0 to 1 and 1 to 0). But this change takes place in rising and falling times that are substantially shorter than

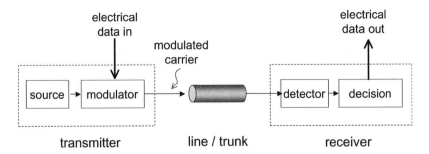

FIGURE 1.22   Generic layout of a transmission system, showing transmitter, line/trunk and receiver constitutive elements.

the bit period, for example, $T_{\text{rise}} = T_{\text{fall}} = T_{\text{bit}}/5$. This means that the modulator's input signal has frequency tones up to $f' = 5B$. Therefore, the modulator's frequency response should not cut off the higher frequency components of the signal. Likewise, bit sequences also exhibit low-frequency tones, which are generated by long strings of 0s or 1s. For instance, an equal-value bit sequence of 100 bits corresponds to a frequency component at $f'' = B/100$. The low-frequency cut-off of the modulator's response should also be adapted to the lowest rate at which long equal-value bit sequences can occur. This illustrates the need for special coding algorithms (e.g., Manchester) where such occurrences can be avoided (see Section 1.2).

The carrier to be encoded is generated by an oscillator or monochromatic *source*. In optical or radio systems, for instance, the carrier source is a *laser* or a *microwave* oscillator. The carrier frequency f_c should be such that $f_c > B$, i.e. individual bits spread over several carrier periods. The most powerful communication systems are optical, since the carrier frequency (at $\lambda = 1.55\,\mu\text{m}$ wavelength) is $f_c = c/\lambda = 193.5\,\text{THz}$ or $193{,}500\,\text{GHz}$ ($c = 3.0 \times 10^8\,\text{m/s}$ being the speed of light). This compares with micro-wave or radio carriers used in mobile and satellite communications, which range from 1 GHz to 6 GHz.

Line or trunk.   The transmission line (trunk) is the physical element that carries with more or less fidelity the bandpass signal to the other end of the system. It can be *free space*, as in mobile or satellite communications, or *guiding*, as in coaxial cables or optical fibers. In either case, several unwanted phenomena serve to degrade the signal quality, namely the intrinsic waveform SNR. The first cause is *attenuation*, or *loss*. A certain fraction of the signal is unavoidably absorbed or scattered by the medium, resulting in power loss. The attenuation rate is expressed in *decibels per kilometer*, as defined by the formula:

$$\text{Loss}_{\text{dB/km}} = \frac{1}{L_{\text{km}}} 10 \log_{10}\left(\frac{P_{\text{out}}}{P_{\text{in}}}\right) \tag{1.31}$$

where L_{km} is the trunk length (in km) and $P_{\text{in,out}}$ are the input/output powers. Because the loss rate can be significant, the trunk length should not exceed some maximum distance. The system transmission distance, or haul, can however be expanded by use of *signal repeaters* along the line, which periodically regenerate, resynchronize and reshape the signal (see the next subsection). Other causes of signal degradation from transmission are: *additive noise, nonlinearity noise* and *waveform dispersion.* Additive noise comes from *in-line signal amplification,* not to be confused with the action of signal repeaters or regenerators. In-line amplification (RF or optical) periodically boosts the signal power to compensate the trunk attenuation. This does not improve the other waveform characteristics. Rather, the process introduces amplification noise, which accumulates at each stage and degrades the SNR. *Nonlinearity* is another source of noise, which is an effect of the signal interaction with the medium, and through the medium's intermediate, with other multiplexed channels. The result of nonlinearity is random power loss or fluctuations, and waveform distortion. Finally, *dispersion* reflects the fact that the different frequency tones making up the signal waveform do not travel at the same velocity. In optical systems, the dispersion, D, is measured in picoseconds per nanometer per kilometer, or ps/nm/km. After propagation over distance L, the time separation experienced by two frequency tones having a wavelength difference $\Delta\lambda$ is given by the relation

$$T_{ps} = D_{ps/nm.km}L_{km}\Delta\lambda_{nm} \tag{1.32}$$

Thus, dispersion causes the bits to spread over time, mingle into each other and eventually vanish. In lightwave systems, this linear dispersion effect can be exactly cancelled by using fiber trunks with alternatively opposing dispersion coefficients D. However, the simultaneous or combined effects of dispersion and nonlinearity cannot be exactly compensated, resulting in unavoidable SNR degradation. Altogether, (dispersion-compensated) communication-systems trunks must be regarded as both *lossy* and *noisy transmission channels.* See the following subsection for loss budget analysis.

Receiver. standard receivers are made of a *detector* and a *binary decision circuit.* The detector converts the modulated signal carrier (bandpass signal) into the original data waveform (baseband signal). This process is called *demodulation,* or *detection,* for short. The detection system performance and complexity depends upon the modulation format type (see following subsection). The decision circuit, as its name indicates, analyzes the detected waveform and makes a decision on the received bit value. The circuit output is the same string of data bits as the one input to the transmission system, except for mistaken bits, called *bit errors.* As we shall see below the *bit-error rate* (BER), is a fundamental measurement of the overall system's transmission quality. See more on the detection and decision principles in the following subsections.

The system's receiver can be directly connected to another transmitter. The combined receiver/transmitter apparatus is called *transceiver* or *regenerator.* If transceivers are periodically placed along a transmission line in order to extend the system's haul, they are called *repeaters* (or *in-line regenerators*). Transmission systems are generally made of two links in order to ensure bi-directional communication, as with telephone signals. Thus, each end of the trunk comprises both transmitter and receiver, and the pairing apparatus is called a *transponder.*

The transmission system can be made to handle several signal channels, each one using a different carrier. In Section 1.1, we described the principle of *frequency-division multiplexing* (FDM) for analog voice channels. The FDM principle can also be applied to binary data. In this case, individual channels are characterized by their own carrier frequencies, the combination of which results in a *frequency comb* of equally spaced bandpass tones. In lightwave systems, the FDM channel separation is greater than a few GHz, corresponding to nanometers or fractions of nanometers (1 nm = 125 GHz at 1.55 μm wavelength). In this case, it is customary to refer to the multichannel approach as *wavelength-division multiplexing*, or WDM. If the WDM channels are as closely spaced as a fraction of a nanometer, the approach is called *dense wavelength-division multiplexing*, or DWDM.

The complete layout of a FDM/WDM, bidirectional transmission system is shown in Figure 1.23. It is simply the combination of many single-channel, bidirectional systems, but with the trunk resource being shared by all the channels. This is done by using multiplexers (MUXs) and demultiplexers (DMUXs) at transmitting and receiving ends, respectively. The total system bit rate, also called *system capacity*, is then given by $N \times B$, where N is the number of channels and B the baseband or channel bit rate. If the system uses cables, as opposed to free space, for the transmission trunks, the total capacity can be further enhanced by combining several trunks (e.g., 2 to 100) into a single cable. The approach, which is extensively used in high-capacity lightwave systems is called *space-division multiplexing* or SDM. For high-capacity systems, a trade-off must be found between SDM (number of parallel trunks) and FDM/WDM (number of frequency/wavelength channels). If cost considerations of transceivers are ignored, the FDM/WDM approach is more economical and bandwidth-efficient. Another possibility is to increase the FDM/WDM bit

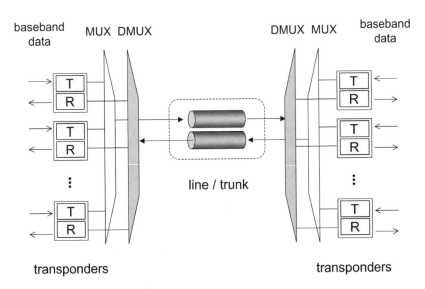

FIGURE 1.23 Generic layout of a multi-channel (FDM or WDM) transmission system, showing transponders, MUX/DMUX devices and line/trunk constitutive elements.

rate by using *electric time-division multiplexing* or TDM. With TDM, electrical data corresponding to different signal channels are interleaved in a bit-to-bit or byte-to-byte fashion, as described in Chapter 2. Thus, hundreds, or hundreds-of-thousands of binary voice channels (64 kbit/s) can be time-multiplexed to form high-bit rate aggregates with rates of up to tens of Gbit/s. The resulting data can then be transmitted by optical or radio systems, each aggregate corresponding to a single frequency of wavelength carrier. Multiplexing these carriers into FDM or WDM aggregates enables one to achieve ten-fold to one hundred-fold capacity increases, leading to terabit (Tbit/s) optical systems (see Desurvire, 2004).

1.5.2 Direct-Detection Binary Receivers

We consider first the so-called *direct-detection* (DD) binary receivers. The DD process can be broken down into three successive stages: signal *detection*, bit *integration* and bit-value *decision*, as summarized in the detailed layout in Figure 1.24. The function of detection is to convert the bandpass signal (modulated carrier) into the baseband signal (original data bits). In DD, the input signal modulates the current of the receiving circuit. In optical systems, for instance, a semiconductor *photodiode* is used as the current modulator. The process is referred to as *square-law detection*, because it is equivalent to an instantaneous amplitude (A) to power (P) conversion (as we have seen in Section 1.1, $P(t)$ is proportional to $A^2(t)$).

The resulting signal current $(I(t) \propto P(t))$ is then passed through an analog pre-amplifier stage. A fraction η of the output (typically 50%) is used to feed a *clock recovery* (CR) circuit. The recovered clock is a sinusoidal signal whose frequency is equal to the baseband frequency or bit rate (B). A simple CR scheme consists of passing the signal $\eta I(t)$ through a bandpass filter centered at $f = B/2$, followed by electronic frequency doubling. This is because NRZ signals do not have frequency components at $f = B$. This is not the case for RZ signals, however, and CR can therefore be done directly at the baseband rate B. Other CR schemes

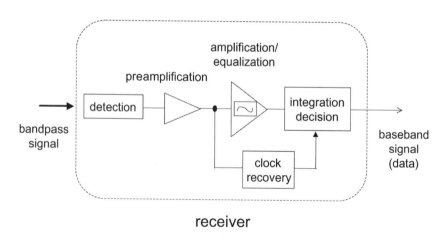

FIGURE 1.24 Principle of direct binary detection.

involve a *phase-locked loop* (PLL). In PLLs, an oscillator is forced to tune to $f = B/2$ or B and exactly synchronize to the bit frequency by means of a feedback signal.

The other fraction of the preamplified signal $((1 - \eta)I(t))$ is passed into a second analog amplification stage. This stage has three functions, which are *power amplification, automatic gain control* (AGC) and *spectral equalization*. Power amplification first brings the signal to the desired processing level. The function of AGC is to compensate for input power fluctuations and deliver the same signal power at the output. Spectral equalization makes it possible to correct any frequency-dependent power loss that has occurred during signal transmission, detection and preamplification. The two analog amplification stages necessarily introduce *electronic noise*, which adds to the noise of the input signal current. It is usually referred to as *Johnson* noise (see Desurvire, 2004).

The output signal is then fed to a circuit for power *integration* and bit-value *decision*. This circuit is driven by the extracted clock. During a full bit period, the bit power is integrated, resulting in a voltage signal which grows as a linear function of the accumulated power. Consider for instance ON/OFF NRZ or RZ modulation. If the received bit is a "1" (signal ON or presence of a pulse), the resulting voltage should be maximum, or $V_{int} = V_{max}$. If the bit is "0" (signal OFF, or absence of a pulse), the voltage should be minimum, $V_{int} = V_{min}$. Why "should be"? Because in the presence of noise, the ON and OFF bit powers are randomly fluctuating about some mean values. Such a fluctuation is reflected in the integrated signal voltage V_{int} which takes therefore V_{max} and V_{min} as average values. After such an integration, comes the time for decision. The decision circuit has for input the analog signal V_{int}, which is approximately defined by the amplitude range V_{max}, V_{min}. The circuit output is a noiseless, binary signal exactly defined by $V_{dec} = 0$ or $V_{dec} = 1$. Such an output is generated by a *flip–flop* device. The flip–flop compares the analog input V_{int} with a *decision* signal level V_d. It switches its output to $V_{dec} = 1$ if $V_{int} \geq V_d$ and to $V_{dec} = 0$ in the contrary case, $V_{int} < V_d$. Typically, the decision level is set to $V_d = (V_{max} - V_{min})/2$. However, if there is more noise associated to one of the bit levels, the decision level must be lowered or raised from the mid value. In any case, it can happen that, because of the noise associated with a bit, the integrated signal V_{int} exceeds or falls below the decision level, while the original bit was a 0 or a 1, respectively. As a result a *decision error* is made. As previously mentioned, the quality rating of transmitted/received signals is the number of such bit errors per number of detected bits, or bit-error-rate (BER). The international standard for so-called *error-free* transmission/reception is BER $\leq 10^{-9}$, meaning that not more than one received bit per billion is erroneous. As will be described in Section 1.6, there are *error-correction codes* (ECC) which bring higher BERs to the error-free performance level (or, more generally, decrease overly high BERs).

The receiver function and signal-processing operations described above are relatively complex, both in principle and in actual hardware implementation. At high bit rates (10–40 Gbit/s), each of the electronic subcircuits must be carefully designed for optimal overall performance. Such an optimization is made relatively difficult by the fact that each of the building blocks has its own finite bandwidth and frequency response.

The DD receiver is of minimal complexity when using a format such as ON/ OFF NRZ or ASK. FSK and PSK can also be used in DD without significant receiver design change. FSK signals can be converted into ON/OFF ASK signals by use of a narrow band-pass filter placed in front of the receiver. The filter center frequency f_0 is made to match the frequency of one of the two FSK tones. The filter's rejection of the unwanted tone must be chosen sufficiently high (e.g., -30 to -40 dB) to ensure minimal power in the "0" bits and optimal eye opening. PSK signals have a constant amplitude and phase jumps $\pm\phi$. The phase jumps can be made equal to $\pm\pi/2$ so that bits of opposite value correspond to out-of-phase carriers. Such a PSK signal is also equivalent to bipolar ASK. The signal can then be converted into ON/OFF (unipolar) ASK if a differential coding is used. In differential coding, the bit value is changed with respect to the previous bit if it is a "0" and is not changed if it is a "1", as shown in Figure 1.25. If bits are re-combined two by two by means of a simple one-bit-delay line, the out-of-phase adjacent bits will destructively interfere (zero amplitude and power) and the in-phase bits will constructively interfere (non-zero amplitude and power), which corresponds to an ON/OFF ASK signal. There are several reasons to use constant-envelope formats such as FSK and PSK/(bipolar-ASK), as opposed to NRZ/(unipolar-ASK). The purpose of constant-envelope modulation is to prevent nonlinear interference between bits and multiplexed channels. For optical signals, however, constant-envelope signals can be unstable in certain dispersion and nonlinearity regimes. The value of PSK is that the signal is made of a single carrier tone, as opposed to FSK (two tones) and NRZ (carrier plus two side-bands). Having a narrow, single-tone spectrum is advantageous for transmission over dispersive trunks, since the effect of time broadening is negligible. These considerations illustrate some of the options offered by binary coding in direct detection. The number of options is actually far greater in the case of *coherent* detection, which is described in the next subsection.

We focus next on the issue of BER characterization in DD receivers. We have seen that bit-errors are caused by fluctuations in the received bit voltages. We have also seen that these fluctuations are characterized by probability distribution functions, or PDFs. Let $x = V_{int}$ be the random variable associated with the integrated voltage, $p_1(x)$ be the PDF associated with the "1" bits and $p_0(x)$ be the PDF associated with the "0" bits. The corresponding mean and standard deviations are

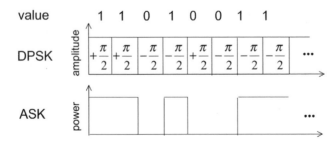

FIGURE 1.25 Top: binary coding with differential phase-shift keying (DPSK). Bottom: conversion into ON/OFF ASK after combining adjacent bits through a one-bit delay line.

$(\langle x_0 \rangle, \sigma_0)$ and $(\langle x_1 \rangle, \sigma_1)$, respectively. Figure 1.26 shows a plot of the two PDFs, which corresponds to the eye diagram in Figure 1.20. The decision level, $x = d$, is set somewhere between the two mean values $\langle x_0 \rangle$, $\langle x_1 \rangle$. It is seen that $p_0(x)$ and $p_1(x)$ overlap to some extent. In terms of probabilities, such an overlap means that it is possible for "0" or "1" bits to yield voltages above or below the decision level, respectively, which in both cases leads to a *decision error*. The probabilities associated with these two possible events, namely $p_0(x > d)$ and $p_1(x < d)$, are given by the shaded *areas* shown in Figure 1.26. The total bit-error probability, or BER, is thus defined by

$$\text{BER}(d) = \frac{1}{2}[p_0(x > d) + p_1(x < d)] \qquad (1.33)$$

In this expression, the factor $\frac{1}{2}$ accounts for the fact that the probabilities for incident bits to take the values "0" or "1" are both equal to $\frac{1}{2}$, assuming sufficiently long NRZ sequences.

Thus the BER is a function of the decision level d, and one must find the value for which the function is minimum, which requires some careful analysis. The analytical definition of the two functions involved in equation (1.33) depends upon the actual PDF types. For Gaussian PDFs, the definition turns out relatively simple. Minimization of the BER then yields the *optimum decision level*:

$$d_{\text{opt}} = \frac{\sigma_1 \langle x_0 \rangle + \sigma_0 \langle x_1 \rangle}{\sigma_1 + \sigma_0} \qquad (1.34)$$

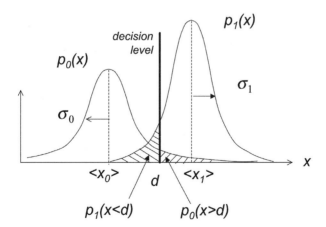

FIGURE 1.26 Plot of the probability distributions $p_0(x)$ and $p_1(x)$ associated with "0" and "1" bits, showing finite overlap in the decision region. The shaded areas correspond to the probability of "1" bit being mistaken for a "0" bit, or $p_1(x < d)$ (left), and the reverse case, or $p_0(x > d)$. By definition, the bit-error rate is given by the half sum of the shaded areas.

The minimum BER is then given by the following expression:

$$\text{BER}_{\text{min}} \equiv \text{BER}(d_{\text{opt}}) \approx \frac{1}{Q\sqrt{2\pi}} \exp\left(-\frac{Q^2}{2}\right) \tag{1.35}$$

where the parameter Q, referred to as the *Q-factor*, is defined by

$$Q = \frac{\langle x_1 \rangle - \langle x_0 \rangle}{\sigma_1 + \sigma_0} \tag{1.36}$$

It is seen from this result that, under the Gaussian-PDF (or Gaussian noise) assumption, the BER lends itself to a remarkably simple definition. The function $\text{BER}_{\text{min}} = f(Q)$ is rapidly decaying with Q. It can easily be checked that the value of $\text{BER} = 10^{-9}$, corresponding to error-free operation, is achieved when $Q \approx 6$. The Q-factor is also defined in decibel units, as follows:

$$Q^2_{\text{dB}} = 10 \log_{10} Q^2_{\text{linear}} \equiv 20 \log_{10} Q_{\text{linear}} \tag{1.37}$$

Thus, error-free operation corresponds to $Q^2_{\text{dB}} = 20 \log_{10}(6) = +15.5\,\text{dB}$.

The above analysis concerned the BER minimization with respect to decision level. As previously mentioned, phase noise will also be present in the received signal waveform. Therefore, the time-phase t_d at which bit-integration starts must also be carefully determined so that minimal BER is achieved. This determination can be done by measuring the BER while the phase t_d is continuously varied over a one-bit period. If the PDF mean and variances are not exactly known (as is generally the case for actual systems), the same operation can be performed with the decision level, d. In the general case, therefore, BER minimization is a two-dimensional optimization problem.

The search for optimum decision and phase levels can actually be performed without going through BER measurements over the full range of associated intervals. This is quite fortunate, since actual transmission systems usually operate at error rates significantly below $\text{BER} = 10^{-9}$. Such two-dimensional measurements could therefore be extremely long and tedious to implement (see exercise at end of chapter). The method for BER and optimal decision/phase evaluation is referred to as *Q-factor extrapolation*. An example of implementation is provided in Figure 1.27. The BER is only measured at relatively high values (e.g., $\text{BER} = 10^{-10} - 10^{-5}$), which is obtained in two regions located below and above the optimum decision level, respectively. The two families of data points are then fitted with an algorithm which assumes a given PDF type (e.g., Gaussian). The crossing of the two fitting curves locates the minimal BER and associated optimal decision level.

1.5.3 Coherent Detection

The principle of *coherent detection* is more complex than that of direct detection. It is because the detection is sensitive to both the amplitude and phase of the received signal. The principle is to mix the incoming signal with a strong, unmodulated wave called the *local oscillator* (LO), prior to detection. The LO frequency can

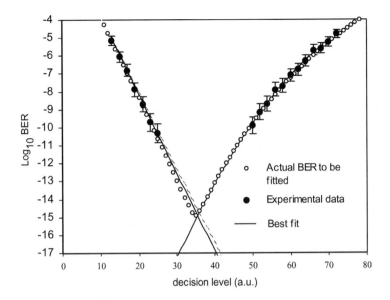

FIGURE 1.27 Bit-error-rate extrapolation method. The actual BER (open symbols) is measured over a limited decision interval (dark symbols). A fitting algorithm is used to extrapolate the BER and determine the minimal BER value, $\text{BER}_{\min}$ with corresponding optimal decision level, d_{opt} (here $\text{BER}_{\min} = 10^{-15}$ and $d_{\text{opt}} = 35$).

be chosen equal to that of the signal carrier. In this case, the scheme is referred to as *homodyne* detection. If the LO frequency is different, this is referred to as *heterodyne* detection.

Consider an amplitude-modulated (ASK) input signal defined as

$$f_s(t) = A_s(t) \sin(\omega_s t + \varphi_s),$$

and an LO wave defined as

$$f_{\text{LO}}(t) = A_{\text{LO}} \sin(\omega_{\text{LO}} t + \varphi_{\text{LO}})$$

where A_{LO} is a constant amplitude. The signal and LO waves are summed and input to a square-law detector. As its name indicates, the detector's current is proportional to the power of the incident wave, as defined by:

$$
\begin{aligned}
P(t) &= [f_s(t) + f_{\text{LO}}(t)]^2 \\
&= [A_s(t) \sin(\omega_s t + \varphi_s) + A_{\text{LO}} \sin(\omega_{\text{LO}} t + \varphi_{\text{LO}})]^2 \\
&= A_s^2(t) \sin^2(\omega_s t + \varphi_s) + A_{\text{LO}}^2 \sin^2(\omega_{\text{LO}} t + \varphi_{\text{LO}}) \\
&\quad + 2A_s(t)A_{\text{LO}} \sin(\omega_s t + \varphi_s) \sin(\omega_{\text{LO}} t + \varphi_{\text{LO}})
\end{aligned}
\tag{1.38}
$$

To further develop the above expression, we use the formulae

$$\sin a \sin b = [\cos(a - b) - \cos(a + b)]/2$$

and

$$\sin^2 a = [1 - \cos 2a]/2,$$

which gives:

$$P(t) = \frac{A_s^2(t)}{2}[1 - \cos(2\omega_s t + 2\varphi_s)] + \frac{A_{LO}^2}{2}[1 - \cos(2\omega_{LO} t + 2\varphi_{LO})]$$

$$+ A_s(t)A_{LO}\cos[(\omega_s - \omega_{LO})t + (\varphi_s - \varphi_{LO})]$$

$$- A_s(t)A_{LO}\cos[(\omega_s + \omega_{LO})t + (\varphi_s + \varphi_{LO})] \tag{1.39}$$

which we rewrite in the form:

$$P(t) = \frac{1}{2}[A_s^2(t) + A_{LO}^2 + 2A_s(t)A_{LO}\cos(\omega_{IF}t + \Delta\varphi)]$$

$$+ g(2\omega_s, 2\omega_{LO}, \omega_s + \omega_{LO}) \tag{1.40}$$

where we have introduced $\omega_{IF} = \omega_s - \omega_{LO}$ and $\Delta\varphi = \varphi_s - \varphi_{LO}$. The function $g(2\omega_s, 2\omega_{LO}, \omega_s + \omega_{LO})$ represents the rest of the oscillating terms.

The result we have obtained shows that the detected wave power has three features:

1. An oscillating term at the signal/LO frequency difference, ω_{IF}, which we call the *intermediate frequency* or IF.

2. A second group of terms oscillating at double frequencies or higher: $2\omega_s, 2\omega_{LO}, \omega_s + \omega_{LO} \equiv 2\omega_{LO} + \omega_{IF}$

3. A time-dependent term of magnitude $A_s^2(t) + A_{LO}^2$

We assume next that the electronic circuit associated with the detector has a limited bandwidth and blocks frequencies above ω_{IF}. We also assume that LO power A_{LO}^2 is substantially greater than the modulated signal power $A_s^2(t)$, so that we can make the approximation $A_s^2(t) + A_{LO}^2 \approx A_{LO}^2$. Under these two assumptions, the detector's current $I(t) \propto P(t)$ is simply defined by

$$I(t) \propto A_{LO}^2 + 2A_s(t)A_{LO}\cos(\omega_{IF}t + \Delta\varphi) \tag{1.41}$$

This result shows that the baseband signal, $A_s(t)$, is carried by the IF tone. The signal can be retrieved if we pass the current through a filter centered at ω_{IF}. There are two possible choices for the IF:

1. With homodyne detection, $\omega_{IF} = 0$ and $\tilde{I}_{ASK}(t) \propto A_s(t)A_{LO}\cos(\Delta\varphi)$. The signal/LO phase difference $\Delta\varphi$ must be chosen so that the cosine term is maximized, $|\cos(\Delta\varphi)| = 1$, or $\Delta\varphi = \varphi_s - \varphi_{LO} = \pm\pi$, which gives $\tilde{I}_{ASK}(t) \propto A_s(t)A_{LO}$. Such a phase adjustment must be precise and constant, which requires the implementation of a *phase-locked loop* (PLL), or

2. With heterodyne detection ($\omega_{IF} \neq 0$) the oscillating signal must be time-averaged, which results in $\tilde{I}_{ASK}(t) \propto A_s(t)A_{LO}\sqrt{\langle\cos^2(\omega_{IF}t + \Delta\varphi)\rangle} = A_s(t)A_{LO}/\sqrt{2}$. This corresponds to a 3 dB difference or "penalty" in signal

current power $\tilde{I}_{\text{ASK}}^2(t)$, as compared to the homodyne case (see the following further discussion)

We note that in both homodyne and heterodyne detections, the LO acts as a signal amplifier, since the output current power is proportional to the LO wave power, A_{LO}^2, which can be made arbitrarily high. Thus, the resulting SNRs are significantly higher than in the direct-detection case.

The same calculation can be carried out with FSK signals. Note that with FSK, only heterodyne detection applies. If the detector current is passed through a narrow-band filter centered about one of the two FSK tones ($f_1 = \omega_s - \delta f, f_2 = \omega_s + \delta f$), it is easily found that the resulting current is of the same form as in the ASK case, with the IF being $\omega_{\text{IF}} = \omega_s \pm \delta f - \omega_{\text{LO}}$. Therefore, the conclusions for heterodyne FSK are the same as for heterodyne ASK.

The same calculation as previously also applies to PSK signals. The input signal is now defined as $f_s(t) = A_s \sin[\omega_s t + \varphi_s(t)]$, and the resulting detector current is

$$I(t) \propto A_{\text{LO}}^2 + 2A_s A_{\text{LO}} \cos[\omega_{\text{IF}} t + \Delta\varphi(t)] \tag{1.42}$$

where $\Delta\varphi(t) = \varphi_s(t) - \varphi_{\text{LO}}$. We then have two possible choices:

1. With homodyne detection ($\omega_{\text{IF}} = 0$), we have $\tilde{I}_{\text{PSK}}(t) \propto A_s A_{\text{LO}} \cos[\varphi_s(t) - \varphi_{\text{LO}}]$. One must chose the LO phase such that the cosine term switches between -1 and $+1$, for example $\varphi_{\text{LO}} = -\pi/2$ with $\varphi_s(t) = \pm\pi/2$. Such a phase adjustment must be precise and constant, which requires the implementation of a PLL.

2. With heterodyne detection ($\omega_{\text{IF}} \neq 0$) the oscillating signal must be time-averaged, which results in $\tilde{I}_{\text{PSK}}(t) \propto A_s(t) A_{\text{LO}} \sqrt{\langle \cos^2[\varphi_s(t) - \varphi_{\text{LO}}]\rangle} = \pm A_s(t) A_{\text{LO}}/\sqrt{2}$. This corresponds to a 3 dB penalty in signal current power $\tilde{I}_{\text{PSK}}^2(t)$, as compared to the case of homodyne PSK (see the following further discussion).

We now compare ASK and PSK in terms of Q-factors. As we have seen earlier, bipolar ASK is the same as PSK. Therefore, one must make the comparison only with unipolar ASK, which corresponds to ON/OFF keying. With ON/OFF keying, the amplitudes of the incident 0 and 1 bits are $A_S(\text{``0''}) = 0$ and $A_S(\text{``1''}) = A_{\text{max}}$, respectively. Since 0 and 1 bits have the same probability of occurrence ($\frac{1}{2}$), the average signal power incident upon the detector is $\langle P \rangle = A_{\text{max}}^2/2$, which gives $A_{\text{max}} = \sqrt{2\langle P \rangle}$. For homodyne coherent ASK(ON/OFF), the Q-factor is thus:

$$Q_{\text{ASK}} = \frac{\langle \tilde{I}(\text{``1''})\rangle_{\text{ASK}} - \langle \tilde{I}(\text{``0''})\rangle_{\text{ASK}}}{\sigma_1 + \sigma_0} \equiv \frac{\sqrt{2\langle P \rangle} A_{\text{LO}}}{\sigma_1 + \sigma_0} \tag{1.43}$$

Consider next PSK. Since the modulation has a constant envelope, the average signal power incident upon the detector is $\langle P' \rangle = A^2$, which gives $A = \sqrt{\langle P' \rangle}$.

For homodyne coherent PSK, the Q-factor is thus:

$$Q_{PSK} = \frac{\langle \tilde{I}("1") \rangle_{PSK} - \langle \tilde{I}("0") \rangle_{PSK}}{\sigma_1 + \sigma_0} \equiv \frac{2\sqrt{\langle P' \rangle} A_{LO}}{\sigma_1 + \sigma_0} \qquad (1.44)$$

Equating the two Q-factors, we obtain

$$Q_{ASK} = Q_{PSK} \Leftrightarrow 2\sqrt{\langle P' \rangle} = \sqrt{2\langle P \rangle} \Leftrightarrow \langle P' \rangle = \langle P \rangle / 2 \qquad (1.45)$$

We see from this last result that in identical noise conditions, *PSK requires half the signal power of ASK (or FSK) to achieve an identical Q-factor (or BER) performance.*

Finally, we compare homodyne and heterodyne detection performance. Let $\langle p \rangle$ and $\langle p' \rangle$ be the mean signal powers used in heterodyne detection of ASK and PSK, respectively, and $\langle P \rangle$, $\langle P' \rangle$ the corresponding mean powers in the homodyne case. The previous results showed that heterodyne detection introduces a $1/\sqrt{2}$ penalty factor in the detector currents. Equating the Q-factors for homodyne/heterodyne detection in ASK and PSK immediately shows that $\langle p \rangle = \langle P \rangle$ and $\langle p' \rangle = \langle P' \rangle$. Thus, it can be concluded that in identical noise conditions, *homodyne detection requires half the signal power as homodyne detection to achieve an identical Q-factor (or BER) performance.*

The mean-power requirements for identical BER in ASK, FSK and PSK are summarized in Table 1.3. Thus, if the mean power P gives a certain BER in homodyne PSK, the required power is two-fold or four-fold in the other approaches or formats. By convention, the input power requirement for achieving a BER of 10^{-9} is called the *receiver sensitivity*.

In light-wave systems, the receiver sensitivity can also be expressed in *photons/bit*. For a bit-average optical power of $\langle P \rangle$ (in Watts), the number of photons/bit is given by

$$\langle n \rangle_{photons/bit} = \frac{\langle P \rangle}{h\nu B} \qquad (1.46)$$

where $h = 6.62 \times 10^{-34}$ (Joules-seconds or Js) is Planck's constant, ν (Hz or s^{-1}) the frequency, and B (bit/s) the bit rate. It can be shown that ideal homodyne PSK receivers have a requirement of *9 photons/bit*. Based on the results in Table 1.3, it is seen that other coherent formats have corresponding sensitivities

TABLE 1.3 **Mean power requirements for achieving identical bit-error-rate performance with homodyne and heterodyne coherent detection using ASK, FSK and PSK signaling**

Format	Homodyne	Heterodyne
ASK	2P	4P
FSK	–	4P
PSK	P	2P

in the 18–36 photons/bit range. In practice, these high sensitivity levels are very difficult to achieve when considering Gbit/s systems This is because some of the electronic components in the receiver (detector, pre- and power amplifiers) introduce noise and power penalties, which cannot be completely removed by use of high-power LOs. Further, some fraction of the received signal is lost because of imperfect signal-to-detector coupling, finite power conversion efficiency, and signal sampling for clock and phase recovery purposes. Each of these losses adds its own penalties, raising the power requirement.

Direct-detection (DD) receivers have sensitivities substantially lower than coherent receivers, because of the same electronic-noise limitations and the absence of a LO as signal booster. In optical systems, a way to overcome this receiver noise is to use *optical preamplification*. Optical preamplification acts in a way similar to LO signal boosting. For this type of DD receiver, the theoretical sensitivity is about 40 photons/bit. In realistic systems, the best achievable sensitivities are in the vicinity of 100 photons/bit.

1.5.4 System Power Budget

As previously described (Figure 1.5a), a communications system is made up of three elementary blocks: transmitter, trunk and receiver. The receiver sensitivity (P^{sens}), sets the minimum signal power which must be available at the receiving end. Since the trunk has a finite transmission loss (T), the power launched at the transmitter end must be set to some value (P^{trans}) determined by both sensitivity and loss. Some extra *power margin* (M) can be given to overcome intrinsic system penalties and temporary/long-term performance degradation causes. The signal power requirement is thus given by the equality, called *power budget*:

$$P^{trans}_{dBm} + T_{dB} = M_{dB} + P^{sens}_{dBm} \tag{1.47}$$

For instance, if the line has a loss of $T = -10\,dB$, the margin is set to $M = +3\,dB$, and the sensitivity is $P^{sens} = -30\,dBm$, the transmitter power must be set to $P^{trans} = +10\,dB + 3\,dB - 30\,dBm = +17\,dBm$.

1.5.5 In-line Regeneration and Amplification

The power budget determines the maximum allowable loss or distance between the transmitter and the receiver. The distance is fixed by the maximum available power at the transmitter end, the best available receiver sensitivity, and the lowest possible trunk loss. In order to cover this distance, there are two possibilities:

1. After detection, a new signal can be re-transmitted from the receiving point. As previously mentioned, the arrangement of a receiver/transmitter in tandem is called a *transceiver* (not to be confused with a *transponder*, which is the pairing of receiver and transmitter ensuring bidirectional communication, see Figure 1.23). This process is called *regeneration*, and the transceiver a *regenerator*. Often, the word *repeater* is equivalently used.

2. The signal can be boosted by an amplifier, without any detection stage. The amplifier gain is set to compensate for the trunk loss, bringing back the signal power to some nominal value. This is called *repowering* or *reamplification*. An amplifier placed on the signal path is referred to as a *line amplifier*. Often, line amplifiers are also called repeaters, but as we shall see, their function is not equivalent to that of a full regenerator.

The basic regenerator layout is shown in Figure 1.28. The output of the decision circuit in the receiver subblock is fed into an amplifying circuit. This circuit drives the modulator in the transmitter subblock. Since the data integrity is completely restored (within a few bit errors), the corresponding output eye diagram is at maximum opening. Intensity and phase noises are minimal, data pulses exhibit no distortion, and the signal power is reset to a nominal level. It is customary to refer to this clean-up process as *3R regeneration*. The three R's stand for repowering (or reamplification), retiming (or resynchronization), and reshaping.

For a bi-directional trunk system, the whole regenerator apparatus must be duplicated to ensure two-way operation. It is clear that because of the amount and sophistication of the electronic components involved, a regenerator is a complex and costly device. Furthermore, it is intrinsically made to process a single baseband signal, as opposed to multiplexed signals. In multiplexed systems, the aggregate signals must first be demultiplexed, passed through their dedicated regenerators, and then remultiplexed. Such a complex apparatus is more appropriate to a system terminal, such as a central node in a communication network. This is because switching and data processing operations can also be made as signals are demultiplexed and regenerated. On the other hand, using such electronic in-line regeneration to extend the system distance is not practical when considering multiplexed signals.

For multiplexed signals, the better approach is in-line amplification. Chapter 2 provides a description concerning amplified analog trunk systems with FDM voice channels. Light-wave systems with WDM signals utilize in-line *optical amplifiers*.

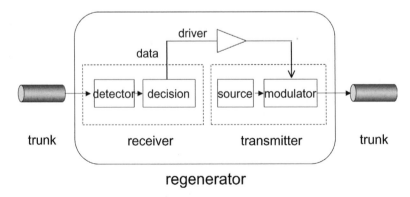

regenerator

FIGURE 1.28 Regenerator layout, showing receiving and transmitting subblocks. The same regenerator apparatus must be duplicated for the other transmission direction.

As we saw above, the single function of a line amplifier is to repower the signals that have been weakened by transmission. Hence, they act as a *1R regenerator*, in contrast with the previous 3R case. It is also customary to refer to (in-line) amplified systems *as unrepeatered*, In a system with k line amplifiers, the power budget is simply:

$$P_{dBm}^{trans} + kT_{dB} + kG_{dB} = M_{dB} + P_{dBm}^{sens} \qquad (1.48)$$

where G is the amplifier gain, set to $G_{dB} = -T_{dB}$ or in linear units $G = 1/T$. The principle of in-line amplification is central to long-haul (LH) and ultra-long-haul (ULH) optical systems. The terms LH and ULH refer to transmission distances in the $500-1{,}000$-km and in the $5{,}000-10{,}000$-km ranges. *Undersea cable systems*, for instance, span unrepeatered distances from 6,500 km to 12,000 km in the Atlantic and Pacific oceans.

1.5.6 Noise Figure of Active/Passive Transmission System Elements

The possibility of periodically repowering signals in a long link does not mean that distance limits have been finally overcome. Indeed, each amplification stage contributes to a small amount of SNR degradation. Each lossy transmission stage also contributes to SNR degradation since the signal is attenuated at constant noise power. The cumulative effect of this active and passive SNR degradation must be taken into account in the system budget. For this, let us define a new parameter called the *noise figure*.

By definition, the noise figure (NF) of a given transmission device or system is the ratio of input to output SNR, i.e.,

$$NF = \frac{SNR^{in}}{SNR^{out}} \qquad (1.49)$$

In decibel units, we have consistently $NF_{dB} = SNR_{dB}^{in} - SNR_{dB}^{out}$. By definition, the NF is greater than unity (dB positive), since the SNR is always reduced or degraded by the passive/active transmission process, giving $SNR_{dB}^{out} \leq SNR_{dB}^{in}$. Thus the NF represents the amount of SNR degradation introduced by devices or transmission elements. Such a degradation is cause by the intrinsic noise they generate and add as a background to the traversing signal. A purely transparent noise-free and loss-free medium has a noise figure of unity, or $NF = 0$ dB. It can be shown that for microwave or optical devices, the NF of passive attenuation is equal to the loss absolute value, i.e.,

$$\begin{cases} NF(loss) = \dfrac{1}{T} \\[2mm] NF_{dB}(loss) \equiv -10\log_{10}\left(\dfrac{1}{T}\right) = -T_{dB} = |T_{dB}| \end{cases} \qquad (1.50)$$

In the case of amplification, all amplifier types (electronic/microwave, optical) generate noise and are therefore characterized by a specific noise figure. The formal definition depends upon the amplifier type.

For microwave amplifiers (or amplifying devices operating in the microwave frequencies domain), the NF is defined through:

$$\left(\begin{array}{l} NF^{RF}(amp) \equiv \dfrac{P_N}{Gk_B T_0 \Delta f} \equiv 1 + \dfrac{T_A}{T_0} \\[12pt] NF^{RF}_{dB}(amp) \equiv 10 \log_{10}\left(\dfrac{P_N}{Gk_B T_0 \Delta f} \right) \equiv 10 \log_{10}\left(1 + \dfrac{T_A}{T_0} \right) \end{array} \right. \tag{1.51}$$

where P_N is the noise power, G is the gain, k_B is the Boltzmann's constant ($k_B = 1.38 \times 10^{-23}$ *Joules/Kelvin* or J/K), T_0 is the absolute "room" temperature ($T_0 = 290$ K) and Δf is the amplifier bandwidth. Defining the *equivalent amplifier temperature* through $T_A = P_N/(Gk_B \Delta f)$, the NF is equal to $1 + T_A/T_0$. For a microwave amplifier whose equivalent temperature is $T_A = T_0$, the NF is equal to 2 dB or 3 dB. Most microwave amplifiers have NFs between 3 dB and 10 dB.

For optical amplifiers (or amplifying devices operating in the optical frequencies domain), the NF has a different definition:

$$\left(\begin{array}{l} NF^{optic}(amp) \equiv \dfrac{1 + 2n_{sp}(G-1)}{G} \\[12pt] NF^{optic}_{dB}(amp) \equiv 10 \log_{10}\left(\dfrac{1 + 2n_{sp}(G-1)}{G} \right) \end{array} \right. \tag{1.52}$$

where G is the gain and n_{sp} is the *spontaneous emission factor*. Ideal optical amplifiers have $n_{sp} = 1$. At high gains $[1 + 2(G-1)]/G \approx 1$, and the NF therefore reduces to 2 dB or 3 dB. Typical optical amplifiers have NFs in the 3 dB to 6 dB range.

Since an amplified transmission line (microwave or optical) is made of a cascade of lossy elements and amplifiers, it is useful to have a definition for the overall line NF. Assume a chain of transmission elements, i, in the sequence order $i = 1, 2, 3 \dots$. Each element is characterized by a net linear transmission (<1) or gain (>1), called H_i, and by a NF, which we call F_i. The net NF of the system is given by the cascading formula:

$$F = F_1 + \frac{F_2 - 1}{H_1} + \frac{F_3 - 1}{H_1 H_2} + \cdots \tag{1.53}$$

This is a very important result. It shows that the overall line noise figure depends on the order in which the different elements are placed. It can be shown that the resulting NF is minimized when the first element of the chain has the lowest NF (see Exercises at the end of the chapter).

Consider finally a *transparent* amplified transmission line. This corresponds to the situation where each amplifier compensates for the loss of the previous (or following) trunk elements. Thus, each element is alternatively defined through $F = $ NF(amp) $\equiv F_{amp}$ or $F = $ NF(loss) $\equiv F_{loss} = 1/T$, with individual transmission $H = G$ (gain) or $H = T$, and with $GT = 1$ (transparency). Assuming for instance that amplification precedes loss, it is easily established from the cascading formula that the NF of a transparent line with k loss-compensated trunks is given by:

$$F = F_{amp} + \frac{F_{loss} - 1}{G} + \frac{F_{amp} - 1}{TG} + \frac{F_{loss} - 1}{GTG} + \cdots$$

$$= kF_{amp} - 1 + k\frac{F_{loss} - 1}{G} \equiv k\left(F_{amp} + 1 - \frac{1}{G}\right) - 1 \qquad (1.54)$$

Expressing the result in terms of decibel SNRs:

$$SNR_{out}^{dB} = SNR_{in}^{dB} - F_{dB}$$

$$= SNR_{in}^{dB} - F_{amp}^{dB} - 10\log_{10} k$$

$$- 10\log_{10}\left[1 + \frac{1}{F_{amp}}\left(1 - \frac{1}{G}\right) - \frac{1}{kF_{amp}}\right]$$

$$\approx SNR_{in}^{dB} - F_{amp}^{dB} - 10\log_{10} k - 10\log_{10}\left(1 + \frac{1}{F_{amp}}\right) \qquad (1.55)$$

In the last equation, the gain, G, and the number of amplified segments, k, are assumed sufficiently high to justify the approximation. From this result, it can be concluded that amplified-line systems degrade the SNR by approximately k times the line-amplifier NF, or equivalently in decibels $\Delta SNR_{dB} = -F_{amp}^{dB} - 10\log_{10} k$. Consider for instance two (microwave or optical) lines of 10 and 100 amplified segments with amplifier noise figure of $F_{amp}^{dB} = 3$ dB. The corresponding SNR penalties are $\Delta SNR_{dB} = -13$ dB and -23 dB, respectively. If the amplified segments are 50 km long, the overall system lengths are 500 km to 5,000 km. This shows that line amplification makes it possible to realize long-haul or very long-haul systems with relatively low SNR degradation.

We conclude this subsection by introducing a useful relationship between the received SNR and the Q-factor. This relationship applies to amplified optical systems, where the received SNR is given by the ratio of the mean incident signal power (under ON/OFF modulation) to the total incident noise power. The two equivalent relationships between SNR and Q-factor are as follows:

$$Q = \sqrt{\frac{2 \cdot SNR}{1 + \frac{1}{2 \cdot SNR} + \sqrt{\frac{1}{2 \cdot SNR}\left(2 + \frac{1}{2 \cdot SNR}\right)}}} \qquad (1.56)$$

or

$$SNR = Q(Q + \sqrt{2})$$

It is seen from the second definition that the SNR is approximately given by Q^2. Since, at low BERs, the Q-factor is not significantly higher than unity, it is better to keep the correct definition, $SNR = Q(Q + \sqrt{2})$. In the case of error-free transmission (BER $= 10^{-9}$), we have $Q = 6$ and the corresponding detection SNR is given by $SNR = 6(6 + \sqrt{2}) = 44.48$ or 16.5 dB. Knowing the received SNR requirement for a given BER performance, it is thus possible to very simply determine the launched power given system parameters such as line transmission loss, distance between amplifiers and amplifier noise figure. These performance estimates assume ideal linear transmission. In practice, dispersion and nonlinearity introduce other penalties, which contribute to a decrease in the BER performance. However, such a basic estimate provides a first and accurate minimum-performance level for system-design reference.

■ 1.6 ERROR-CORRECTION CODING

This section provides a basic introduction to the principles of *error-correction coding* (ECC). The following subsections are extracted from Appendix H in *Desurvire et al. (2002)*, see Bibliography at the end of this book.

Error correction is one of the most intriguing feature of binary communications. The idea is that one can, to some extent, correct the bit-errors made at the receiver end. This requires the transmission, with the signal data (payload) of a certain number of *control bits*, from which errors become detectable and fixed. Alternatively, both payload and control bits can be encoded together, so that they form an ensemble with self-correcting potential. The intriguing feature of ECC is that one can *reverse* the effect of bit-value uncertainty in the noisy detected signals. A good analogy is restoring a bad TV picture to a crisp one. The price to pay for this is extra bandwidth, namely the higher bit rate required to transmit the full ECC signal.

The first approach of the ECC technique is to periodically append to the signal data stream a certain number of *redundancy* bits which, for the same signal payload rate, results in a higher bit-rate signal. The decoder uses the redundancy bits to decide whether signal bits have been properly detected. A most obvious redundancy scheme is to repeat the information a certain number of times, N, increasing the bit rate by the same factor. For instance, one can transmit 1111 for "1" symbols and 0000 for "0" symbols. According to this code, the received sequence

   1011 0010 0111 1111

is immediately interpreted as very likely to represent the initial word

   1111 0000 1111 1111

which was restored following some form of *majority logic*. However, there is no absolute certainty in such a decoding. Indeed if two errors occur in the 4-bit sequence (however less likely the event), receiving 1100 or any other permutation thereof could incorrectly be decoded as 1111 or 0000, while the correct word could have been either 0000 or 1111. In such a case, majority logic fails to correct errors, unless a larger number of redundancy bits is used. It is clear that decoding reliability (BER → 0) grows with increasing redundancy, but at the expense of large bandwidth waste. It is like exchanging phone numbers in a building under construction: repeating the numbers two or three times may not prove to be completely reliable! The difference between human beings and machines is that we have ingrained and powerful ECC faculties. For instance, if during a December evening you read a display in the streets stating

HAMPY HXLIDEYS

you will instantaneously correct the 3 wrong letters, because of some long-educated algorithms of automatic association and inference. But this example is easy because the information is somewhat expected. The same message received in June with some more mistaken letters will not be so easily correctable. In most practical cases, we must use minimal redundancy power to make a noisy communication get through. Usually a single repeat of the message is sufficient to achieve absolute accuracy. (But to get the correct spelling of an unusual last name, this may yet require getting letters one by one!). We shall now examine better and more efficient ways to overcome the situation.

1.6.1 Linear Block Codes

Based on the intuitive arguments previously discussed, we can chose to transmit our information by successive blocks of n bits. The first part of this block is made of k *message bits* which represent the signal payload (or information), corresponding to 2^k possible messages. The second part is made of $m = n - k$ *parity bits* that will be used for coding purposes. The (n, k) bit sequence is referred to as a *linear block code*. We define the *code rate* as the ratio $R = k/n$, which represents the proportion of message bits actually transmitted in the block ($1 - R$ representing the redundancy proportion). In order to provide decoded/corrected data at the initial signal rate the encoded-block bit rate should be increased by the factor $1/R$, corresponding to a percentage additional bandwidth of $1/R - 1$, which is referred to as the percentage *bandwidth expansion factor*. One must then find an appropriate coding for the parity bits which will make it possible, upon decoding, to detect and to correct errors in the received messsage. For this sequential reason, the approach is called *forward error correction,* or FEC. The following description will require some familiarity with basic matrix-vector operations.

How are linear block codes generated? Define the input message bits by the k-vector $X = (x_1 \ldots x_k)$, and the block-code (encoder output) by the n-vector $Y = (y_1 \ldots y_n)$. The block-code bits y_i are calculated from the linear combinations:

$$y_i = g_{1i}x_1 + g_{2i}x_2 + \cdots g_{ki}x_k \qquad (1.52)$$

where g_{lm} ($l = 1 \ldots k$, $m = 1 \ldots n$) are binary-number coefficients. This definition can be put into a vector$-$matrix product, $Y = X\tilde{G}$, or explicitly

$$Y = (y_1 \ldots y_n) = (x_1 \quad x_2 \quad \cdots \quad x_k) \begin{pmatrix} g_{11} & g_{12} & \cdots & g_{1n} \\ g_{21} & g_{22} & & \vdots \\ \vdots & & & \vdots \\ g_{k1} & \cdots & \cdots & g_{kn} \end{pmatrix} \equiv X\tilde{G} \qquad (1.53)$$

The matrix $\tilde{G}$ defined is called the *generator matrix*. It is always possible to rewrite the generator matrix under the so-called *systematic form*:

$$\tilde{G} = [I_k|P] = \begin{pmatrix} 1 & \cdots & \cdots & 0 & p_{11} & p_{21} & \cdots & p_{1m} \\ 0 & 1 & & \vdots & p_{12} & p_{22} & & \vdots \\ \vdots & & & \vdots & \vdots & & & \vdots \\ 0 & \cdots & \cdots & 1 & p_{1k} & \cdots & \cdots & p_{km} \end{pmatrix} \tag{1.54}$$

where I_k is the $k \times k$ identity matrix (which leaves the k message bits unchanged) and P is a $m \times k$ matrix which determines the redundant parity bits. Thus the encoder output $Y = X\tilde{G} = X[I_k|P]$ is a message block of k bits followed by a parity block of $m = n - k$ bits. This arrangement where the message bit-sequence is unchanged by the encoder is called a *systematic code*.

For future use, we define the $n \times m$ *parity-check matrix*:

$$\tilde{H} = [P^T|I_m] = \begin{pmatrix} p_{11} & p_{12} & \cdots & p_{1k} & 1 & \cdots & \cdots & 0 \\ p_{21} & p_{22} & & \vdots & 0 & 1 & & \vdots \\ \vdots & & & \vdots & \vdots & & & \vdots \\ p_{1m} & \cdots & \cdots & p_{mk} & 0 & \cdots & \cdots & 1 \end{pmatrix} \tag{1.55}$$

where P^T is the transposed matrix of P (note how their coefficients are symmetrically permuted about the diagonal elements). We note then the important property

$$HG^T = GH^T = 0 \tag{1.56}$$

which stems from

$$HG^T = [P^T|I_m] \cdot \begin{bmatrix} I_k \\ P^T \end{bmatrix} = P^T + P^T \equiv 0 = (HG^T)^T = GH^T \tag{1.57}$$

In this expression, we have used the property that for binary numbers x, the sum $x + x$ is identical to zero ($0 + 0 = 1 + 1 \equiv 0$). From the property in equation (1.56), we also have

$$Y\tilde{H}^T = 0 \tag{1.58}$$

which stems from $Y\tilde{H}^T \equiv X\tilde{G}\tilde{H}^T = X(\tilde{G}\tilde{H}^T) = 0$.

We consider now the received block code, which we define as the n-vector Z. Since the received block code is a "contaminated" version of the original block code Y, we can write Z in the form:

$$Z = Y + E \tag{1.59}$$

where E is an error vector whose i^{th} coordinate is 0 if there is no error and 1 otherwise. Post-multiplying equation (1.59) by $\tilde{H}^T$ and using property from equation

(1.58), we obtain

$$S = Z\tilde{H}^{\mathrm{T}} = (Y + E)\tilde{H}^{\mathrm{T}} = E\tilde{H}^{\mathrm{T}} \tag{1.60}$$

The *m*-vector *S* is called the *syndrome*. The term syndrome is used to designate information helping to diagnose a "disease," which here is the error contamination of the signal. A zero syndrome vector means that the block contains no errors $(E = \vec{0})$. As the example described thereafter illustrates, single-error occurences (*E* has only one nonzero coordinate) are in one-to-one correspondence with a given syndrome *S*. Thus when computing the syndrome $S = Z\tilde{H}^{\mathrm{T}}$ at the receiver end, we immediately know two things:

1. Whether there are errors, as indicated by $S \neq \vec{0}$,

2. Assuming that it is a single-error occurence, where it is located.

It is useful at this point to illustrate the process of error detection and correction through a basic example. We consider the block code $(n, k) = (7, 4)$, which has $k = 4$ message bits and $m = 3$ parity bits. Such a block which is of the form $n = 2^m - 1$ with $m \geq 3$ is referred to as a *Hamming code*. According to the definition in equation (1.54), we define the following generator matrix (out of many other possibilities)

$$\tilde{G} = \begin{pmatrix} 1 & 0 & 0 & 0 & 1 & 1 & 0 \\ 0 & 1 & 0 & 0 & 1 & 0 & 1 \\ 0 & 0 & 1 & 0 & 0 & 1 & 1 \\ 0 & 0 & 0 & 1 & 1 & 1 & 1 \end{pmatrix} \tag{1.61}$$

which gives, from equation (1.55), the parity-check matrix

$$\tilde{H} = \begin{pmatrix} 1 & 1 & 0 & 1 & 1 & 0 & 0 \\ 1 & 0 & 1 & 1 & 0 & 1 & 0 \\ 0 & 1 & 1 & 1 & 0 & 0 & 1 \end{pmatrix} \tag{1.62}$$

and its transposed version

$$\tilde{H}^T = \begin{pmatrix} 1 & 1 & 0 \\ 1 & 0 & 1 \\ 0 & 1 & 1 \\ \hline 1 & 1 & 1 \\ \hline 1 & 0 & 0 \\ 0 & 1 & 0 \\ 0 & 0 & 1 \end{pmatrix} \tag{1.63}$$

The $2^k = 16$ possible message words *X* and their corresponding block codes *Y*, as calculated from equations (1.53) and (1.61), are listed in Table 1.4. For clarity, the parity bits are shown in bold numbers. As expected, the block codes are made of a first sequence of 4 (original) messages bits followed by a second sequence of 3 parity bits. For instance, the message word 0111 is encoded into the block code 0111 001, where the three bits on the right are the parity bits.

TABLE 1.4 Message words and corresponding block codes in an example of Hamming code (7,4)

Message Word X	Block Code Y
0000	0000 **000**
0001	0001 **111**
0010	0010 **011**
0011	0011 **100**
0100	0100 **101**
0101	0101 **010**
0110	0110 **110**
0111	0111 **001**
1000	1000 **110**
1001	1001 **001**
1010	1010 **101**
1011	1011 **010**
1100	1100 **011**
1101	1101 **100**
1110	1110 **000**
1111	1111 **111**

For future use, we define the following quantities in ECC formalism:

Hamming weight: number of "1" bits in a given code word;

Hamming distance between two code words: number of bit positions in which the two codes differ (consequently, the Hamming weight is the distance between a given code and the all-zero code word;

Minimum Hamming distance d_{min}: minimum number of "1" bits in the code word (excluding the all-zero code word). It is easily checked from Table 1.4 that $d_{min} = 3$, which is a specific property of Hamming codes.

Assume now that the received block Z contains a *single bit error,* whose position in the block is unknown. If the error concerns the first bit of $Z = Y + E$, this means that the value of the first bit of Y has been increased (or decreased) by 1, corresponding to the error vector $E = (1,0,0,0,0,0,0)$. For an error occuring in the second bit, we have $E = (0,1,0,0,0,0,0)$, and so on to bit seven. For each single-error occurence, we can then calculate the corresponding syndrome vector using the relation $S = E\tilde{H}^T$ in equation (1.60). The result of the computation is shown in Table 1.5. It is seen that to each error occurence corresponds to a unique syndrome. For instance, if the syndrome is $S = (0,1,1)$, we know that a bit error occurred in the third position of the block sequence, and correction is made by adding 1 to the bit (equivalently, by switching its polarity or by adding E to Z). Thus any single-error occurence can be readily identified and corrected by this method.

What happens if there is more than one error in the received block? Consider the case of double errors in our $(n,k) = (7,4)$ example. The possibilities of single and

**TABLE 1.5 Syndrome vector
S = (s₁,s₂,s₃) associated with
single-error patterns
E = (e₁,e₂,e₃,e₄,e₅,e₆,e₇) in the Hamming
block code (7,4) defined in Table 1.4**

Error Pattern E	Syndrome S
0000000	000
1000000	110
0100000	101
0010000	011
0001000	111
0000100	100
0000010	010
0000001	001

double errors in the block code are 7 for single, and $C_7^2 = 7!/[2!(7-2)!] = 21$ for double, representing 28 different error patterns. Since the syndrome is only 3 bits, it can only take $2^3 - 1 = 7$ configurations. Thus, the syndrome is no longer associated with a unique error pattern. It is an interesting computing exercise to determine the probabilities associated with each pattern. The result shows for instance that $S = (0,0,1)$ and $S = (0,1,1)$ are associated with 3 and 5 error patterns, respectively. All other configurations correspond to 4 error patterns. Thus, syndrome decoding can only detect the presence of errors, but cannot locate them with 100% accuracy since they are associated with more than one error pattern. Correcting the block from this information has a 1/3 to 1/5 probability of being right, which represents only minor improvement. It is clear that with Hamming codes having a greater number of parity bits, this imperfect correction improves, as the mapping $E \rightarrow S$ becomes less redundant. The approach is however costly in bandwidth use, and error correction is not absolutely efficient. This situation illustrates the fact that in most cases, syndrome decoding cannot determine the exact error pattern E, but only the one that is most likely to correspond to the syndrome, E_0. The Hamming distance between E and E_0 is minimized but is nonzero. Therefore, the substitution $Z \rightarrow Z + E_0$ represents a best-choice correction rather than an exact corrective operation. This is referred to as *maximum-likelihood decoding*.

The above limitations can be rigorously explained by a fundamental property of linear block codes. The property states that the code has the power to correct *any* error patterns of Hamming weigth, w provided that

$$w \leq \left\{ \frac{d_{\min} - 1}{2} \right\} \tag{1.64}$$

where $d_{\min}$ is the *minimum Hamming distance* and where the brackets indicate the highest integer no greater than the number contained. In the case of Hamming codes, we have seen that $d_{min} = 3$, which gives $w = 1$ as the maximum number of errors that can be corrected in Hamming block codes, independently of the block-code size.

1.6.2 Cyclic Codes

Cyclic codes represent a subset of linear block codes which obey two essential properties:

   1. the sum of two code words is also a code word;
   2. any cyclic permutation of the code word is also a code word.

Cyclic codes corresponding to (n,k) block codes can be generated by polynomials $g(p)$ of degree $n - k$. We shall briefly show the principle of this encoding method. Assume a block code $Y = (y_0 \ldots y_{n-1})$ of n bits y_i, which are now labelled from 0 to $n - 1$. From this code word, we can generate a polynomial

$$Y(p) = y_0 + y_1 p + y_2 p^2 + \cdots y_{n-1} p^{n-1} \tag{1.65}$$

where p is a real variable. We multiply $Y(p)$ by p and perform the following term re-arrangements:

$$pY(p) = y_0 p + y_1 p^2 + y_2 p^3 + \cdots + y_{n-1} p^n$$

$$= y_0 p + y_1 p^2 + y_2 p^3 + \cdots + y_{n-1}(p^n + 1) - y_{n-1}$$

$$= \left[\frac{y_{n-1} + y_0 p + y_1 p^2 + y_2 p^3 + \cdots + y_{n-2} p^{n-1}}{p^n + 1} + y_{n-1} \right](p^n + 1)$$

$$\tag{1.66}$$

To derive the above result, we have ued the property that for binary numbers, subtraction is the same as addition. Equation (1.66) can then be put into the form:

$$\frac{pY(p)}{p^n + 1} = y_{n-1} + \frac{Y_1(p)}{p^n + 1} \tag{1.67}$$

where

$$Y_1(p) = y_{n-1} + y_0 p + y_1 p^2 + y_2 p^3 + \cdots + y_{n-2} p^{n-1} \tag{1.68}$$

is the reminder of the division of $pY(p)$ by $M = p^n + 1$, or

$$Y_1(p) = pY(p) \bmod[M] \tag{1.69}$$

where mod[M] stands for "modulo M" (the same way one writes $6 : 3 = 0 \bmod[3]$ or $8 : 3 = 2 \bmod[3]$). It is seen from the definition in equation 1.68 that $Y_1(p)$ is a code-word polynomial representing a cyclicly shifted version of $Y(p)$, in equation 1.65. Likewise, the polynomials $Y_m(p) = p^m Y(p) \bmod[M]$ are all cyclicly shifted versions of the code $Y(p)$. It is then possible to use the $Y(p)$ polynomials as a new way of encoding messages, as we show next.

Define $g(p)$ as a *generator polynomial* of degree $n - k$ which *divides* $p^n + 1$. As an example for $n = 7$, we have the irreducible-polynomial factorization:

$$p^7 + 1 = (p + 1)(p^3 + p^2 + 1)(p^3 + p + 1) \equiv (p + 1)g_1(p)g_2(p) \qquad (1.70)$$

showing that $g_1(p)$, $g_2(p)$, $(p + 1)g_1(p)$ and $(p + 1)g_2(p)$ are possible generator polynomials for $k = 4$ (first two) and $k = 3$ (last two). Next define the message polynomial $X(p)$ of degree $k - 1$:

$$X(p) = x_0 + x_1 p + x_2 p^2 + \cdots x_{n-1} p^{k-1} \qquad (1.71)$$

We can now construct a cyclic code for the 2^k possible messages $X(p)$ according to the definition:

$$Y(p) = X(p)g(p) \qquad (1.72)$$

where $g(p)$ is a generator polynomial. Equation (1.72) represents a polynomial version of the previous matrix definition in equation (1.53), i.e. $Y = XG$. Although we will not make use of it here, one can define the *parity-check polynomial* $h(p)$ from the relation

$$g(p)h(p) = 1 + p^n \qquad (1.73)$$

or equivalently, $g(p)h(p) = 0 \bmod[1 + p^n]$, which is the polynomial version of the matrix equation (1.56) i.e., $HG^T = GH^T = 0$.

We illustrate next the polynomial encoding through the example of the $(n,k) = (7,4)$ Hamming block code. From equation (1.70), we can choose $g(p) = p^3 + p + 1$ as the generator polynomial (which divides $p^7 + 1$). The coefficients of the polynoms $Y(p)$, as calculated from equation (1.72) for all possible message polynoms $X(p)$, are listed in Table 1.6. It is seen that the obtained code representation is not systematic, i.e., the message bits no longer represent a separate word at the beginning or end of the code, unlike in linear block-codes previously seen (Table 1.4). For instance, as Table 1.6 shows, the original 4-bit message 0111 is encoded into 01100011.

Detecting and correcting errors in cyclic codes can be performed as follows. Assume that $Z(p)$ is the result of transmitting $Y(p)$ through a noisy channel. We divide the result by $g(p)$ and put it in the form:

$$Z(p) = q(p)g(p) + s(p) \qquad (1.74)$$

where $q(p)$ and $s(p)$ are the quotient and the remainder of the division, respectively. If there were no errors, we would have $Z(p) = Y(p) = X(p)g(p)$, meaning that the quotient would be the original message, $q(p) = X(p)$, and the remainder would be zero, $s(p) = 0$. As previously, we call $s(p)$ the *syndrome* polynomial. A non-zero syndrome means that errors are present in the received code. As previously shown for the vectors E,S, it is possible to map the error polynomial $e(p)$ into

TABLE 1.6   Message polynomial coefficients and corresponding cyclic code-word polynomial coefficients in the (7,4) block code example, as corresponding to the generator polynomial $g(p) = p^3 + p + 1$

Message Word $X(p)$ $x_4\, x_3\, x_2\, x_1\, x_0$	Cyclic Code $Y(p)$ $y_7\, y_6\, y_5\, y_4\, y_3\, y_2\, y_1\, y_0$
0000	0000000
0001	0001011
0010	0010110
0011	0011101
0100	0101100
0101	0100111
0110	0111010
0111	0110001
1000	1011000
1001	1010011
1010	1001110
1011	1000101
1100	1110100
1101	1111011
1110	1100010
1111	1101001

$s(p)$, which makes it possible to associate syndrome and error patterns. Looking at Table 1.6, we observe that the minimum Hamming distance for this cyclic code is $d_{min} = 3$. According to equation (1.64), the Hamming weight of error patterns that can be corrected is $w \le \{(d_{min} - 1)/2\} = 1$, meaning that only single errors can be corrected. As previously discussed, this is a general property of Hamming codes, which is not affected by the choice of cyclic coding. The error-detection and correction capability of various cyclic codes is discussed in the next subsection.

1.6.3 Types of Error-Correcting Codes

The previous description provided a number of conceptual tools as well as a first introduction to the properties of ECCs. We shall now complete this presentation by mentioning different types of binary codes and their correcting capabilities without going into details.

• *Cyclic Redundancy Check (CRC) Codes*: this is the name given to any cyclic code used for error detection. Binary CRC codes (n,k) can *detect* error bursts of length $\le n - k$ and various other error-burst patterns, such as combinations of $\le d_{min} - 1$ errors, for instance. CRC codes can *correct* all error patterns of odd Hamming weigth (odd number of bit errors) when the generator polynomial has an even number of nonzero coefficients.

- **Golay Code**: (23,12) cyclic code generated by either of the two polynomials $g(p) = 1 + p^2 + p^4 + p^5 + p^6 + p^{10} + p^{11}$ or $g(p) = 1 + p + p^5 + p^6 + p^7 + p^9 + p^{11}$, which are both dividers of $1 + p^{23}$. The corresponding minimum distance is $d_{min} = 7$, corresponding to 3-bit error-correction capability.

- **Maximum-Length Shift-Register Codes**: cyclic code of the form $(2^m - 1, m)$ with $m \geq 3$, which are generated by polynomials of the form $g(p) = (1 + p^n)/h(p)$, where $h(p)$ is a primitive polynomial of degree m (meaning an irreducible polynomial dividing $1 + p^q$ with $q = 2^m - 1$ being the smallest possible integer). The minimum distance for this code is $d_{min} = 2^{m-1}$, indicating the possibility of correcting a maximum of $w = \{2^{m-2} - 1/2\}$ simultaneous errors. For instance, the code (15,4) has a 3-error correction capability. The code rate (number of message bits divided by the block length) is $R = 4/15 = 26\%$, which is relatively poor in terms of bandwidth use. The maximum-length code sequences are labeled as "pseudo-noise," owing to their autocorrelation properties which closely emulate white noise.

- **Bose–Chaudhuri–Hocquenghem (BCH) Codes**: cyclic codes of the form $(2^m - 1, k)$ with $m \geq 3$, $k \geq n - mt$, t integer, and generated by polynoms dividing $1 + p^{2m-1}$. Extended lists of such polynoms are provided in standard tables. The first of these is $g(p) = 1 + p + p^3$, corresponding to the block (7,4) with $t = 1$. The minimum distance for these codes is $d_{min} = 2t + 1 = 3, 5, 7, \ldots$ corresponding to error-correcting capabilities of $w = t = 1, 2, 3, \ldots$, respectively. For instance, the block (31,11) with $t = 5$ can be corrected for number of errors corresponding to almost half of the 11 message bits, while the code rate is $R = 11/31 = 35.5\%$.

- **Reed–Solomon (RS) Codes**: these are BCH codes based on a specific arrangement, noted RS(N,K). The code block is made of N symbols comprising K message symbols and N-K parity symbols. The message/parity symbols of length m are nonbinary (e.g., $m = 8$ for *byte* symbols). The total block length is thus Nm. The RS code format is $RS(N = 2^m - 1, K = N - 2t)$. Its minimum distance is $d_{min} = N - K + 1$, corresponding to a *symbol* error-correcting capability of $t = (N - K)/2$. The code rate is $R = (N - 2t)/N = 1 - 2t/N$. For instance a RS code with $N = 255$ symbols ($m = 8$) and $t = 12$ ($K = 231$) corresponds to a code rate of 90% (or a bandwidth expansion factor of $N/K - 1 = 10.4\%$. Thus symbol-errors can be corrected up to about $1/20$ of the block length with only $N - K = 24$ symbols, representing an ECC overhead of nearly 10%. This example illustrates the power of RS codes and justifies their widespread use in optical telecommunications.

- **Concatenated Block Codes**: it is possible to concatenate two separate ECCs, usually a nonbinary (outer) code, which is labeled (N,K) and a binary (inner) code, which is labeled (n,k). The message coding begins with the outer code and ends with the inner code, yielding a block of the form (nN,kK), meaning that each of the K nonbinary symbols are encoded into k binary symbols, and similarly for the parity bits. At the receiving end, the block successively passes through the inner decoder first and the outer decoder second. The corresponding code rate and minimum Hamming distance are given by $R' = kK/nN$ and $d'_{min} = d_{min}D_{min}$, meaning that both parameters are given by the products of their counterparts for each code. Since the code rates k/n and K/N are less than unity, the resulting code rate is subtantially reduced. However, the error-correction capability $w = \{(d_{min}D_{min} - 1)/2\}$ is subtantially increased, approximately like the square of that of the individual codes. Using for instance the concatenation of two RS codes of the previous example, $RS(255,231)$, the code rate is 81% (bandwidth expansion factor 22%) while the correction capability is 71 bit errors, representing 31% of the initial message block length.

Other ECCs techniques with high signal-correcting efficiencies, but which we shall overlook here, are *convolutional codes* and *treillis codes*. An example of convolutional code, CSOC, is dicussed in Desurvive (2004).

Since ECCs can only correct error patterns having up to $w = \{(d_{min} - 1)/2\}$ bit errors, it is clear that in the general case the BER is never zero. We shall not detail the optimum receiver *decision* techniques, usually referred to *as hard-decision* and *soft-decision decodings* respectively. Suffice it to state that in the soft-decision decoding approach, the receiver decision is optimized for each symbol type, which eventually minimizes symbol errors. On the other hand, hard-decision decoding consists of making a single choice between "1" and "0" values for each of the received symbol bits, as we have described in Section 1.5.

The result of the operations of decoding and error-correction is a significant BER decrease. Such a decrease is associated with a Q-factor improvement. One can then define the *coding gain* as the decibel ratio

$$\gamma = 10 \log_{10}[Q_c/Q_{unc}] \tag{1.75}$$

where Q_c and Q_{unc} represent the Q-factors for the corrected and uncorrected signals, respectively. Assume for instance that the Q-factor is doubled from $Q_{unc} = 3$ to $Q_c = 6$. The coding gain is therefore $\gamma = 10 \log_{10}[6/3] = 3\,dB$. The corrected signal is so-called "error-free" since $Q = 6$ corresponds to BER $= 10^{-9}$. On the other hand, the uncorrected signal with $Q = 3$ corresponds to BER $= 1.5 \times 10^{-3}$. Thus, we see that a relatively small coding gain of 3 dB is extremely powerful in reducing the BER (by about six orders of magnitude!) and restoring signal integrity. As we have seen, such a performance improvement only requires a small bandwidth overhead (typically 5–15%). One drawback of ECC (yet not sufficient to give up its advantages) is the increase in terminal complexity due to the need for complex encoder/decoder circuits. The ECC processing is made by integrated circuits operating at comparatively low-bit-rates (e.g., 622 Mbit/s). For high-bit-rate signals (e.g., 40 Gbit/s) the implementation of ECC require complex intermediate frequency conversions through $1:4/(4:1)$ then $1:16/(16:1)$ demultiplexing and remultiplexing operations.

■ 1.7 CHANNEL INFORMATION CAPACITY

In previous sections, the words *data, symbols, message* and *information* were used in several places with only minimum definitions but straightforward understanding. A communication system is a physical channel which has the function of conveying, from one location to another, all of the above contents (data, symbols, messages, information.) with highest possible fidelity or accuracy. One can conceive of information as being the most meaningful part of the contents being communicated. On the other hand, 0 and 1 data bits are the least meaningful or intelligible of the contents. The previous sections showed that in practical communication channels, a fraction of the data is invariably lost (or equivalently, some of the bits mistaken). But the message integrity can be recovered to some large extent, depending upon the power of the error-correction coding. While we know how to characterize with accuracy how data can be conveyed through a communication channel, it is more difficult to make the same quantitative assessment concerning information.

Newspapers, radio and television represent some of the most familiar communication channels for information. They operate in a *broadcast* mode, meaning that the information contents are conveyed to a relatively large group of receiving (listening, seeing) entities in a one-way fashion. Since in all these communication media the information/messages are being conveyed/retrieved at a symbol rate corresponding to a few hundred words per minute, one could say that they have very limited *bandwidth*. Such a view is correct if one only considers symbol rate. But imagine the case where an unexpected event of serious implications happens. Very high information content can in fact be encapsulated into very few words or symbols (e.g., "President Smith died today"). On the other hand, very low information content can be associated with extensive communications, such as certain political speeches or corporate reports.

These examples illustrate that *information* means more than data/symbol transmission or interpretation rate. Information can be very weak with lots of data, or very powerful with minimum data. As Claude Shannon established in his *information theory*, the information content in a given message, i.e., what receiving parties actually value the most, is in inverse proportion to the *probability* of this message being communicated, or its degree of expectation. If we know more or less in advance what the message holds, there is nothing much we learn by its being communicated to us. Likewise, a broadcast channel transmitting the same message all the time (e.g., a fixed text or TV image), regardless of its transmission bit rate (extensive document or high-resolution movie clip), would be rated as virtually informationless.[1] It seems indeed that information and uncertainty are closely associated.

1.7.1 Channel Information and Entropy

To formalize the above concept, we define the measure of information of a given message "n" of probability p_n according to the expression:

$$I_n = \log_2 \frac{1}{p_n} = -\log_2 p_n \qquad (1.76)$$

---

[1]What is true of world news may not be of commercials. It is indeed a paradox that the same commercials with fixed advertisement patterns (e.g., neon lights on top of a building, posters in a stadium) convey real information. As a tentative explanation, one can say that the commercial's information, already stored in the recipient's mind, is just recalled for message reactivation. Under these specific conditions, communication channels with seemingly zero information rate truly exist.

where $\log_2(x) = \log(x)/\log(2)$ is the base-two logarithm. The quantity I_n is referred to as the message's *self-information*. If the message content is unique or perfecly defined, thus $p_n = 1$ and consistently, $I_n = 0$. In this case, there is no information. The unit of information is called the *bit*. Such a definition can be clarified as follows.

Assume then that there are two possible messages to be conveyed. Call these messages YES or NO. If they have equal probabilities ($p_1 = p_2 = 1/2$), their self-informations are also equal, namely $I_1 = I_2 = -\log_2(1/2) = 1$. The *bit* is therefore the information unit we gain when one out of two equally possible messages is received.

Next, assume that YES is a more likely message, for instance with $p_1(\text{YES}) = 3/4$ and $p_1(\text{NO}) = 1/4$. The corresponding informations are $I_1(\text{YES}) = -\log_2(3/4) = 0.41\,\text{bit}$ and $I_2(\text{NO}) = -\log(1/4) = 2\,\text{bits}$. This shows that there is substantially more information in the answer being NO than being YES. This is because NO being the least expected, we gain more information from the message, consistently with the considerations made at the beginning of the section.

When there are several possible messages with different probabilities of occurrence, one should not just add up the corresponding informations. They are not likely to be equally communicated. As we have seen, messages with high information are the least likely, and the reverse is also true. Thus how can one characterize the actual information flow? A way to do this is to take the *statistical mean* of the information, instead of its absolute value. Such a mean takes the form of the weighted sum:

$$H = p_1 \log_2 \frac{1}{p_1} + p_2 \log_2 \frac{1}{p_2} + \cdots + p_n \log_2 \frac{1}{p_n} = -\sum_n p_n \log_2 p_n$$

$$\equiv -\langle \log_2 p \rangle \tag{1.77}$$

The statistically averaged measurement of information, H, is called the source *entropy*, by analogy with physics.[2] Consistently, an absolutely certain message with probability $p = 1$ is associated with zero entropy or zero information.

Consider next the previous YES/NO examples. We have $H = -p_1 \log_2 p_1 - p_2 \log_2 p_2$ and we get $H = -2(1/2)\log_2(1/2) = 1\,\text{bit}$ for equally probable messages and $H = -(1/4)\log_2(1/4) - (3/4)\log_2(3/4) = 0.807\,\text{bits}$ for the second case. This result brings a different picture. We see that the entropy is highest when the messages are equally probable, and is lower otherwise. More generally, it can be shown that the highest entropy corresponds to the case where all possible messages have equal probability. This is explained by the fact that uncertainty is maximized. If YES and NO are equally probable, we are left with maximum doubt about the outcome. In the second example (YES being more probable), we trust one answer

---

[2]Readers with a physics background know that entropy is a measure of disorder, or more accurately, of the number of different states W_n a physical system can randomly occupy. Entropy is then defined as $S = -k_B \langle \log W \rangle$, the brackets having the meaning of statistical average and W the meaning of a probability distribution (k_B = Boltzmann's constant). The same concept applies for the channel messages. High entropy means a large amount of possible and unexpected information.

more than the other, and our uncertainty is reduced. This shows why entropy is a correct way of defining the net channel information, or equivalently, the average amount of message uncertainty.

1.7.2 Coding Efficiency

The messages to be conveyed through the channel takes the form of symbols, which belong to some reference source *alphabet,* which we call X. Since the symbols have various probabilities of occurrence, p_n, the source alphabet is characterized by an entropy, $H(X) = -\langle \log_2 p \rangle$. Each of the symbols is represented by a *code word.* For instance, transmitting the symbol/letter A consists of sending the code word 1000001 if one uses the ASCII code (Section 1.1). In ASCII all symbols have the same 7-bit length, but it is not the case in Morse code, for instance. Attributing the same code-word length to all symbols of the alphabet may not be the most efficient way to use the channel, since some symbols are more frequent or likely than others. If each symbol is given a bit length l_n, generally different from one another, we can define an *average symbol length*:

$$\langle l \rangle = \sum_n l_n p_n \qquad (1.78)$$

While there is an infinite number of possibilities to attribute word lengths to the symbols, there exists a minimum value which we call $\langle l \rangle_{min}$. We can define the alphabet's *coding efficiency* as the ratio $\eta = \langle l \rangle_{min}/\langle l \rangle$, which has a maximum of unity. The *source-coding theorem,* demonstrated by C. Shannon, states that the minimum code word, $\langle l \rangle_{min}$, is equal to the alphabet's entropy, H. The coding efficiency is thus given by

$$\eta = \frac{H(X)}{\langle l \rangle} \qquad (1.79)$$

How can one code the alphabet symbols in order to approach 100% efficiency? The answer is given by *Huffman coding.* The coding algorithm consists of several steps

- Listing the symbols in decreasing order of probability;
- Attributing a 0 and a 1 bit to the lowest two symbols;
- Summing their probabilities, making of the pair a single symbol and reordering the list (in the event of equal probabilities, always move the pair to the highest position);
- Restart from step one, until there is only one pair left.

An example of Huffman-coding implementation is provided in Figure 1.29. The alphabet is $X = (x_1, x_2, x_3, x_4)$ with respective probabilities 0.5, 0.2, 0.2, and 0.1. We find that the average word length is $\langle l \rangle = 0.5 \times 1 + 0.2 \times 2 + 0.2 \times 3 + 0.1 \times 3 = 1.8$ bit/word. On the other hand, the alphabet entropy is $H = -0.5 \log_2 0.5 - 2 \times 0.2 \log_2 0.2 - 0.1 \log_2 0.1 = 1.76$. The coding efficiency is therefore $\eta = H/\langle l \rangle = 1.76/1.8 = 97.8\%$. If we had attributed to each of these symbols the same length of 2 bits, we would have $\langle l \rangle = l = 2$ and $\eta = 1.76/2 = 88\%$.

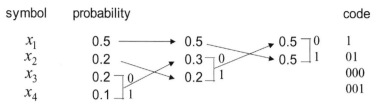

FIGURE 1.29 Example of Huffman coding, showing algorithm to attribute minimal average word length to a symbols alphabet $X = (x_1, x_2, x_3, x_4)$ with respective probabilities 0.5, 0.2, 0.2 and 0.1.

This example illustrates that Huffman coding brings the coding efficiency very close to the 100% limit.

What is the coding efficiency if the N symbols of an alphabet have the same probability? The answer is straightforward. Since all probabilities are equal to $p_n = 1/N$, the entropy of this alphabet is $H = \log_2 N$. According to Huffman algorithm, the symbol lengths must be all the same, namely $\langle l \rangle = l = k = \log_2 N$. The efficiency is thus $H/\langle l \rangle = 100\%$.

1.7.3 Mutual Information, Equivocation and Channel Capacity

The concept of *mutual information* involves further abstraction. But the exercise is well worth it, because it leads to the understanding of how, in the case of non-ideal or noisy channels, entropy and ultimate *channel capacity*, are in fact related. The following represents a very minimal introduction, in the interest of easier familiarization. For this reason, only the important definitions and immediately expoitable results will be discussed, overlooking lots of (otherwise) interesting properties.

We assume that due to some internal noise process, the communication channel imperfectly transmits the symbols, which leads to transmission errors. Let the received symbol alphabet be $Y = (y_1 \ldots y_n)$ and the original (source) alphabet be $X = (x_1 \ldots x_n)$. The associated symbol probabilities are noted $p(y_j)$ and $p(x_i)$. By definition, the quantity $p(y_j|x_i)$, called the *transition probability*, is the probability of receiving the symbol y_j when the source sent the symbol x_i into the channel. Assume for simplicity that the two alphabets have a one-to-one index correspondence, which does not affect generality. If the channel had no noise, then the transition probabilities $p(y_{j \neq i}|x_i)$ would all be identically zero, while $p(y_{j=i}|x_i)$ would be unity. The transition probabilities thus define the channel properties when using a specific source alphabet distribution, $p(x_i)$.

We define the *conditional entropy associated with the output symbol* y_j when all possible x_i are sent according to the formula:

$$H(y_j|X) = -\sum_i p(y_j|x_i) \log_2 p(y_j|x_i) \tag{1.80}$$

The quantity $H(y_j|X)$ is a random variable which has an associated probability $p(y_j)$. Its value when averaged over all possibilities, which we call $H(Y|X)$, is thus given by

$$H(Y|X) = -\sum_j \sum_i p(y_j)p(y_j|x_i) \log_2 p(y_j|x_i) \qquad (1.81)$$

The resulting conditional entropy $H(Y|X)$, which is called *equivocation*, represents the symbol uncertainty obtained *after* signal generation, transmission and detection. In contrast, the entropies $H(X)$ and $H(Y)$ represent the symbol uncertainties at the source (without knowing about the transmitted/detected symbols), and at the detector after transmission (without knowing about the source symbols). As we have seen, $H(X)$ is the full measure of the original information. The noisy channel adds some uncertainty, but this uncertainty is not information. As a result, the detected entropy $H(Y)$ contains some amount of information-less or extra uncertainty. This extra uncertainty is precisely the conditional entropy, $H(Y|X)$. It makes sense then to substract $H(Y|X)$ from $H(Y)$, in order to get a measurement of the actually received information, as we shall do next.

We define the *mutual information* $I(X, Y)$ as the entropy difference

$$\begin{aligned} I(X,Y) &= H(Y) - H(Y|X) \\ &\equiv H(X) - H(X|Y) \end{aligned} \qquad (1.82)$$

the second equality ($\equiv$) being implied without demonstration. In Shannon's theory, the ultimate *channel capacity* is defined as the maximum value of $I(X,Y)$ that one could obtain by a proper choice of the input distribution $p(x_i)$. The unit of channel capacity is *bit per channel use*. We shall illustrate this concept through a basic example.

Assume a two-symbol source alphabet $X = (x_1, x_2)$, with equal symbol probability, $p(x_1) = p(x_2) = 1/2$. The output alphabet is $Y = (y_1, y_2)$. The transition probabilities, $p(y_j|x_i)$ are shown in Figure 1.30. We chose $p(y_i|x_i) = \varepsilon$ and $p(y_{j \neq i}|x_i) = 1 - \varepsilon$, with $0 \leq \varepsilon \leq 1$ being a free real parameter. By definition of conditional probabilities, $p(y_j) = \sum_i p(y_j|x_i)p(x_i)$, which gives equiprobable output symbols $p(y_1) = p(y_2) = 1/2$. As we have seen earlier, equiprobable symbol distributions correspond to entropies of $H = 1$, or $H(X) = H(Y) = 1$. But the fact that the output and input alphabet entropies are equal does not mean that the channel is ideal! Indeed, we find that the equivocation

$$H(Y|X) = -\sum_j \sum_i p(y_j)p(y_j|x_i) \log_2 p(y_j|x_i)$$

$$= -\left[\begin{array}{l} p(y_1)p(y_1|x_1) \log_2 p(y_1|x_1) + p(y_1)p(y_1|x_2) \log_2 p(y_1|x_2) \\ +p(y_2)p(y_2|x_1) \log_2 p(y_2|x_1) + p(y_2)p(y_2|x_2) \log_2 p(y_2|x_2) \end{array} \right]$$

$$= -\frac{1}{2}\left[\begin{array}{l} \varepsilon \log_2 \varepsilon + (1 - \varepsilon) \log_2(1 - \varepsilon) \\ +(1 - \varepsilon) \log_2(1 - \varepsilon) + \varepsilon \log_2 \varepsilon \end{array} \right]$$

$$= -[\varepsilon \log_2 \varepsilon + (1 - \varepsilon) \log_2(1 - \varepsilon)] = f(\varepsilon) \qquad (1.83)$$

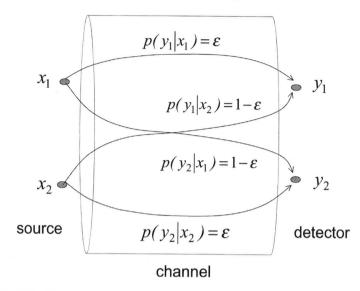

source   detector

channel

FIGURE 1.30 Noisy channel with two-symbol input (x_1, x_2) and output (y_1, y_2) alphabets. The corresponding transition or conditional probabilities, $p(y_j|x_i) = f(\varepsilon)$, where $0 \le \varepsilon \le 1$ is a free real parameter, are indicated.

is, in the general case, not identically zero. Figure 1.31 shows plots of the equivocation entropy $H(Y|X) = f(\varepsilon)$ and the mutual information $I(X,Y) = H(Y) - H(Y|X) \equiv 1 - H(X|Y) = 1 - f(\varepsilon)$. Here, we can consider that this mutual information is maximized and is equal to the *channel capacity*, since the choice of equi-probable input symbol yields maximum input entropy $H(X)$. It is seen that the equivocation is reaching a maximum of unity when $\varepsilon = 0.5$, corresponding to zero mutual information. In this case, the channel has *zero capacity*, and it can be therefore called a *useless channel*. This corresponds to a situation where the transition probabilities $p(y_j|x_i)$ are all equal to $\frac{1}{2}$. Thus, the received symbols have random values that are fully uncorrelated to the source symbols, hence the channel uselessness. In the opposite case where either $\varepsilon = 0$ or $\varepsilon = 1$, the equivocation vanishes and the channel capacity is maximized to the value $I(X,Y) = H(X) = H(Y) = 1$. This corresponds to the situation where the transition probabilities $p(y_{j \ne j}|x_i)$ are identically zero and $p(y_i|x_i)$ are identically unity. There is full correlation between the source and the received symbols, which corresponds to the case of a *noiseless* or ideal channel.

The intermediate cases where $0 < \varepsilon < 0.5$ and $0.5 < \varepsilon < 1$ correspond to non-optimized channel capacities. This example has shown that the output entropy may be equal to the input entropy, but in situations covering anything from a useless to a noiseless/ideal channel. Second, the input symbol distribution may be such that the mutual information equals the channel capacity, but the channel capacity is not necessarily maximum.

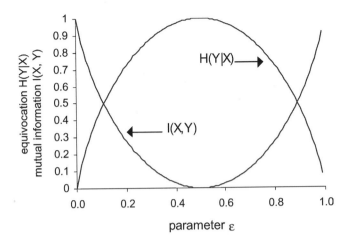

FIGURE 1.31 Equivocation entropy $H(Y|X)$ and mutual information $I(X, Y)$, plotted as functions of the parameter ε. The case $\varepsilon = 0.5$ corresponds to a useless channel ($H(Y|X) = 1$) and $\varepsilon = 0$ to a noiseless channel ($H(Y|X) = 0$).

To finish this example, it is interesting to analyze the BER properties of this specific channel. The fact that the channel capacity is increasing symetrically as one moves away from $\varepsilon = 0.5$ suggests identical channel noise in the two regions. In the first region A, $0 < \varepsilon < 0.5$, for which the channel is ideal at $\varepsilon = 0$, the BER is given by $\mathrm{BER_A} = p(x_1)p(y_1|x_1) + p(x_2)p(y_2|x_2)$ or $\mathrm{BER_A} = \varepsilon$. In the second region B, $0.5 < \varepsilon < 1$, for which the channel is ideal at $\varepsilon = 1$, the BER is given by $\mathrm{BER_B} = p(x_1)p(y_2|x_1) + p(x_2)p(y_1|x_2)$ or $\mathrm{BER_B} = 1 - \varepsilon$. When $\varepsilon \to 0.5$, the BER approaches the $\frac{1}{2}$ limit, corresponding to the useless channel. Taking for instance $\varepsilon = 1 \times 10^{-9}$ or $\varepsilon = 1 - 1 \times 10^{-9}$ gives identical BER values, i.e., $\mathrm{BER_A} = \mathrm{BER_B} = 1 \times 10^{-9}$, corresponding to an acceptable error level, or to the regime of (so-called) error-free transmission.

1.7.4 Shannon–Hartley Law

Further developing the analysis, we assume that the source alphabet X is *continuous*, with a frequency spectrum confined in the region $f < B$. The symbols $x_1 \ldots x_i$ now represent *samples* of the continuous variable x, which is a waveform amplitude. The sampling rate, $2B$ samples per second, is the Nyquist rate, and the channel is used for duration T. The source symbol probability $p(x)$ is given by the normal or Gaussian distribution, which is defined by

$$p(x) = \frac{1}{\sigma_{\mathrm{in}}\sqrt{2\pi}}\exp\left(-\frac{x^2}{2\sigma_{\mathrm{in}}^2}\right) \tag{1.84}$$

where $\sigma_{\mathrm{in}}^2 = \langle x^2 \rangle = P_S$ is the symbol variance, and P_S the average power of the signal waveform. By definition, the source entropy corresponding to discrete samples x_i is

$$H(X) = - \sum_i p(x_i) \log_2 p(x_i) \tag{1.85}$$

which using the average definition of $p(x)$ and the property $\log_2[\exp(u)] = u \log_2(e)$ can be developed into

$$H(X) = - \sum_i \frac{1}{\sigma_{in}\sqrt{2\pi}} \exp\left(-\frac{x_i^2}{2\sigma_{in}^2}\right) \left[\log_2\left(\frac{1}{\sqrt{\sigma_{in}^2 2\pi}}\right) - \frac{x_i^2}{2\sigma_{in}^2}\log_2(e)\right]$$

$$= \frac{1}{2}\log_2\left(2\pi\sigma_{in}^2\right) \sum_i p(x_i) + \frac{1}{2\sigma_{in}^2}\log_2(e) \sum_i x_i^2 p(x_i) \tag{1.86}$$

The samples x_i can be chosen sufficiently close for the following approximations to be valid: $\sum_i p(x_i) = 1$, and $\sum_i x_i^2 p(x_i) = <x^2> = \sigma_{in}^2$ (the same result would be obtained by replacing the discrete summations by *integrals*). Hence the source entropy takes the form:

$$H(X) = \frac{1}{2}\log_2\left(2\pi\sigma_{in}^2\right) + \frac{1}{2\sigma_{in}^2}\log_2(e) \equiv \frac{1}{2}\log_2\left(2\pi e\sigma_{in}^2\right) \tag{1.87}$$

Consider next the effect of channel noise, which we assume to be characterized by a Gaussian PDF with variance $\sigma_{ch}^2 = P_N$, where P_N is the channel noise power. If the channel noise is *additive*, it is not correlated with the input signal uncertainty. As a result, the total uncertainty measured at the detector is $\sigma_{out}^2 = \sigma_{in}^2 + \sigma_{ch}^2$, as associated with a Gaussian PDF defined as $p(z)$. The output entropy $H(Y)$ can be defined the same way as done previously for $H(X)$, this time using σ_{out}^2 instead of σ_{in}^2. Consistently, we obtain:

$$H(Y) = \frac{1}{2}\log_2\left(2\pi e\sigma_{out}^2\right) \tag{1.88}$$

The next step is to find the *equivocation entropy* $H(Y|X)$. The received waveform has a random amplitude given by $y_j = x_i + z_i$, where z_i is the random variable associated with the channel noise with power $<z^2> = \sigma_{ch}^2 = P_N$. The probability of measuring y when x has been (exactly) sent is therefore $p(y|x) = p(z)$. We then apply the definition

$$H(Y|X) = - \sum_j \sum_i p(y_j)p(y_j|x_i) \log_2 p(y_j|x_i)$$

$$= - \sum_j \sum_i p(y_j)p(z_i) \log_2 p(z_i)$$

$$= - \sum_j p(y_j) \sum_i p(z_i) \log_2 p(z_i) \equiv H(Z) \equiv \frac{1}{2}\log_2\left(2\pi e\sigma_{ch}^2\right) \tag{1.89}$$

In this derivation, we have used the property $\sum_j p(x_j) = 1$, the entropy definition for the channel noise, $H(Z) \equiv \sum_i p(z_i) \log_2 p(z_i)$, and used the result corresponding to a Gaussian PDF of variance σ_{ch}^2.

By definition, the *channel capacity* is given by the input distribution which maximizes the mutual information $I(X,Y) = H(Y) - H(Y|X)$. Using the previous results for $H(Y)$ and $H(Y|X)$, we obtain:

$$I(X,Y) = H(Y) - H(Y|X) = \frac{1}{2}\log_2(2\pi e \sigma_{out}^2) - \frac{1}{2}\log_2(2\pi e \sigma_{ch}^2)$$

$$= \frac{1}{2}\log_2\left(\frac{\sigma_{out}^2}{\sigma_{ch}^2}\right) = \frac{1}{2}\log_2\left(\frac{\sigma_{in}^2 + \sigma_{ch}^2}{\sigma_{ch}^2}\right) \qquad (1.90)$$

and using $\sigma_{out}^2 = \sigma_{in}^2 + \sigma_{ch}^2, \sigma_{in}^2 = P_S, \sigma_{ch}^2 = P_N$:

$$I(X, Y) \equiv \frac{1}{2}\log_2\left(1 + \frac{P_S}{P_N}\right) \qquad (1.91)$$

Does this mutual information correspond to the channel capacity? We shall use here, as a postulate (which is easily demontrated), that given a signal power constraint (P_S), for continuous symbol alphabets the entropy is maximized with the Gaussian distribution, called here $p(x)$. It is therefore sensible that the maximum mutual information associated with the noisy channel corresponds to $p(x)$ as a choice for input distribution. A Gaussian distribution for channel noise was also assumed, which given the noise power constraint P_N therefore maximizes the equivocation $H(Y|X) = H(Z)$. Based upon these two observations, we can state that the mutual information is maximized (given signal and noise power constraints) and is therefore equal to the channel capacity, C.

Remember that the channel capacity is expressed in bits per channel use. At the Nyquist sampling rate $2B$ per second, the channel capacity per unit time is, then, $C_{bit/s} = 2BC_{bit}$. From the previous result, we get

$$C_{bit/s} \equiv B\log_2\left(1 + \frac{P_S}{P_N}\right) \qquad (1.92)$$

or equivalently in dimensionless units of bit s^{-1} Hz^{-1}:

$$C_{bit\,s^{-1}\,Hz^{-1}} \equiv \log_2\left(1 + \frac{P_S}{P_N}\right) \qquad (1.93)$$

The above result is known both as Shannon's channel-capacity theorem and as the Shannon–Hartley law, based on the earlier contribution by Hartley to Shannon's analysis. The bit s^{-1} Hz^{-1} figure is also called *information spectral density*, or ISD.

The ISD, as defined by the ratio of the occupied signal bandwidth (Hz) to the signal bit rate (bit/s) is often called *spectral efficiency*. It would be a correct appellation if the ISD were bounded to unity, corresponding to a 100% efficiency, or 1 bit s^{-1} Hz^{-1}. But the Shannon–Hartley law shows that the ISD is in fact unbounded. At high signal-to-noise ratios SNR $= P_S/P_N$, we have $C \approx \log_2$ (SNR), meaning that every 3-dB increase of SNR corresponds to an ISD increase of one additional bit $s^{-1}\,Hz^{-1}$. For instance, $SNR = 2^{10} = 1024 = +30.1dB$ corresponds, by definition, to an ISD of 10 bit $s^{-1}\,Hz^{-1}$. But what about NRZ signals for which the bit rate and occupied bandwidth are equal, giving a maximum ISD of 1 bit $s^{-1}\,Hz^{-1}$? The answer is that the capacity result $C = \log_2(1 + SNR)$ is most general with respect to any coding possibilities. If one restricts the coding to formats such as NRZ, PSK or M-ary QAM, the result is quite different, because we *specify* what the source entropy is. Consider indeed NRZ. The bit-symbol distribution $p(x_i)$ is uniform, with value $\frac{1}{2}$. We have seen earlier that in this case, the entropy is $H(X) = 1$. According to equation (1.82), the channel capacity is given by $C = H(Y) - H(Y|X) = H(X) - H(X|Y) \leq H(X) = 1$. For NRZ signals, the capacity is therefore bounded by $C = 1$ bit per channel use $(H(X|Y) = 0)$, or 1 bit $s^{-1}\,Hz^{-1}$ since the bit rate and channel use rate are identical, and it is less in the case of a noisy channel $(H(X|Y) \neq 0)$. Likewise, for PSK and M-ary QAM, the bit-symbol distribution is also uniform but equal to $p(x_i) = 1/M$, the source entropy is therefore $H(X) = \log_2 M$ and the maximum capacity is $C = \log_2 M$. This shows that both coding scheme and channel noise define channel capacity. Without any coding consideration, on the other hand, channel capacity is defined with respect to the received SNR, but how to achieve the coding (leading to optimum mutual information) is not specified.

1.7.5 Bandwidth Efficiency

At this point, we must replace the Shannon–Hartley law in the more elaborate information theory developed by Shannon. Two of the results, (a) the *channel-coding theorem*, and (b) the *channel capacity theorem*, can be stated in the form:

1. With an alphabet source of entropy $H(X)$ producing symbols at rate B, an encoder producing coded blocks at coding rate $B' \leq B$ and a noisy channel of capacity C used at the coding rate B', if the condition

$$H(X) \leq \frac{B'}{B}C$$

   is verified, then there exists a coding scheme for which information can be transmitted with an arbitrary small probability of error.

2. For a noisy channel used at the code rate B', under the constraints of signal power P_S and additive Gaussian noise power P_N, the bit/s capacity is equal to

$$C_{bit/s} = B'_{Hz} \log_2\left(1 + \frac{P_S}{P_N}\right)$$

   and through appropriate coding schemes, information can be transmitted at this rate with an arbitrary small probability of error.

We shall see now how to interpret the meanings of these two theorems. The concept of error-rate mimimization to arbitrary low levels is associated with error-correction coding (ECC), namely the generation of complex code-words from the source alphabet words. In Section 1.6, we saw that for *linear block codes*, coding k message bits involves the use of $n - k$ parity bits, resulting in code words of n bits. The relative coding rate is therefore $B'/B = k/n < 1$, corresponding to a block length $T' = (n/k)T_{\text{bit}}$ longer than the message word length kT_{bit}. (Note that the source bit rate B can be increased by the factor n/k so that the message through-put rate, from uncoded to decoded information remains unaffected; but should the bandwidth increase requirement n/k become significant, such a bit-rate compensation cannot be implemented.)

The first theorem provides a condition for minimal-error transmission, given the source entropy H and the channel capacity C. Taking our previous example of a noisy channel, Figure 1.30, and on the assumption of equiprobable alphabet ($H(X) = 1$), we have obtained the channel capacity $C = I(X|Y) = 1 + \varepsilon \log_2 \varepsilon + (1 - \varepsilon) \log_2 (1 - \varepsilon)$ as a function of the parameter ε, which is plotted in Figure 1.31. For instance, the capacity is about $C = 0.3$ bits^{-1} Hz^{-1} for $\varepsilon = 0.2$ or 0.8. The theorem condition $H(X) \leq (B'/B)C$ is thus equivalent to $(B'/B) \geq C/H(X) = 0.3$, or $B' = 0.3B$, or $T' = (1/0.3)T = 3.33T$. This result shows that the block code must be at least 3.33 times the uncoded message length. This is a big penalty in terms of required code-word length, but consider that the uncorrected BER corresponding to either $\varepsilon = 0.2$ values is BER $= 2 \times 10^{-1}$ (see previous subsection for details). If we took instead $\varepsilon = 1 \times 10^{-4}$ or $\varepsilon = 1 - 1 \times 10^{-4}$ to give BER $= 1 \times 10^{-4}$, the channel capacity becomes $C \approx 0.998$, and the theorem condition is $(B'/B) \geq C/H(X) = 0.998$, or $B' = 0.998B$, corresponding to a code word of $T' = (1/0.998)T = 1.002T$. This result means that at least two parity bits per 1,000 message bits would be required. The important conclusion is that within the theorem inequality condition, block codes exists (or any other code type) for which transmission through this noisy channel can be made with arbitrary small error rate, and the minimal code length is provided by the said inequality.

In the second theorem, signal power and channel noise constraints, and hence the received SNR, do define the channel capacity. It just states that *there exist* block codes (or any other code type) for which transmission through this noisy channel can be made with arbitrary small error rate. There are no indications as to what the code words and lengths could be. We can however get some intuitive idea of the requirements by comparing the channel capacity obtained at given SNR with those achievable by usual (noncoded) M-ary formats if given SNR and BER.

First define the signal and noise energies per symbol as $E_S = P_S/B$ and $E_N = P_N/B$, respectively. Assuming an ISD of $C_{\text{bit/s}}$ the signal power can also be expressed as $P_S = E_S C_{\text{bit/s}}$. Inserting these definitions into equation (1.92), we obtain

$$C_{\text{bit/s}} = B \log_2 \left(1 + \frac{E_S}{E_N B} C_{\text{bit/s}} \right) \tag{1.94}$$

or equivalently

$$C_{\text{bits}^{-1} \text{Hz}^{-1}} = \log_2 \left(1 + \frac{P_S}{P_N} C_{\text{bit/s/Hz}} \right) \tag{1.95}$$

The previous equation can be reversed to

$$SNR = \frac{P_S}{P_N} = \frac{2^{C_{\text{bit/s/H}}} - 1}{C_{\text{bits}^{-1}\,\text{Hz}^{-1}}} \qquad (1.96)$$

We can then plot the resulting law in a *bandwidth-efficiency diagram*, SNR = $f(C_{\text{bits}^{-1}\,\text{Hz}^{-1}})$, or equivalently $C_{\text{bits}^{-1}\,\text{Hz}^{-1}} = f^{-1}(\text{SNR})$, which is shown in Figure 1.32. We first observe from the figure that at low SNR, the ISD asymptotically vanishes ($\log_{10} C \to -\infty$). From equation (1.96), this value is calculated to be SNR = $\log(2) = 0.693 \approx -1.6$ dB. This low SNR bound, called the *Shannon limit*, defines a region where no transmission can be achieved with indefinitely low

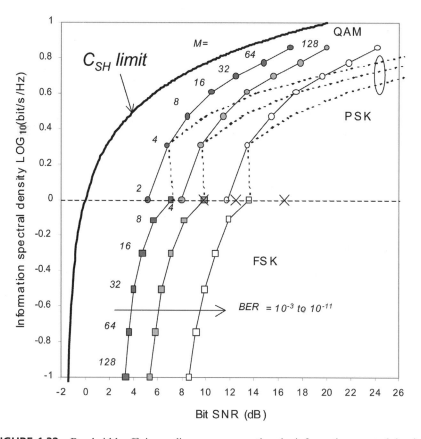

FIGURE 1.32 Bandwidth-efficiency diagram, representing the information spectral density $\log_{10}[C_{\text{bit/s/Hz}}]$ as a function of the bit SNR (dB). The thick line correspond to ISDs bounded by the Shannon–Hartley law. The families of points correspond to ISDs of the M-ary signal modulation formats FSK, PSK and QAM, at constant bit-eror rate ($BER = 10^{-3} - 10^{-11}$). The two crosses on the $C = 1$ bits^{-1} Hz^{-1} axis correspond to the same BER range for ON/OFF, NRZ signals. [After E. Desurvire et al., 2002. See Bibliography at end of this book.]

error. For a given SNR above the Shannon limit, the region located under the curve $C_{bits^{-1} Hz^{-1}} = f^{-1}(SNR)$ defines operating conditions where error correction can be made up to arbitrarily low error levels. In order to visualize the type of codes required, the figure also shows plots of ISDs obtained with M-ary signal modulation formats such as FSK, PSK, and QAM. The families of points linking each format type with increasing M, were computed at constant bit-error rate (BER $= 10^{-3}–10^{-11}$). The two crosses on the axis $\log_{10} C = 0\,dB$ or $C = 1\,bits^{-1}\,Hz^{-1}$ correspond to the same BER range for ON/OFF, NRZ signals. Note that the iso-BER curves move away from the limiting curve $C_{bits^{-1} Hz^{-1}} = f^{-1}(SNR)$ as the BER increases.

We can then interpret Figure 1.32 as follows. At constant SNR, a given M-ary format has a well-defined BER and ISD. Decreasing or increasing the SNR results in higher or lower BER, respectively, as is well known. However, as long as the point (SNR,ISD) remains under the limiting curve $C_{bits^{-1} Hz^{-1}} = f^{-1}(SNR)$, a means of coding these formats exists for which the BER can be made arbitrarily small. Consider for instance 128-QAM, which has an ISD of $\log_2 128 = 7\,bits^{-1}\,Hz^{-1}$. A SNR of $+15$ dB yields BER $= 10^{-3}$. The theorem states that there is a code (at least as complex) which can yield any smaller BER for the same SNR and ISD. It is seen from the figure that as the SNR is increased, the iso-BER points of M-ary QAM asymptotically move towards the limiting curve. But the BER remains unchanged, and therefore, this convergence does not suggest that M-ary QAM is the code answer. Rather, this illustrates that classical M-ary formats can approach the channel capacity as the number of coding levels M is increased, but always with finite error. On the other hand, such formats can be encoded with complex algorithms in such a way as to achive the channel capacity *and* the lowest transmission error.

✅ EXERCISES

Section 1.1
1.1.1 Calculate how much energy is contained in the wave generated by the pebble's fall into the pond, assuming that the pebble mass is $m = 100$ g and that it is launched from height of $h = 1$ m (elementary physics background required).

1.1.2 Assume the wave defined by equation (1.1) to be a light wave with $\lambda = 1.5\,\mu m$ wavelength. What are the frequency and the period of the wave? (speed of light $c = 3 \times 10^8$ m/s)

1.1.3 Calculate the power density in an optical fiber whose core diameter is 4 μm, assuming that the light signal power is 1 mW. Convert the result to W/cm^2. What can be concluded?

1.1.4 Assume a waveform defined by equation (1.10). What is the modulation index m for which the total power in the side-bands is equal to that in the carrier tone?

1.1.5 Without (!) a calculator, convert the following powers into dBm: $P_1 = 10\,nW$, $P_2 = 4\,\mu W$, $P_3 = 500\,\mu W$ and $P_4 = 25$ mW.

Section 1.2
1.2.1 Make the table of binary numbers ranging from 0 to 15.

1.2.2 Convert your year of birth into the binary system, and show the result in two octets (clue: decompose the number in powers of two, starting from the highest).

1.2.3 Draw the voltage diagram of the bit string 1001 1000 1101 first in Manchester then differential Manchester code, assuming in the second case that the bit preceding the string is 1 in low state.

1.2.4 Draw the voltage diagram of the bit string 1001 1000 1101 first in AMI then in HDB2 code, assuming that the bit preceding the string is 1 in low state.

1.2.5 Generate a Gray-code 4-bit symbol correspondence for 8-ary PAM with the same coding rule as in the quaternary space diagram shown in Figure 1.12.

Section 1.3
1.3.1 Calculate the nonlinear quantization increments in the case of μ-law with $\mu = 255$ and 1-byte sample coding, assuming a waveform with amplitude range $[-1, +1]$.

Section 1.4
1.4.1 Determine the mean and standard deviation of the following number set: $\{-1, 9, 5, 2, -8, 10, -2, -3, 7, 1, 2\}$.

1.4.2 Assume a normal distribution $p(x)$ with zero mean and standard deviation σ. Determine the abscissa x for which the distribution is equal to 3/4, 1/2, 1/2, 1/4 and 1/8 of its peak value. Make an approximate plot of the normal distribution, as based upon these values, while assuming $\sigma = 1$.

Section 1.5
1.5.1 Assume a transmission system which is $L = 100$ km long. The power received at the system end represents 1% of the input. What is the trunk loss rate, in dB/km? What is the percentage of received power in a system having the same trunk loss but a length of $L = 150$ km?

1.5.2 Assume a light wave at carrier frequency $f_c = 200$ THz, whose spectrum is 120 GHz wide. What is the time broadening experienced by the waveform after 100 km propagation in a medium having a dispersion coefficient of 17 ps nm^{-1} km^{-1} and carrier velocity of $c = 3 \times 10^8$ m/s?

1.5.3 A 10-Gbit/s system continuously transmits data with a bit error rate of BER $= 10^{-12}$. How many errors are made over the course of a single day?

1.5.4 What is the Q-factor requirement for achieving a bit-error rate very near 10^{-12}?

1.5.5 How long would it take to make a measurement of BER $= 10^{-14}$ in a 10-Gbit/s system, assuming that at least 10 error counts would be required?

1.5.6 What are the ideal receiver sensitivities in photons/bit for homodyne and heterodyne ASK signals? Express the figures in dBm, assuming a 2.5-Gbit/s bit rate and $\lambda = 1.55$ μm wavelength.

1.5.7 A lightwave receiver has a sensitivity of $P^{sens} = -40$ dBm. The transmission line has a loss coefficient of -0.2 dB/km. The available transmitter power is $P^{trans} = +15$ dBm. Assuming a 3-dB system margin, what is the maximum system length permitting error-free transmission?

1.5.8 Assume for a transmission line a loss coefficient of -0.3 dB/km. What would be the transmitter powers necessary to obtain a signal power of 1 mW at distances $L = 100$ km, 250 km and 500 km? How would these figures change with in-line amplifiers placed every 50 km? What can be concluded from the result?

1.5.9 Calculate the noise figure of a 100-km-long microwave transmission line with loss of 1 dB/km and amplifiers placed every 10 km (amplification preceding loss), assuming an amplifier temperature of $T_A = 626.5$ K.

1.5.10 Assume you want to realize a high-gain amplifier made by cascading two smaller amplifiers of gains G_1 and G_2 with corresponding noise figures F_1 and F_2. Consider the following cases:

$$\text{(a) } G_1 > G_2, F_1 = F_2 = F$$
$$\text{(b) } G_1 = G_2 = G, F_1 > F_2$$

In which order would you place the amplifiers in each case to obtain the smallest possible noise figure for the twin-amplifier system? What do you conclude from the results?

1.5.11 Provide the noise figure definitions for a transparent trunk segment where: (a) amplification precedes loss; and (b) amplification follows loss.

1.5.12 An amplified system with $k = 100$ transparent segments of identical noise figures $F_{seg} = 3.16$ ($F_{seg}^{dB} = +5$ dB) is operating error-free. What is the decibel SNR used at transmitter level?

Section 1.6

1.6.1 What is the bit-error rate of the message HAMPY_HXLIDEYS, assuming that each letter or symbol takes one byte and that the message is communicated within a single second?

1.6.2 Define a syndrome table corresponding to the linear block code (6,3), assuming that the parity matrix is equal to $P = \begin{pmatrix} 1 & 1 & 0 \\ 0 & 1 & 1 \\ 1 & 1 & 1 \end{pmatrix}$. Then detect possible single-bit error in the three received code words $Y_1 = (010100)$, $Y_2 = (001111)$ and $Y_3 = (110001)$ and identify the original/corrected message words.

1.6.3 An error-correcting code makes it possible to bring the bit-error-rate from BER $= 5 \times 10^{-4}$ to BER $= 1.0 \times 10^{-12}$. What is the corresponding coding gain? (Clues: (a) use $Q = 7.0$ as the Q-factor for BER $= 10^{-12}$ and (b) find the uncorrected Q-factor by progressive approximations.)

Section 1.7

1.7.1 A movie has 99% chances of being played in a theater on a certain date. What is the information of its being cancelled, compared to its being played?

1.7.2 Preparing his/her vacations, an individual knows that the probability that

(a) the car will have a mechanical problem is 0.1%;

(b) one of the participants be sick is 10%

(c) serious business calls will be received is 40%

Give the entropy associated with the vacation events, assuming that all three cases are uncorrelated possibilities. How far is the result from maximum entropy?

1.7.3 An 8-letter alphabet (A,B,C,D,E,F,G,H) has the following symbol probabilities: $p(A) = 8u$, $p(B) = 7u$, $p(C) = 6u$, $p(D) = 5u$, $p(E) = 4u$, $p(F) = 3u$, $p(G) = 2u$ and $p(H) = u$, where u is a strictly positive real number. Determine a possible Huffman code for this alphabet and provide the corresponding coding efficiency.

a MY VOCABULARY

Can you briefly explain any of these words or acronyms in the digital signal and telephony context?

3R regeneration
A-law
ADPCM
AGC
Alphabet
AM
AMI
Amplitude
Analog
ASBC
ASK
Attenuation
Bandpass filter
Bandpass signal
Bandwidth
Baseband (signal)
Baud rate
Baudot (code)
BCH (code)
BER
Binary
Bipolar
  (modulation)
Bit
Bit period
Bit rate
Block code
Boltzmann's
  (constant)
Broadcast
Budget (power)
Capacity (system)
Carrier (wave)
Channel
Clipping
  (competitive)
Clipping
  (connection)
Clock recovery
CMI
Code rate
Coding gain
CR
CRC (code)
Coding
Coding efficiency
Coherent detection
Companding
Conditional entropy
Constellation

CVSD
Data
dBm
Decibel
Decimal
Decision
Delta modulation
Demodulation
De-multiplexer
De-multiplexing
Detection
Detector
Dibit
Digital
Direct detection
Dispersion
DPCM
DPSK
DSI
Duplex
DWDM
ECC
Electromagnetic
Energy
Entropy
Equivocation
Error (bit)
Error correction
Eye diagram
Eye closure
FEC
FFT
Flip-flop
FM
Free space
Fourier transform
Frequency
Frequency domain
FSK
Gaussian
  distribution
Generator (matrix)
Generator
  (polynomial)
Golay code
Gray code
Hamming weight
Hamming distance
Hard-decision
  decoding

HDBn
Heterodyne
  (detection)
Hexadecimal
  format
Homodyne
  (detection)
Huffman code
IA2 code
IF
Information (signal)
Integration (bit)
IM
Information
Intensity
Intermediate
  frequency
ISD
Johnson noise
LH (system)
Light
Line
LO
Loss (signal)
LPC
M-ary (modulation)
Manchester (code)
Margin (system)
Mean
Message bits
Miller (code)
Modem
Modulation format
Modulation index
Monochromatic
Morse (code)
Multiplexer
Multiplexing
Mutual information
μ-law
NF
Noise
Noise figure
Noiseless channel
Nonlinearity
Normal distribution
NRZ
ON/OFF
OOK
PAM

Payload (signal)
PCM
PDF
Period
Parity bits
Photodiode
Photon
Planck's (constant)
PLL
PM
Power (wave)
Power density
Power spectrum
Preamplification
PSK
Pulsation
Q-factor
QAM
Quantization
Quantization noise
Quat
Reamplification
Repowering
Receiver
Reed-Solomon code
Redundancy bits
Regenerator
Repeater
RF
RS code
RZ
SDM
Self-information
Sensitivity
  (receiver)
Shannon limit
Side-bands
Sinusoid
Soft-decision
  decoding
Spectral efficiency
Spectrum
Standard deviation
Symbol
Syndrome
Systematic code
TASI
TDM
Temperature
  (amplifier)

Time domain

Tone (frequency)

Transceiver

Transmitter

Transponder

Trunk

ULH (system)

Unipolar
(modulation)

Unrepeatered
(system)

Useless channel

Variance

Velocity

Voice

Wave packet

Waveform

Waveguide

Wavelength

WDM

Word (binary)

2B1Q (code)

4-PAM

CHAPTER 2

Telephony and Data Networking

In this chapter, we shall revisit the plain old telephone system, its basic infrastructure and ancestral traffic-aggregation hierarchy. Computer local-area networks, and their different network access protocols, are also described.

■ 2.1 PUBLIC SWITCHED TELEPHONE NETWORKS (PSTN) AND SERVICES

Some readers may recollect from their early childhood those telephones hanging on the wall, where one would have to turn a crank to call a remote operator. One would have to tell that operator the number and location of the person to be called. The operator would then call back with the other person manually switched onto the line. This procedure was a serious ritual as telephones were still rare and for exceptional use. The operator could also hear the entire conversation ... What a change in home telephony, since! The most lucky users have a wireless, digital telephone handset with error correction, which they can carry anywhere on their premises without change in signal quality or loss of connection. These phone appliances come up with scores of programmable functions, the most popular ones being voice messaging and automatic dial-up and call-back. Not to mention the other services offered by the operator. Wireless phones, with similar features and services following you almost anywhere in the world, complete the picture of our new telephonic environment.

The instantaneity and quality of the telephone connection makes many forget the highly complex reality of the network, as described in the Introduction. Before entering into the details of this complex communication infrastructure,

Wiley Survival Guide in Global Telecommunications: Signaling Principles, Network Protocols, and Wireless Systems, by E. Desurvire
ISBN 0-471-44608-4 © 2004 John Wiley & Sons, Inc.

it is worth recalling the basics of the public switched telephone network, also referred to as *PSTN*. In small towns, residential or rural areas, the PSTN is formed by those tiny cables supported by telephone poles disposed along streets and roads. In most towns and areas, however, the telephone cables are buried. It is not only for esthetic purposes, but also for robustness, as external wires are exposed to a variety of hazards. The most lethal is icing, which makes the cables break under the ice's weight. Storms, falling trees, and hunters' shots complete the list of threats.

2.1.1 PSTN Topology

The standard layout or topology of a PSTN is shown in Figure 2.1. It consists in an array of *central offices* (CO) or *exchanges* linked by *trunks* or *junctions*. Each CO manages its own set of local lines (referred to as the local *loop*). In modern networks, the trunks carry digital signals, while the loop carries analog signals, as supported by the old-fashioned *twisted copper-wire pairs*. The typical distance between a customer phone terminal and the CO is 12,000 feet or 3.6 km. More remote terminals require in-line *repeaters*, which regenerate the signals weakened by transmission loss through the cable.

The twisted pairs coming from phone terminals (homes, buildings) are bundled together into telephone cables of increasing size as the CO location is progressively approached. In the CO basement, telephone cables can have as many as 4,000 twisted pairs, which provides an idea of the actual physical system. As the number of subscribers grows, it becomes therefore impractical to assign a dedicated twisted-pair line to each (a principle familiarly referred to as "one dog, one bone"). The problem is solved by grouping together a subset of subscriber lines. This is done by first converting the voice signals from analog to digital, then by digital channel *multiplexing*. Multiplexing means that several voice channels are being transmitted simultaneously through a single wire pair, but at a higher overall frequency.

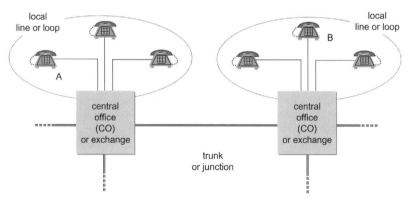

FIGURE 2.1 Basic layout of public telephone switched network (PSTN), showing local loops connected by central offices (CO) or exchanges.

See Section 2.3 for further description. It is shown there that for instance, nearly 100 phone terminals can be connected to a given CO through 20 wire pairs, instead of 2×100 wires. This illustrates the wiring economy that can be achieved by multiplexing, and such a basic concept remains valid at any network scale.

The function of the PSTN is to connect together the different telephone subscribers on a one-to-one basis. This is achieved by creating a physical link, or *connection circuit* between them, upon request of the calling party. For the local loop, one could conceive of a permanent, dedicated network of wires connecting each subscriber to all the others, as shown in Figure 2.2. It is easily established that for N subscribers, the number of permanent connections is $N(N-1)/2$. Thus, the number of required connections grows as the square of the number of users, which is wholly impractical. Instead, visualize that the dashed lines in the figure represent temporary connections, or *switched circuits*. Note that there are not as many physical switches in the CO as there are possible connection configurations, since it is unlikely that all subscribers would use their phones at the same time. If the call concerns a subscriber outside the local loop, then a long-distance circuit connection is generated between the two COs involved.

How does the CO identify the number requested for connection? Forget the times when one had to tell the number to the operator (although we still have to do this in case of connection problems). We are familiar with rotating dials and touch-tone dials, which have replaced them. For each number selected by the finger $(1,2 \ldots 9,0)$, the rotating dial generates a clicking signal whose duration

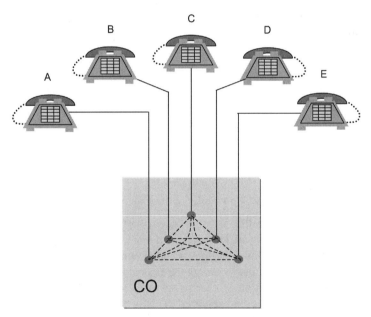

FIGURE 2.2 Connecting users within a local loop: dashed lines represent connections or switched circuits that are either permanent (impractical approach) or temporary (practical approach).

corresponds to the number position (zero being the longest). The sequence of clicks, if patiently and properly dialed (Remember this?) is then identified by the CO. The more practical and rapid "touch-tone" dialing was introduced by AT&T in 1963. It is based upon a different principle, *called dual-tone multi-frequency* (DTMF). Each number is identified by a pair of closely-spaced tones (hence sounding dissonant) which we hear like a little song as dialed in sequence. The tones frequency pairs have been chosen in order not to be mistaken for the human voice. Compared to the rotary technique, dialing is ten times quicker, i.e., about one second for CO acquisition. DTMF can also be used to handle servicing through the phone connection, for instance to interrogate your bank account with your confidential code, or to be guided through the directory of services in a public institution.

2.1.2 Making a Phone Connection

A circuit-switched connection enables two phone terminals to be in real-time communication, referred to as *duplex* (*simplex* referring to one-way communication). How is such a circuit connection actually initiated and established between the two phone terminals? Consider the case of person A located in a first CO loop (CO1) calling person B in a second CO loop (CO2), as shown in Figure 2.1. Here are the successive stages:

1. Person A picks up the handset.
2. CO1 senses the "off-hook" mode and sends *dial tone* to A.
3. Person A dials the number.
4. CO1 identifies this number as located in the CO2 loop, and sends request to CO2.
5. CO2 acknowledges to CO1, sends ringing signal to B, while CO1 sends a *ringing tone* to A.
6. Person B picks up the handset, which is sensed by CO2.
7. The voice circuit connection between A and B is established.
8. At the end of the conversation, the two persons hang up their handsets and CO1/CO2 disconnect the circuit.
9. CO1 records the date, time and length of the phone call for billing.

This list looks very familiar, since we know all of these signaling tones so well. The intriguing fact is that we take it for granted that a given CO is capable of handling hundreds of thousands of call requests and circuit connections simultaneously, from local to long-distance, and recording the associated history, in addition to many other service functions. Yet on Mother's Day or New Year's Eve, even this enormous handling capacity may be exhausted! This is because the CO does not have as many circuit switches as there are subscribers, as it is unlikely that *all* subscribers would place a call request simultaneously (yet such a situation might be approached on these special occasions).

2.1.3 Interoffice Trunking and PSTN Environment

The different COs in an operator's network are connected by trunks. For analog interoffice connections, the trunks are usually made of 4-wire, two-pair cables, with analog repeaters located every 10–15 km at most. The reason for using two pairs instead of one comes from circuit stability issues when repeaters are on the path. Figure 2.3 shows the repeater configuration in a 2-wire trunk. The repeater boosts the incoming signals that have been attenuated by transmission loss. It also has re-shaping and equalizing functions to compensate for various distortion causes. Signal boosting is done through analog, unidirectional amplifiers. The 2-wire line must therefore be split into two one-way lines, which is achieved through hybrid transformer circuits, or *hybrids*. The gain of each of the amplifiers is set to the formula

$$G_{dB} = T_{dB} + 2H_{dB} + E_{dB} \tag{2.1}$$

where all terms are expressed in *decibels* and in absolute value (see Chapter 1). The term T accounts for the line transmission loss, corresponding to the previous trunk segment (e.g., 1 dB). The term H is the throughput loss introduced by the hybrid on each splitting branch (e.g., 3 dB). The term E is the excess signal loss occurring between the two hybrids in either branch (e.g., 4 dB). So in the example we have $T + 2H + E = 11$ dB, thus he amplifier gain is set to $G = 11$ dB. Unfortunately, there is signal leakage in the hybrids. Some of the signal power of the upper branch makes it to the lower branch, and conversely. As can be shown through an exercise (see Exercises at the end of the chapter), all the conditions are met for ring oscillations, which saturate the amplifiers. This phenomenon is called *repeater instability*. The solution to this problem is shown in Figure 2.4. It consists in using two dedicated wires per path (referred to for clarity as "eastbound" or "westbound"), each path having its own amplifier chain. In this case, the net transmission loss is reduced (no hybrid loss at each amplifier stage) and the line does not suffer from instability.

With optical interoffice trunking, the links connecting COs are made of single optical fiber pairs, with optical amplifiers every 50–100 km. Each fiber carries

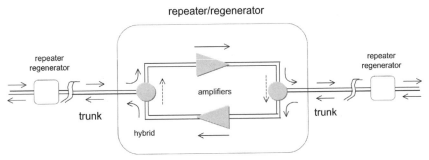

FIGURE 2.3 Basic layout of a 2-wire analog repeater. Arrows show signal propagation directions. Hybrids are used to split the 2-wire trunk into a 4-wire system, with separate amplification branches. Dashed arrows represent signal leakage through the hybrids.

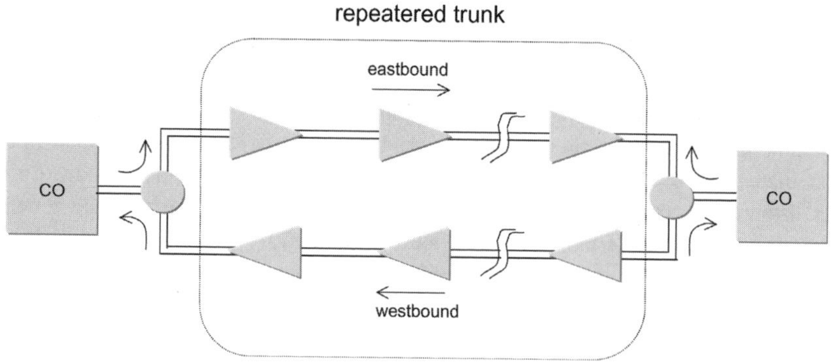

repeatered trunk

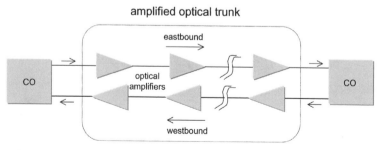

FIGURE 2.4 Basic layout of a 4-wire interoffice trunk with analog repeaters.

several wavelength channels, in the form of a single wavelength-multiplexed optical signal. Optical amplifiers operate in a way similar to analog amplifiers, with the exception that the wavelength-multiplexed signal is boosted at once, without channel-by-channel demultiplexing. They are also unidirectional, although some specialized configurations make possible two-way, bidirectional amplification. The trunk apparatus is shown in Figure 2.5. Optical amplifiers are often designated as *optical repeaters*, which is not accurate but accepted. Their only function is to boost the signals, although filtering features for wavelength equalization purposes can be included. Since optical fibers have the lowest possible loss (0.2 dB/km at 1.55 μm wavelength), optical amplifiers/repeaters can be spaced as far apart as 100 miles (160 km), the average distance between major cities in the United States. Another difference with the analog trunk cable is that optical repeaters can be powered directly from the CO stations, instead of locally, at each repeater stage. The trunk then includes concentric copper wires for the electrical feed. This feature is implemented in submarine cable systems, which can contain up to 12 active fiber pairs with hundreds of wavelength channels each. Interestingly, these cable types have only one copper wire, since the sea is used as the return or ground. It is also worth mentioning that inter-CO optical trunks can be implemented without optical line repeaters, owing to the comparatively low transmission loss of fibers, and the high optical receiver sensitivities that can be achieved.

amplified optical trunk

FIGURE 2.5 Basic layout of a single fiber pair interoffice trunk with optical amplifiers.

Several COs can be connected to form a network. A preferred network topology is the *ring*, Figure 2.6. As the figure shows, one or several COs can be also connected to the outside, meaning to a wide-area network (WAN), via interexchange carriers (IXC) and/or local ISPs. The physical location where outside entities such as IXCs and ISPs provide their access to the PSTN are called *points of presence*, or POP. Such a hierarchy constitutes the framework in which PSTNs operate. Because of its historical longevity and high popularity, the PSTN commodity is also referred to as *plain old telephone service*, or POTS. Here, the world *service* corresponds to the entity providing a response to a given connection request or transmission need. It should not be confused with the *application*, i.e., the type of connection or transmission. For instance, a facsimile transmission or a connection to the Internet are both applications. The service is first provided by the local, then the long-distance operator. With the Internet, the connection request is passed from the PSTN to the *Internet service providers* (ISP), for which the PSTN acts as an *Internet gateway*. Therefore, the PSTN is a local multiservice or multivendor entity, with a wide range of applications of which *voice-circuit switching* is the most elementary and revenue-profitable. This does not preclude that the same circuit-switching infrastructure and twisted copper wires be used to carry signals other than voice,

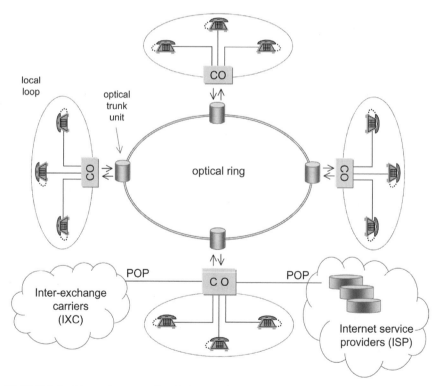

FIGURE 2.6 Network connecting independent central offices (CO) via an optical ring. Connection to a wider network is made through interexchange carriers (IXC) or Internet service providers (ISP). POP: points of presence.

namely video and (IP-based) data, which is extensively described in the other chapters.

2.1.4 Private Branch Exchanges (PBX) and Centrexes

The conceptual derivative of the central office (CO) in the PSTN is the *private (automatic) branch exchange*, or PBX (PABX). The PBX (or PABX), is a privately owned telephone network which operates on local premises. It is also connected to the PSTN for outside calls. It is installed in all major businesses, as the central node for all telephones wires. Colleagues can call each other at no charge (and using simpler phone numbers called *extensions*), while the company is billed at once by the CO for outside calls. Telephone terminals can be activated at will by the PBX for outside, or long-distance or international calling, according to employees' needs.

One of the key applications of PBX is voice messaging. DTMF dialing is extensively used to access *voice mail* through a confidential code, to store, delete or replay messages, and for many other functions such as caller identification on liquid-crystal displays (LCD), *audio-conferencing* (several phone lines connected together) and *call forwarding* to a collaborator when one must be temporarily out of office.

The *Centrex* offers a service similar to the PBX, i.e., a centralized and dedicated phone system for the needs of a single business or institution, except that the switching is made at the CO level. The centrex offers the same functionality as with the PBX, but it operates in multiple business or institution sites (e.g., bank branches, large business centers, university campuses, administrations). The advantage is that there is no need for installing or duplicating PBXs on each premises. A second advantage is centralized billing and discounts based on global subscription packages.

2.1.5 Integrated Services Digital Networks (ISDN)

With the development of digital communications (see Section 2.3), new networking services could be offered through PSTN owners, including *nonvoice* applications such as *facsimile, telex, video-telephony/conferencing*, remote-sensing/control systems, and other computer data links. Hence the name of *integrated service digital networks*, or ISDN, which appeared in the late '70s. Because of its slow and erratic beginnings, ISDN has often be described with humor as meaning "Innovation Subscribers Don't Need," or "the technology which took 15 years to become an overnight success."

The ISDN break down into *bearer services, supplementary services* and *teleservices*. The first represent the minimum provision of a analog/digital phone connection at 3.1 kHz/64 kbit/s, but also a nonvoice, 64 kbit/s data connection. Supplementary services concern improvements in telephone functionality, some of which were already available in the earlier PSTN, but with a wider range of features. To quote only a few of these functionalities:

- Call diversion (automatic call routing to another party);
- Automatic call-back when called-party line is available (known as *call connection to busy subscriber*, or CCBS);

- Direct dialing of PBX numbers without PBX operator mediation;
- Identification of the name/number of the calling party before picking up the call;
- Multipoint conference call from desktop, including associated computer connections;
- Real-time billing information, etc.

The teleservices essentially concern telex, high-quality facsimile transmission, video-telephoning/conferencing, remote sensing (telemetry), electronic/radio messaging and many other on-line service possibilities. In these times of Internet and 3G mobile communications, some of these ISDN features have lost some impact, but they may be viewed as complementary services, when not the only option available within a given performance range or geographical location.

ISDN can be implemented via two types of interface: the *basic rate interface/access* (BRI or BRA) and the *primary rate interface/access* (PRI or PRA), which we describe next.

The BRI connects directly to the ISDN through an ordinary telephone line, or alternatively via a PBX or a centrex. To connect with basic twisted-pair wires, the distance between the customer premises and the ISDN exchange must not exceed 7 km. Elementary implementation of BRI in the customer premises requires *network terminating equipment* referred to as *NT1*. The NT1 physically converts the two-wire twisted-pair into a 4-wire or two-channel circuit, plus a single-wire data/signaling circuit, as illustrated in Figure 2.7. The two channels are called *B-lines* (or *B-channels)* and operate at 64 kbit/s. The data or signaling circuit is called a *D-line* (or *D-channel*) and operates at 9.6–16 kbit/s. This is referred to as a 2B + D layout.

The customer is thus equipped with a total potential capacity of $2B + D = 2 \times 64$ kbit/s $+ 16$ kbit/s $= 144$ kbit/s. The two B-channels can be used at the same time for instance for 128 kbit/s video-conferencing. Alternatively, the two B-channels can be simultaneously used for separate purposes, such as digital

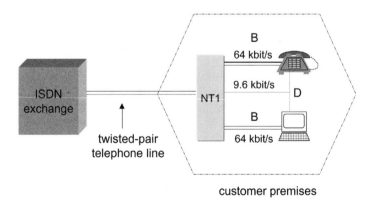

customer premises

FIGURE 2.7 Basic rate interface (BRI) for ISDN connection, showing network terminal equipment (NT1) which divides the incoming telephone line into two B-channels at 64 kbit/s and one D-channel at up to 16 kbit/s.

voice (ISDN telephone) and data communication (personal computer), with the C-channel as additional signaling/data link. The actual bit-rate specification of the interface is 192 kbit/s, which comprises the 144 kbit/s of customer-use capacity and 48 kbit/s of control overhead. This control bandwidth makes it possible to implement a *local bus network*, with up to eight terminals and up to 500 m length. The D channel ensures that the proper terminal picks up the incoming calls, which can be of different types (voice, fax, data or video).

The PRI represents an upgrade of the BRI, see Figure 2.8. It uses a second terminal equipment called *NT2*, and is referred to as a 23B + D layout (North America and Japan) or 30B + D layout (Europe). As the name indicates, the NT2 makes it possible to split the incoming bandwidth into 23/30 B-channels at 64 kbit/s, with the signaling D channel now operating at 64 kbit/s. Typically, the NT2 is a PBX, whose functions are to manage call concentration and routing. As with the BRI, the calls handled by the NT2/PBX may concern any types of digital signals, namely voice, fax, data, or video. As with the BRI, the total capacity can be assigned to a single, high-bandwidth service, but this time there is more bandwidth available and there are more options to provision it. The different bandwidth options, which are labeled *H-channels*, are summarized in Table 2.1. The maximum capacity is achieved with the H12-channel, namely 30×64 kbit/s $= 1.92$ Mbit/s. As shown in Section 2.3, the 23B + D granularity (24×64 kbit/s) makes the approach compliant with the first digital hierarchy level used in North America and Japan for voice multiplexing (as referred to as T1 or J1, respectively).

Broadband ISDN, called B-ISDN, represents the bandwidth extension of ISDN made possible by using several 64 kbit/s lines in parallel. The high-speed aggregates are first demultiplexed, then transmitted through separate paths, then remultiplexed after proper sequencing. At high bit rates, such a transport requires highly-reliable packet-switching networks and routing protocols, for which ATM (*asynchronous transfer mode*, see Chapter 3) represents key technology.

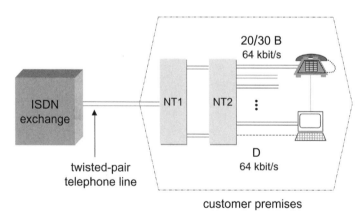

FIGURE 2.8 Primary rate interface (PRI) for ISDN connection, showing network terminal equipments (NT1 and NT2) which divide the incoming telephone line into two 23/30 B-channels at 64 kbit/s and one D-channel at 64 kbit/s.

TABLE 2.1   **H-channels used in PRI ISDN**

H-Channel	Number of B-Lines	Capacity
H0	6	384 kbit/s
H10	23	1.472 Mbit/s
H11	24	1.536 Mbit/s
H12	30	1.920 Mbit/s

The extended services offered by B-ISDN concern fields as varied as high-resolution video-conferencing, multimedia networking, teleshopping, teleworking, tele-education and telemedicine. Such services complete, and customize in a different way, similar potentials now offered by the Internet.

■ 2.2 ANALOG FREQUENCY-DIVISION MULTIPLEXING

Analog voice channels occupy a relatively limited bandwidth of 300 Hz – 3.5 kHz. It is therefore sensible to combine several voice channels into an aggregate signal, through frequency-division multiplexing (FDM). The principle of FDM is detailed in Chapter 1, Section 1.1. Basically, the frequency spectrum is sliced into a group of 4-kHz bands, each of which carries a baseband voice signal, or tributary channel. Thus tributary 1 is found in the 0–4-kHz band centered at $f_1 = 2$ kHz, tributary 2 in the 4–8-kHz band $f_2 = 6$ kHz, and so on. Voice channel to band attribution is made randomly, on a band-availability basis. The voice signals are used to modulate the available carriers, which generates two symmetrical side-bands having the same information, as shown in Figure 1.7. In order to save spectral space, only one of the side-bands is selected by use of a bandpass filter. The FDM signal is thus made of the analog superposition of all the frequency-shifted, single-side-band spectra, as illustrated in Figure 1.7. At the other end of the transmission line, demultiplexing is achieved by passing the FDM signal through a comb of bandpass filters centered at frequencies $f_1, f_2 \dots f_n$. The output of each filter is a baseband voice tributary.

The early telephone systems used FDM with up to 24 channels, resulting in 24×4 kHz = 94 kHz aggregates. Analog-FDM prevailed until the mid-70s, when it was overtaken by digital coding (PAM and PCM) and time-division multiplexing (TDM) (see Chapter 1, Section 1.3). Analog transmission in wired telephone networks had several drawbacks. The wires would act as antennas and effectively pick up radio noise falling in the multiplex band. Also, line amplifiers used periodically to repower the signal contributed to their own frequency noise, which accumulated in proportion with the line distance. Finally, FDM signals could leak between adjacent wires, resulting in a parasitic voice-background noise appropriately called cross-talk. For long-distance calls, the result was a poor quality signal where voice was heard over a loud "shhhhhhhh" background noise, sometimes along with other conversations.

2.2.1 FDM Hierarchy

Yet analog FDM remains a key technique in multiple-access radio systems (Chapter 4). Apart from the better transmission quality of radio waves, one of the reasons for choosing FDM is the versatility of frequency-band allocation, permitting multiple users located at different points in space to share the frequency spectrum effectively. A specific FDM hierarchy had to be created, based upon various combinations of submultiplexes. Figure 2.9 shows the different multiplexing stages.

The starting level of the hierarchy is a 12-channel aggregate, called a *group*. The FDM operation is achieved with *channel-translating equipment*, or CTE (the name comes from the fact that each analog signal must be frequency shifted before the MUX operation, as we have seen). In the CTE, the carrier frequencies for channel k are $f_{c,k}$ (kHz) $= 60 + k \times 4$, meaning that for 12 channels the carriers range from $f_{c,1} = 64$ kHz to $f_{c,12} = 108$ kHz. Carrier modulation by the 4-kHz voice signals produces two side-bands, which occupy the frequency range $[f_{c,k} - 4$ kHz $f_{c,k} + 4$ kHz$]$. Selecting only the lower side-band (by convention), each of the channels occupies the range $[f_{c,k} - 4$ kHz$, f_{c,k}]$, namely 60–64 kHz for channel 12 to 104–108 kHz for channel 1. Thus, the group bandwidth spreads from 60 kHz to 108 kHz (channel 1 to channel 12) and is therefore 48 kHz. Before the FDM signal is sent through the 4-wire line, it is economically frequency down-shifted by mixing with a $f_c = 120$ kHz carrier. After lowpass filtering, this results in a single-side-band signal spreading from $[f_{min} = f_c - 108$ kHz$, f_{max} = f_{min} + 48$ kHz$] = 12$ kHz–60 kHz.

The next stage combines five groups coming from five 4-wire lines through a *group-translating equipment* or GTE. The 60-channel GTE output is called a *supergroup*. The supergroup bandwidth is then 60×4 kHz $= 240$ kHz. Since the five incoming groups are in 12 kHz–60 kHz bands, they must be frequency-shifted

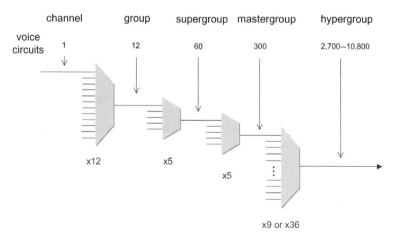

FIGURE 2.9 Hierarchy for analog voice channel frequency-division multiplexing, from channel, group, supergroup, mastergroup and hypergroup. Frequency conversion and multiplexing are realized by channel-translating, group-translating and supergroup-translating terminal equipments (CTE, GTE and STE).

according to the same principle as in the previous stage. The groups k are mixed with carriers at frequencies $f_{c,K}$ (kHz) $= 312 + k \times 48$, such that for 5 groups the carriers range from $f_{c,1} = 360$ kHz to $f_{c,12} = 552$ kHz. The lower side-band in each group occupies the spectrum $[f_{c,K} - 48$ kHz, $f_{c,k}]$. Thus, the resulting super-group signal ranges from 312 kHz to 552 kHz.

The five-fold multiplexing of supergroups coming from five 4-wire lines, via *supergroup-translating equipment* (STE), results in a $5 \times 60 = 300$-channel *mastergroup*. Frequency-shifting of the supergroups is done in the same way as previously described, with a resulting mastergroup spectrum of bandwidth 300×4 kHz $= 1.2$ MHz ranging from 312 kHz to 1548 kHz. Two other alternatives are $15\times$ and $16\times$ multiplexing of supergroups, as referred to as "basic hypergroups."

The following *hypergroup* level can be either 9 mastergoups (2,700 channels) or 36 mastergroups (10,800 channels). The two mastergroup types have different bandwidths and frequency ranges. The first has a bandwidth $2,700 \times 4$ kHz ≈ 12 MHz, ranging from 312 kHz to 12,300 kHz, corresponding to 1,330 kHz per hypergroup band. The second begins at frequency 4,400 kHz and ends at 59,580 kHz, corresponding to bands 1,530 kHz per hypergroup band. The two hypergroup types are referred to as "12 MHz" and "60 MHz," respectively.

The different bands occupied by FDM channels, groups, supergroups, mastergroups and hypergroups are shown in Figure 2.10.

■ 2.3 PLESIOSYNCHRONOUS MULTIPLEXING

In the previous section, it was shown that *multiplexing* voice channels into a single signal makes it possible to simplify the wiring and reduce the number of physical connections. The two main techniques are frequency-domain multiplexing (FDM) and time-domain multiplexing (TDM). The principle of FDM is described in Chapter 1. Here, we shall focus on TDM, which requires preliminary voice-signal

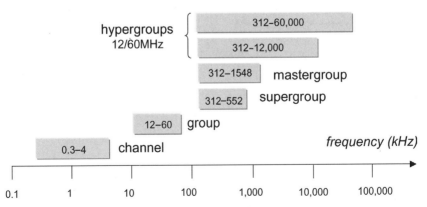

FIGURE 2.10 Frequency bands allocated to channels, groups, supergroups, mastergroups and hypergroups (two types).

conversion from analog to digital (referred to as *pulse-code modulation* or PCM). In Chapter 1, it is shown that analog signals are a superimposition of modulated waves spreading over a certain frequency range. For voice signals, this range is approximately 300 Hz–3.5 kHz (the full hearing range being actually 20 Hz–20 kHz for human beings). The *Nyquist theorem* states that for proper digital conversion, analog signals must be sampled at twice the highest frequency component. For voice channels, the *sampling rate* has therefore been set to 2×4 kHz = 8 kHz, corresponding to 8,000 samplings per second. Since each digital sampling is coded into a single *byte* (8 bits), the resulting data rate is $8,000/s \times 8$ bits = 64 kbit/s. Therefore, 64 kbit/s is the elementary bit rate for voice channels, from which we can build the multiplexing hierarchy.

2.3.1 T-Span Multiplexing and Framing

Let's consider an example to illustrate the wiring simplification introduced by multiplexing. Assume for instance a local cluster of 96 phone subscribers. This represents no less than 96 twisted-wire pairs (192 wires total) to draw to the CO. Instead, let's multiplex them together to form 4 subgroups of 24 channels, i.e., 4 subgroups at 24×64 kbit/s = 1.536 Mbit/s. Call any subgroup "T1" (or T-span). We can include an extra T1 as a spare. So we now have five T1s to transmit to the CO through four-wire connections (see Section 2.1), representing 20 wires total, instead of 192. Note that the trunk is based on 4-wire connections instead of 2-wire connections for each T1 multiplex. It is also important to note that the line bit rate is 24 times higher, which implies time compressing the 64 kbit/s channels by 24-fold.

The TDM operation is achieved by *queuing* together the voice channels in a byte-by-byte fashion, as shown in Figure 2.11, in a basic example involving four channels. The device is commonly referred to as MUX. The MUX output consists of a stream of 4-byte frames having a four-fold bit rate. In order to be able to identify where the frames start, a *framing bit* is added at the beginning of each frame. In this example, the frames have $(4 \times 8) + 1 = 33$ bits and occupy the same time duration as the original channel byte (8 bits). The actual bit-rate increase is therefore $33/8 = 4.125$. Interestingly, we note that within each frame, the voice-channel rate (at which a given channel byte is transmitted) remains the same. The byte compression is just the way to make enough room for the other channels to be transmitted within the same overall duration. Put simply, the resulting TDM frame rate is the same as the input voice-signal rate.

The reverse operation of MUX is called *demultiplexing*, or DMUX. It can be interpreted in the same way as in Figure 2.11, but read from right to left instead. Since the trunk is bidirectional, the actual TDM device includes both MUX and DMUX functions. MUX is for the outbound traffic (towards CO), and DMUX is for the inbound traffic (from the CO). We refer to the lowest bit-rate channels as *tributaries*, by analogy to rivers.

Consider now *T1 multiplexing* (also called *T-span*). We combine together 24 channels at 64 kbit/s. Each TDM frame has $(24 \times 8) + 1 = 193$ bits, including the extra frame bit. Since the original frame rate is 8,000 times per second

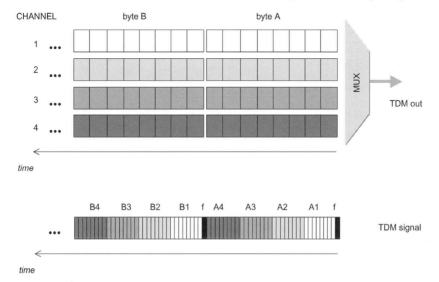

FIGURE 2.11 Byte-to-byte time-domain multiplexing of four channels (top) and resulting output (bottom). Note inclusion of a frame bit (f) at the beginning of each 4-byte frame multiplex. Demuliplexing is the same operation as read from right to left.

(or 8 kHz), and remains unchanged by TDM (as previously seen) the resulting TDM bit rate is 193 bits $\times$ 8,000/s = 1,544,000 bit/s or 1.544 Mbit/s.

2.3.2 Plesiosynchronous Digital Hierarchy

The same operation as T1 multiplexing can be repeated at higher TDM levels. The resulting succession of increasing TDM rates is referred to as *plesiosynchronous digital hierarchy*, or PDH. The word plesiosynchronous refers to a network where each MUX/DMUX station operates at the same clock rate, but the clocks are not locked to each other. Thus resulting TDM signals have the same nominal bit rates, but they can be slightly out of synchronization over time because their respective clock rates are not strictly identical. Let's now construct the PDH scale as implemented in North America, and as defined by the International Telecommunications Union (ITU).

We call DS0 (for digital signal zero) the elementary 64 bit/s voice channel. As previously seen, the next level is T1, which we also call DS1, and which corresponds to 24 $\times$ DS0. We then define T2 (or DS2) as 4 $\times$ DS1, T3 (or DS3) as 7 $\times$ DS2, and DS4 as 6 $\times$ DS3. Figure 2.12 shows the resulting hierarchy. As previously explained, the DS1 rate is slightly higher (1.544 Mbit/s) than 24 times 64 kbit/s (1.536 Mbit/s), because of the inclusion of framing bits. As the TDM process is upgraded to successive stages, clock deviations from each tributary become more important, and therefore extra bits must be included, which are called *stuffing* (or *justification*) bits. Take for instance the case of T2 (6.312 Mbit/s) which corresponds to four T1s (1.544 Mbit/s). The T2 bit rate is slightly higher than four times

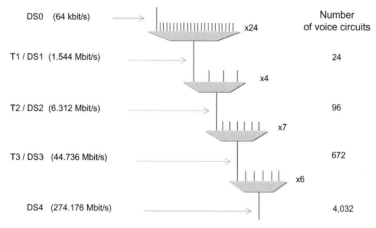

FIGURE 2.12 Plesiosynchronous digital hierarchy, as used in North America, showing successive TDM stages from original T1 voice channels. The resulting TDM bit rates, which include extra framing and stuffing bits, and the corresponding number of voice circuits are shown for each TDM stage.

the T1 rate, which is explained by the adjunction of the framing and stuffing bits (see Exercises at end of this chapter).

To make the PDH picture a bit more complex, Europe and Japan have developed their own systems, as based upon different E1, E2 ... and J1, J2 ... hierarchies. Fortunately, they offer some mutual correspondence with the T1, T2 ... of the North America system, as illustrated by the data summarized in Table 2.2.

TABLE 2.2 Plesiosynchronous hierarchies in North America, Europe and Japan, with level definitions (K = kbit/s, M = Mbit/s)

Level Definition	North America	Europe	Japan	Number of Circuits
DS0	64 K	64 K	64 K	1
DS1/T1/J1	1.544 M	–	1.544 M	24 (24 × DS0)
E1	–	2.048 M	–	32 (32 × DS0)
DS1-c	3.152 M	–	3.152 M	48 (48 × DS0)
DS2/T2/J2	6.312 M		6.312 M	96 (4 × DS1)
E2	–	8.448 M	–	128 (4 × E1)
J3	–	–	32.064	480 (5 × DS2/J2)
E3	–	34.368 M	–	512 (4 × E2)
DS3/T3	44.736 M			672 (7 × DS2)
DS3-c	91.053 M	–	–	1344 (14 × DS2)
J4	–	–	97.728	1440 (3 × J3)
E4	–	139.264 M	–	2048 (4 × E3)
DS4	274.176		–	4032 (6 × DS3)
	–	–	397.2 M	5760 (4 × J4)
E5	–	564.922 M	–	8192 (4 × E4)

We note that the hierarchy level increases differently in the three regions, with only a few matching occurrences. At the fourth multiplexing level, J4, E4, and DS4 aggregate between 1,440 and 4,032 voice circuits. The bit rates increase accordingly, but the requirement for more framing/stuffing bits with increasing hierarchy level in fact decreases the proportion of transmitted *payload*, i.e., the number of bits actually representing the voice channels. For instance, E5 corresponds to 8,192 voice channels (Table 2.2), or $8,192 \times 64$ kbit/s = 524.288 Mbit/s of payload. The actual E5 rate being 564.922 Mbit/s, the payload efficiency is $524.288/564.922 = 93\%$. In comparison, the payload efficiency of E2 is 97%. In reality, the efficiencies are a little smaller due to the fact that some of the bytes of the E1 ... E5 frames are used for framing and signaling.

In T1 systems, the 193-bit TDM frames can be grouped together by packets of 12 (*superframe*, referred to as D4) or 24 (*extended superframe* or ESF). If we put these T1 frames on top of each other, like a matrix with 193 columns and 12 or 24 rows, we get the result shown in Figure 2.13. This is where the approach is interesting. Indeed, one does not need 12 or 24 bits for framing reference. Therefore, the extra frame bits can be reallocated to other useful signaling purposes. One signaling function, for instance, concerns *terminal synchronization*: alternate "frame" bits (6 for D4, 12 for ESF) are set to the same "1" value, which provides a 4-KHz clock frequency. Another example: in the ESF, frame bits 2, 6, 10, 14, 18 and 22 are used for *error-correction*, which makes it possible, as the name indicates, to detect and restore incorrectly received bits (see Chapter 1).

In E1 systems (Europe) the TDM framing principle is different. First, as Table 2.2 shows, it concerns a 32-circuit multiplex. Second, as illustrated in Figure 2.14,

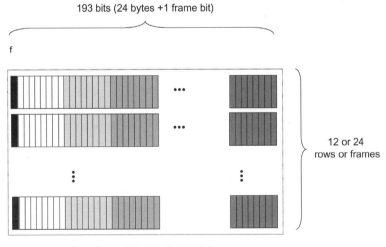

193 bits (24 bytes +1 frame bit)

f

12 or 24 rows or frames

Superframe (12x193=2,316 bits), or extended superframe (24x193=4,632 bits)

FIGURE 2.13 Superframe (or extended superframe) format based upon T1 transmission level, as represented by a matrix of 193 bit-columns and 12 (or 24) rows. The dark bits (f) are T1 framing bits.

256 bits (32 bytes)

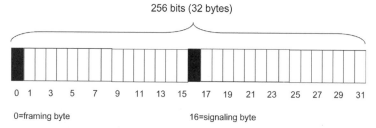

0 1   3   5   7   9   11 13 15   17 19 21 23 25 27 29 31

0=framing byte   16=signaling byte

FIGURE 2.14 A 32-Byte frame format corresponding to E1 transmission.

the first and the 16th byte of the frame are used for framing and signaling, respectively. This leaves only 30 bytes of voice payload, or 31 bytes if signaling is not required. The framing byte makes it possible to achieve synchronization as clock rates may vary slightly from one exchange to the other. The inherent default of synchronization results in underflow (clock too slow) or overflow (clock too fast) of bits. As previously mentioned a few paragraphs above, such under/overflow is compensated by the use of extra *stuffing* or *justification bits*. These extra bits are set to dummy values. If the bit rate turns out to be above the nominal value, the resulting extra bits are substituted to as many justification bits. If the bit rate is under, more justification bits are present. The stuffing/justification bits are removed upon demultiplexing. Signaling through byte 16 is not always mandatory, which leaves room for an extra voice channel. Interestingly, signaling can also be made by "stealing" the lowest-value bits belonging to some voice channels. The loss of these bits results in a very slight decrease of voice quality which is not audible. Such a method of *robbed-bit signaling* only applies to voice transmission and is not allowed for data transmission.

The above has shown that North America/Japan and Europe have two different low-level PDH bit rates that are T1/J1 (1.544 Mbit/s) and E1 (2.048 Mbit/s). A phone call between the two region types thus requires conversions from one to another. It turns out that E1 has been kept as the international standard and T1/J1 operators must make the conversion from E1 or to E1 for incoming or outgoing international calls. The conversion requires fitting the 1.5 Mbit/s traffic into a 2 Mbit/s traffic or the reverse. In order to minimize the waste of bytes (also referred to as *time slots*) and payload, the frames must be split according to an optimized pattern. For instance, E1 (32 slots) can carry 24 slots of a first T1 and 8 slots of a second T1. The remaining 16 slots of the second T1 are then carried by a second E1, along with 16 slots of a third T1, etc. In this approach, each E1 carries two portions of two different T1 frames. This already does not look too simple, but the worst is that the voice bytes from each system do not correspond to the same speech-coding principles! These codes are referred to μ-law (T1) and A-law (E1); see Chapter 1. Therefore, digital-to-digital conversion from one speech coding to another is also required. In the case where data are transmitted instead of voice, the speech-code conversion must be turned off. To simplify the management of both voice and data transmission types, dedicated channels can be assigned for exclusive voice or data transmission.

We now move up to the next PDH level, considering T2 or DS2. Things would be so simple if we could just repeat the same multiplexing and framing operation as

in T1. Oddly enough, the DS2 framing standard consists in *bit-interleaving* the four
T1/DS1 frames, as opposed to byte-interleaving as in the previous multiplexing
level. Such an operation is illustrated in Figure 2.15. In addition, a control bit
(C-bit) is inserted every 48 bits of payload. We thus have four times the sequence
f + 48 + C + 48 + C + 48 + C + 48 + C = 197 bits, or 788 bits total. Adding to
this an extra stuff bit gives 789 bits, corresponding to the nominal rate of
789 bits × 8 kHz = 6.312 Mbit/s (Table 2.1). This whole operation, the conversion
of DS1 to DS2, is called *M12 multiplexing* (12 meaning one towards two). The conversion of DS2 to DS3 (7 × DS2), consistently called *M23 multiplexing*, follows
similar bit-interleaving rules, except that the C-bit is inserted every 84 payload
bits. The overall process of DS1 to DS3 conversion is then called *M13 multiplexing.*

As the reader is now fully aware, the PDH system is far from handy. Since the
tributaries are so intricately mixed together, it is in fact a tedious task to retrieve
them from the higher multiplex aggregates. One important network function consists
of extracting a given tributary from the multiplex and inserting another one in substitution. This process is called *add-drop multiplexing* or ADM. To achieve ADM
implies demultiplexing the tributary of interest all the way down to its hierarchy
level. An example showing ADM with T1 from a T3 signal is given in
Figure 2.16. The complexity of such a basic operation as ADM makes PDH
heavy to handle, difficult to manage and expensive to implement at high bit rates.
The answer to this dilemma is the *synchronous digital hierarchy* (SDH), which is
described in detail in the next chapter. In short, SDH is a more fancy way to aggregate and disassemble the tributaries. In particular, ADM at any sublevel is a straightforward operation, just like dropping and inserting boxes from the side of a pick-up
truck, without touching the rest of the truck payload.

■ 2.4 PACKET-SWITCHED NETWORKS

The description in previous sections concerned the multiplexing and transport of
digital voice signals between central offices (CO) within a PSTN. As we have

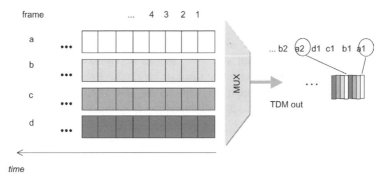

FIGURE 2.15 Principle of bit-to-bit time-division multiplexing, involving four input
tributary frames.

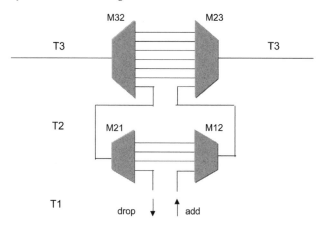

FIGURE 2.16 Add-drop multiplexing of tributary T1 from T3 signal at intermediate network exchange.

seen, data signals (as opposed to voice) can also be mixed up with this traffic. The essence of the PSTN is to be a *circuit-switched* network. Thus a given connection between two users requires mobilizing a given physical link or set of wires for as long as the connection is to be maintained. Since the CO switch can only connect a subset of all possible users at a given time, traffic saturation is possible. Fortunately for telephone subscribers, the network is not generally used at full capacity. But this does also imply that such a network is intrinsically inefficient for extensive data communications. The alternative approach is given by *packet switching*.

The technique of packet switching, developed in the 1970s, consists of breaking down the payloads into many individual *packets* (or *cells)*. The packets are then transmitted separately, sometimes along different networks. In addition to their respective payloads, the packets are labeled with a position number and a destination address, along with other useful signaling information. This series of data is confined into a packet *header*, which is appended to the payload. Generally, the non-payload data are referred to as *overhead*, regardless of their packet position. The overhead can represent a substantial fraction of the total packet, but it represents vital information without which the packet would be unmanageable to the network. A sequence of packets belonging to the same logical message is referred to by names such as *virtual circuits* or *virtual calls* (VC), or more generally, *logical channels*. It is thus seen that packet switching makes it possible to handle a certain number of different virtual circuits through a single transmission link. If *N* packets corresponding to different destinations (or logical channels) are in transit through the link, this link is thus actually handling *N* virtual connections or circuits simultaneously.

Upon reaching a given network node, the packets are first stored, then loaded into a *packet switch*. The switch reads the packet's destination address from the header and forwards the packet according to the optimal route and other traffic priority criteria. The optimal route corresponds to minimizing the number of switching nodes that the packet will have to traverse to reach its destination. But if a neighboring switch is momentarily busy, the route can be changed accordingly, through a longer path. Such a switching process is called *store-and-forward*. The function of packet storage, or *buffering* is essential to solve the problem of simultaneous packet arrivals at switch level. The incoming packets are generally forwarded according to a *first-in first-out* (FIFO) rule, unless certain priorities be pre-assigned to some of the logical channels. Figure 2.17 shows the principle of packet-switched networks. As illustrated in the figure, one important feature of the packet-switched network is that packets may arrive at their destination in random order, due to different route and storage delays. Buffering in the terminal makes it possible to rearrange them into the initial logical-channel sequence.

An important feature of packet-switching is that it becomes less efficient as the packet size increases. Indeed, long packets keep the switch in a given circuit configuration for a certain amount of time, during which other packets (of any size) may have to be buffered. Ideally, the packets should be short enough so that the number of switching operations per unit time is optimized. Clearly a tradeoff must be made with packet size, since the packets also contain overhead information. The smallest size would just be that given by overhead information, corresponding to zero payload. A switching network working only with "overhead information" would be useless. At the other extreme, if the payload is too significant, a given connection path is pre-empted for long periods of time, preventing its use by other network clients.

The configuration where packets from the same logical channel progress through different paths corresponds to *datagram routing*. In this case, some packets can be lost in the process, due to switching overload or routing errors. At a higher level of protocol sophistication, it is possible from the destination user to request the missing packets to be resent. An alternative approach consists of allocating the same route to the packet sequence associated with a given logical channel. The packets then reach the destination together, in the right sequence order. Such a configuration is referred to as *path-oriented routing*. It should not be confused with circuit switching. This is because the path is attached to the packet sequence, not to the switching configuration, which randomly changes as soon as the sequence has been fed into the line. The connection between two neighboring switches may then be either interrupted as the packets travel, or maintained to feed another

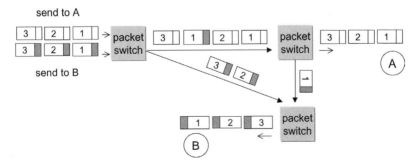

FIGURE 2.17 Principle of packet-switched networks.

sequence from a different logical channel. A drawback of path-oriented routing is that the network resources may not be fully utilized. Indeed, breaking up packet sequences into different paths makes it possible to reduce buffering delays, better share the switch configurations between the channels, and load the network pipes more fully.

Packet-switched networks can be used for both voice and data transmission, starting from limited-size, *local-area networks* (LAN), to larger-sizes *metropolitan* and *wide-area networks* (MAN, WAN), as discussed in other chapters. The most basic LANs concerns local data networks for connecting computer terminals. Specific *protocols* must then be defined to provide the rules whereby calls may be initiated, logical connections may be established and data may be exchanged between such entities. Two widely used packet-switched protocols for data communications are referred to as *X.25* and *frame relay*. Before entering, even superficially, into the details of X.25 and frame relay, it is necessary to define the different layers into which such protocols must be decomposed.

2.4.1 The Open Systems Interconnection (OSI) Model

The *open system interconnection* (reference) model, or OSI, defines networks through a standard set of layers. These layers, which number up to seven, are shown in Figure 2.18. They provide a model according to which different and modular network functions can be implemented in the proper sequence and hierarchy. From top to bottom, the *application layer* is the closest one to the user, since it deals with the perceived contents of the communication. The bottom *physical layer*, where the data are mechanically and electrically generated is the most remote from the user. Each layer performs its own task in order to qualify the data for the next one. The sending terminal proceeds from top to bottom, i.e. from the

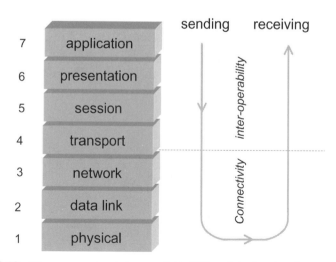

FIGURE 2.18 The seven network layers of the OSI model, showing the conceptual path from sending to receiving.

application (such as typing an e-mail) to the physical emission of "0s" and "1s". The receiving terminal proceeds from bottom to top, i.e., detecting the "0s" and "1s" to make sense of them all the way up to the application. Clearly, the physical layer is the one evolving the most slowly because its function is pretty much well defined or can be rapidly customized. On the other hand, the application layer has the highest potential for evolution or growth, because application needs are rapidly changing. This suggests that the greatest revenues, or market opportunities are to be found at this level, explaining the trend of telecom services. This remains true as long as the physical layer offers *infinite bandwidth* (unlimited bit-rate potential) which we know not to be the case (see a detailed discussion on Moore's law in Desurvire, 2004).

Let us now have a closer look at the different layers of the OSI model, which are also referred to as *layer 1* to *layer 7*. The function of the three bottom layers (1–3) is to ensure *network connectivity*. Connectivity is the series of functions that keeps the communication extant and maintained with adequate quality. It can be lost due to a variety of network failures found at these levels.

• **Physical layer 1** sets the electrical/optical data into a standard form, with the appropriate voltages, frequencies and waveforms. For layer-1 communications, there are numerous ITU standards. In this chapter, we have already seen a few that are used for voice and data: E1, T1 and PDH, and evocated the SDH protocol to be described in Chapter 3. Other protocols such as ADSL and WDM belong to layer 1.

• **Layer two (data link)** ensures end to end synchronization, data flow control, error detection and correction and multiplexing logical channels. The packets are wrapped into a frame. While the packets contain the final destination address, the layer 2 frame includes the *routing address* of the neighboring switch to which it must be sent first. The frames are surrounded at both ends by *flag* bytes, which are used as frame delimiters. The flag bytes have a fixed pattern, namely 0111 1110, which must NOT be found inside the rest of the frame. How can this be achieved, since such a pattern has a finite probability of occuring? The trick is the following: whenever a sequence of 0 followed by five 1s is detected, an extra 0 bit is inserted right after. So XXX111110 is forcibly changed into XXX0111110, whenever it occurs. Then the real closing flag is appended. At the receiving end, the processor detects the first flag pattern, then moves on with the other frame bits. Whenever it reads 111110, it "knows" that if the next bit (at left) is 0, it must be removed, while if it is a 1, it is the closing flag. It results from this reading/uncoding operation that (1) the flag pattern is never found between the two real flags, and (2) data integrity is unchanged. Protocols such as Ethernet, Token-passing, FDDI (see Section 2.5) and ATM (Chapter 3) belong to layer 2.

• **Layer 3 (network)** is concerned with the functions of routing (choice of fastest path) and congestion control (minimization of buffering delays). Routing can be *static* or *dynamic*. Static routing corresponds to a minimum number of hops approach, while dynamic routing chooses the best path according to actual traffic conditions. This can be achieved by analyzing the *tables* that switches exchange with each other. Such tables provide information about the traffic status at each switching node, the ports availability and possible switch congestions. Examples of protocols for layer 3 are X.25 (see Section 2.4) and IP (Chapter 3).

As shown in Figure 2.18, the second group of OSI layers (4–7) ensure *interoperability*. At this point, it is more logical to start from layer 7 down to layer 4, under which the connectivity layers are met.

• **Layer 7 (application)** has to do with the *meaning* of the data received by, or sent to the computer terminal. The word meaning is chosen by analogy with the use of a common and universal language that can be understood by different types of computer machines. Good examples of these "meaningful" applications are the connection to internet websites (*http://www . . .*) and the popular *e-mail* (see Chapter 3). The reader must have seen acronyms such as HTTP (*hypertext file transfer protocol*) and SMTP (*simple mail transfer protocol*). These make it possible to unambiguously send messages throughout the entire world network without being concerned by anything else (at least if the computer is properly configured and connected to some plug or antenna!). Other well-known layer 7 applications include *corporate intranets*, remote access to *databases* (e.g., X.500), and for system administrators, *network management*.

• **Layer 6 (presentation)** concerns the format into which the layer 7 data should be encoded (sending mode) or decoded (receiving mode). Anything we type on the computer keyboard at layer 7 must make sense in the presentation layer 6. The most well-known code is ASCII (American standard code for information exchange). It converts all keyboard characters (plus a set of other useful character functions) into seven-bit numbers (see conversion table in Chapter 1). An extended version of ASCII, used by IBM-compatible machines, is the EBCDIC (extended binary coded decimal interchange code) which encodes characters and functions over single bytes. Data *encryption* can also be performed at this level for confidentiality or security purposes. Layer 6 also enables *data compression* algorithms. For pictures and videos, the most popular ones are JPEG (joint photographic experts group) and MPEG (moving pictures experts group).

• **Layer 5 (session)** controls the opening, unfolding and closing of a given *session*. The concept of session is most familiar to PC users, since it corresponds to whatever happens between the calling and the closing of a given program. The most tangible aspect of a session is a connection to the Internet, during which all kinds of requests (web surfing, file downloading, message posting. . .) can be satisfied.

• **Layer 4 (transport)** takes care of signal quality, error correction, and overall traffic flow. It is also able to choose between different network possibilities (LAN, WAN, PSTN. . .) or to adapt these functionalities between different network types to be traversed, making use of *network addresses*. It is able to follow the request from layer 5 to handle simultaneous sessions, or to have a single session handled simultaneously by different network types.

This description of the OSI model is only introductory. The model is in fact very complex to comprehend in its full scope, and the task is hopeless without some close familiarity with the different standards involved in a each layer.

2.4.2 X.25 and Frame Relay

The two most used standards in packet-switched public networks are *X.25* (after the name of an ITU recommendation) and its simplified later version, *frame relay*. We review next their underlying principles and functionality.

Protocol X.25 is a means to connect the *data terminal equipment* (DTE) belonging to local customer premises (or station) to a *data circuit-terminating equipment* (DCE) which is a gateway (or node) of a public-switched network. The DCE belongs to a *public-switched exchange*, or PSE. The different network elements are shown in Figure 2.19.

The purpose of the (so-called) "X.25 network" is to establish, by means of packet-witching, virtual connections between DTEs. Connection between a given DTE station and DCE node is made through the lowest three OSI layers, namely physical (1), data link (2) and network (3). The protocol defining the access procedure at physical and data link layers 1–2 is called *link access procedure* (LAP), better known under its evolved form *LAP-B*, with 'B' for 'balanced'. The frames exchanged between DTE and DCE can be of variable lengths, but are typically up to 128–256 bytes.

The X.25 frame format is shown in Figure 2.20. Each frame is delimited with a *flag byte* at the beginning (see previous subsection, layer 2).

Then comes a 2-byte header made of an *address* byte and a *control* byte. Unlike its name seems to indicate, the address byte specifies whether the frame is a DCE-to-DTE (or DTE-to-DCE) command or a response from either. This information makes it possible to interpret the control field coming next. The 1-byte control field

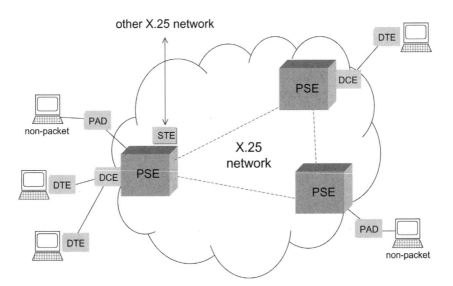

FIGURE 2.19 Elements of a X.25 data network. PSE = public-switching exchange, STE = signaling terminal equipment, DCE = data circuit-terminating equipment, DTE = data terminal equipment.

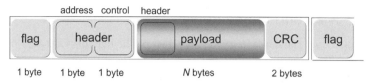

FIGURE 2.20 Frame format of X.25 LAP-B data packet, showing flag, address, control and payload fields.

describes the type of command/responses, and defines three type of frames: *information* (I), *numbered supervisory function* (S) or *unnumbered control function* (U):

1. The mode I (command only) enables the frames to carry information data from one station to another; these frames are referred to as *LAP-B I* frames.

2. The mode S (command only) defines supervisory frames, with instructions for station/node readiness. This includes (command only) the *receiver-ready* (RR) and *receiver not-ready* (RNR) information.

3. The mode U (command/response) characterizes frames having other additional link control function, for instance the instruction for disconnection.

The following payload field (*N* bytes) communicates data for use at network level 3. It comprises a 3-byte header. The header includes the *logical channel number* (LCN), which serves as an identification of the virtual DTE/DTE circuit to which the packet belongs. The LCN occupies 12 bits, corresponding to the possibility for a DTE to access $2^{12} = 4,096$ possible virtual circuits. Note that the LCN is handled at network layer 3, since it is not part of the X.25 frame header. The rest of the payload header includes (among other features) diagnostics, *calling* DTE and *called* DTE addresses (one or several), and requests for specific X.25 network *facilities*. These facilities include retransmission requests, setting one-way channel connections, charging/billing options, dialing connection to PSTN, etc.). The rest consists of the real data payload, whose maximum size is 1,204 bytes.

The last 2-byte field (also called *framed check sequence* or FCS) concerns the function of error correction of the frame's contents (excluding flag) through a *cyclic redundancy code* (CRC), see Chapter 1, Section 1.6. In X.25 frames, the CRC generator polynomial is $g(p) = p^{16} + p^{12} + p^5 + 1$. The highest degree of the polynomial indicates that the code can detect error bursts of lengths up to 16 and other specific patterns.

The X.25 protocol operates in a *synchronous mode*, meaning that the DTE must be synchronously communicating to the DCE. The DTE is referred to as a *packet-mode* device. It is however possible for slower, *nonpacket-mode* devices to connect asynchronously to an X.25 network. This requires the use of a special network interface called a *packet assembler/disassembler*, or PAD. As the name indicates, the PAD prepares and undoes the X.25 frames from/to asynchronous terminals, see Figure 2.19. The protocol under which asynchronous terminals can send control and packet/data ASCII characters to the PAD is referred to as X.28. The protocol interfacing PAD and a remote DTE is called X.29. The main PSE of an X.25 network can also be connected to another X.25 PSE. The required interface is signaling terminal equipment, or STE, as illustrated in Figure 2.19. The interface between two X.25 networks (STE-STE link at layers 1–3) is provided by the X.75 protocol.

The main advantage of X.25 is its ability to connect a relatively large number of data terminals by sharing network resources at kbit/s or tens of kbit/s line rates. Even lines of poor quality can be used, since the protocol has inherent error-correction features. Packet-switching offers the possibility of multiple virtual connections, or channels, between the terminals. But this is possible only if the data traffic remains limited and does not cause packet congestion. Remember that packet-switching can handle traffic more efficiently if the packet size (hence the amount of data/packet) is small, and that the occurrence of bursts of long packets results in switching congestion, delays and possibly packet loss. The main drawback of X.25 is its relative slowness. Indeed, a 128-byte packet at 64 kbit/s rate represents $128 \times 8/64 = 16$ ms. Assuming zero transmission delay, an exchange of only 50 packets between two network terminals takes practically one second! Such a feature is aggravated by the nature of the X.25 protocol, which requires multiple commands, responses, requests and acknowledgments. Bursts of traffic can also congest the network buffering and switching capabilities, resulting in packet-rate slow down and the need for packet re-emission due to loss.

The solution to the above problems is provided by *frame relay*. Initially, the frame relay network resembled a mesh of lines connecting different LANs' central stations. The lines could operate at rates between 9.6 kbit/s and 2.048 Mbit/s, with a capability of $2^{10} = 1,024$ virtual circuits or data links between remote users. The device used to connect to this network is called *universal network interface*, or UNI. The connection between two frame-relay networks is called *network-network interface*, or NNI. The layout of a frame-relay network is shown in Figure 2.21.

Because the frame-relay network resource was available any time, the data links corresponded to a *permanent virtual circuit* (PVC). Since the network is used only when frames need to be *relayed* through the link, such a PVC configuration is inefficient. Rather, the network could be used by a larger number of users if data links could be switched on or off on the actual user's demand, in a "pay-as-you-go" mode. Hence the concept of *switched virtual circuits* (SVC). Thus frame relay corresponds to a subscriber need of relatively low daily usage but relatively high bit rate and important payload, unlike X.25 where all the contrary applies. Therefore, frame relay appeared as the first solution for cost-effective, high-bandwidth, data-only private lines.

The basic frame partitioning in fields for frame relay is identical to that in X.25, as shown in Figure 2.19. The noteworthy differences with X.25 frames are as follows:

- The address field (following the flag) is two-bytes long. It carries the *data-link connection identifier* (DLCI, pronounce 'delsie'), which is equivalent to the logical channel number (LCN) in the X.25 payload header (see above).

- The control field includes (like X.25) the RR/RNR information. In addition, it provides the frame length.

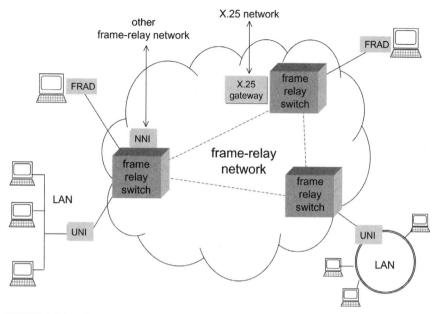

FIGURE 2.21 Elements of frame-relay data network. UNI = universal network interface, NNI = network-network interface, LAN = local-area network.

- The payload field is 100% information, without overhead, unlike in the X.25 payload field. It can be made of from 1,610 bytes up to $2^{16} = 65,536$ bytes, to compare with less than $128-256$ bytes for X.25.

- The full protocol implementation (frame checking and routing) is limited to data link layer 2. This is illustrated by the fact that the data link connection identifier (DLCI) is located directly into the frame header, unlike X.25 whose corresponding LCN is in the frame payload.

Frame relay networks offer the additional service of a guaranteed minimum bit rate during the time of (SVC) connection, referred to as *committed information rate* (CIR). In the absence of congestion, a higher information rate (*excess information rate* or EIR) can be made available to the user, the resulting average rate being between EIR and CIR. Note that in case of congestion, some frames must be discarded. In anticipation of such a risk, the user may initially label some selected frames as "eligible for discard." In order to set priorities for different data types (e.g., voice and video), *frame-relay access devices* (FRAD) were developed. As opposed to UNI devices, however the FRAD are of proprietary design and therefore not mutually compatible.

■ 2.5 LOCAL AREA NETWORKS

Local area networks (LAN) are privately owned networks which appeared in the 1980s. The function is to connect a limited number of computers, terminals, workstations (and other shared appliances) within a delimited area, such as a building, a business center or a university campus, for instance. The physical link can be made of twisted-pair, coaxial cable or radio. The maximum station-to-station distance being of the order of 1–10 kilometers, relatively high bit rates (1–10 Mbit/s) are available, providing (in principle) high network connectivity and reactivity. With optical fiber links, as in FDDI networks (see below), the capacity and transmission distance can even be extended to 100 Mbit/s and 100 km.

The LAN requires an *access protocol*, which can be of either a *random* or *deterministic* nature. We shall overlook deterministic access which is of limited functionality. Random access makes it possible for a group of users to spontaneously connect to the network at any time. This randomness can provoke spurious *collisions*, which are events where two or more terminal users simultaneously call for the use of a common resource (e.g., a printer) or simply attempt to connect to the same end terminal. The situation where data packets of different origins reach the same network point is called a *contention*. The LAN must be able to detect (or sense) collision events, and possibly to resolve contentions by some set of priorities and rules. The access protocols are run at layers 1 (physical) and 2 (data link) of the OSI model. As will be further detailed below, layer 2 is split into two parts: the *medium-access control* (MAC) sublayer, which is specific to the LAN type, and the *logical-link control* (LLC) sublayer. Before entering into the LAN specifics, it is useful to describe the main types of network *topologies*.

2.5.1 Network Topology and Connectivity

The network topology defines the physical connection layout between terminals. We shall discuss topologies which concern all network types, from LANs to wide area networks (WAN). There are four basic topology types, called *bus*, *ring*, *star* and *mesh*, as illustrated in Figure 2.22. Each topology has its own advantages and drawbacks in respect of cabling requirement and evolution potential.

The *bus topology* is the simplest configuration possible. A single cable is laid into the premises (e.g., building floors) and stations can be hooked up where the cable passes by. The network is easily upgraded. The path to connect two given stations is uniquely defined. A drawback is that in case of cable cut (reparation, accident, tampering...) the network fails to operate (network *outage*), or operates only in limited segments.

The *ring topology* is equivalent to a bus topology with the cable forming a closed loop. There are two possible paths to connect to a given station, i.e., clockwise or counter-clockwise. The ring has the same exposure to cable break, but because of the two path possibilities it is possible to implement a redundancy ring (see FDDI below; also see Chapter 2; Desurvire, 2004). Switching the traffic into the redundancy ring makes it possible to by-pass the break point and therefore to prevent network outage. Network upgradability is as straightforward as in the bus case.

The *star topology* is based upon the principle of connectivity to a common *central node*, or network *hub*. The advantage is that the network can be centrally controlled from a single intelligent point. Unless the failure comes from the hub

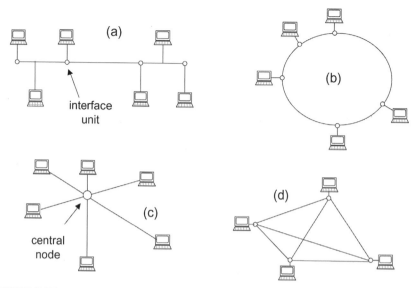

FIGURE 2.22 Different types of local-area network (LAN) topology: (a) bus, (b) ring, (c) star and (d) mesh.

itself, the network is protected against outage. Upgrading the start network with new stations requires pulling more cables towards the hub, and additionally upgrading the hub itself in order to accommodate the new connections.

As we have seen in Section 2.1, the *mesh topology* requires $N(N-1)/2$ wires to connect N stations together. The connection requirement thus grows quadratically with the number of stations and the wiring therefore becomes increasingly complex as the network evolves to accomodate more end users. While the mesh topology is limited to a small number of stations, the advantage is that each station has a unique connection path with the others, which prevents collisions (but not contentions). In case of a cable break between two given stations, the network can still operate through the $(N(N-1)/2)-1$ remaining paths. The mesh topology is advantageous in the context of larger networks (WAN) because it offers the possibility of several *routing paths*.

All topologies can in fact coexist or capitalize on each other's advantages, as the LAN is upgraded. Figure 2.23 shows an evolutive LAN topology made up of a combination of bus and ring subnetworks, connected together in a star arrangement. Figure 2.23 also provides an example of a wide PSTN network based on a ring topology. Internetworking with different topology types offers an infinite variety of complexity, cost and upgradability options, and therefore requires careful optimization.

To illustrate the concept of topology optimization, consider the way major airports are designed. Airports are made of a complex superposition of subnetworks for private cars, taxis, buses, shuttles, pedestrians and airplanes. Vehicles usually access the airport buildings through interleaved and bidirectional rings. The airport building infrastructure is generally configured as a star, at the center of which one gets information and booking services. Each branch of the star then leads to a terminal or

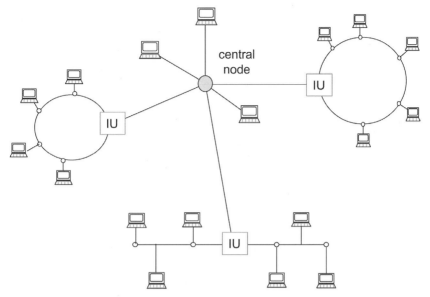

FIGURE 2.23 Evolved topology of a large-size LAN, with star topology and ring or bus subnetworks. IU = subnetwork interconnection unit.

a sub-star of terminals linked by a ring shuttle. The terminals usually form self-contained rings where passengers find their gate and finally access the airplane. Everyone has experienced that sometimes the routing path leading to the gate can be tricky because of all these topology changes! From the airline perspective, flight paths are defined as a mesh of airport-to-airport connections, with major routes converging into hub terminals. At regional level, flights follow a straight, bus-like pattern, with multiple hops. This example illustrates that network topology is more or less optimized for accessing the maximum number of points with the highest performance in (a) transit time, (b) path complexity, (c) link implementation, (d) cost and (e) future upgradability. Another example is the *highway* network, whose topology is a complex mix of star, mesh and loop, in a spider-web fashion. Everyone has experienced that in some big metropolitan areas, the optimum network requirements (a)–(e) are not always met! See also the Exercise at the end of the chapter to get a further appreciation of topology optimization.

In bus and ring topologies, traffic pervades the entire link, while only a limited number of stations are in connection at a time, which represents an inefficient way of sharing bandwidth. It is therefore sensible to segment the network into smaller subnetworks, which is the function of network *bridges*. The bridge acts as a filter to isolate different subgroups of stations, or create a dedicated logical path between them. This decreases traffic congestion and access/connection time, locally increases link bandwidth, and connects subnetworks having different cable or media types (radio, coaxial, fiber . . .). Bridges can also link together two LANs with fixed logical-path attribution (e.g., station A of LAN 1 only communicates to stations D and F of LAN 2, and the reverse). The bridge is essentially a *frame switch*, and it operates at OSI layer 2 (data link), in contrast with an electro-mechanical switch which would relate to OSI layer 1 (physical). Network *switches* provide more sophisticated functions than bridges. Switches also act as network bridging elements but have the ability to store and forward packets/frames, to forward them dynamically according to network addresses, to solve contentions, and to create virtual circuits (see Section 2.4) or virtual LANs. They are also able to ensure full duplex connection between stations, unlike bridges. Depending upon switching protocols (e.g., X.25, frame relay or ATM), the switch operates at OSI layer 2 (data link layer) or layer 3 (network layer). The functional upgrade of a switch is the *router*, which works at OSI layer 3. The router knows all the network addresses, and the status of the network traffic in each node, and can therefore determine the optimal path for transit time and cost savings. Routers can also define traffic priorities, prevent network congestion, and act as control/safety barriers. A higher type of network interconnection is the *gateway*. Gateways are routers which operate at OSI layer 4 (transport) and above. As their name indicates, they permit the transport of data between network types having completely different protocols, for instance X.25 with frame-relay or PSTN, PSTN with Internet, etc.

As previously mentioned in this subsection, the data link layer 2 is split into MAC and LLC sublayers, the MAC being network-specific and the LLC general to all LAN types.

2.5.2 Ethernet

Initially developed by Xerox, the *Ethernet* was the first commercial random-access LAN based on bus topology. The concept relies upon a smart protocol called *carrier-sense, multiple-access with collision detection*, or CSMA/CD. As its name indicates, such a protocol ensures that only one out of multiple users can emit/send data at a time, while possible collision events can be monitored and managed. With carrier-sensing each station always "listens" to whether the bus is already in use by any transmitting station, waiting to transmit (as in a well-behaved conversation). Once the carrier is seen to vanish, the stations can take their turn, while continuing to "listen." If two stations then happen to transmit at the same time, a message collision occurs. The collision event is not detected at the same time by the conflicting stations because of the differences in initiation and transit time of the corresponding messages. In well-behaved conversations, the two people immediately stop talking (at least one generally does it first). In the Ethernet, the first station to detect the collision immediately stops transmitting data and instead generates an alarm or jamming signal, which will be sensed by all stations. As in a conversation, the conflicting parties having detected the alarm signal will hesitate with different reaction times, before trying again. In the Ethernet, each station is able to calculate its own reaction times, based on a pseudo-random algorithm, which makes it possible to minimize collision repetitions.

The early Ethernet generations, known under the names of *10base2*, *10base5* and *10base-T*, operate at 10 Mbit/s and use *unshielded twisted-pairs* (UTP) for the network link. As shown in Figure 2.24, the Ethernet frame is made of an address field (destination and source addresses, 6-bytes each), a 2-byte field indicating the

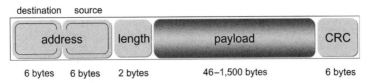

destination   source

| address | length | payload | CRC |

6 bytes   6 bytes   2 bytes   46–1,500 bytes   6 bytes

FIGURE 2.24 Ethernet frame structure, showing address, payload length, payload and error-correction (CRC) fields, with corresponding byte sizes.

length of the payload data, followed by a 64–1,500-byte payload and a 4-byte CRC field for error correction, corresponding to a maximum allowable length of 1,818 bytes. If there are fewer bytes to be transmitted, dummy or "padding" bytes are added to make the payload length up to the 64-byte minimum. Such a minimum length this important for the device to be able to detect collisions in the course of transmitting even very short frames.

A second Ethernet generation, called 100base-T *fast Ethernet*, was made to operate at 100 Mbit/s over up to 100-meter distances. The more recent third generation was called *gigabit Ethernet* (GigE) and operates at 1–10 Gbit/s over multi-kilometer distances. This capacity upgrade is made possible by use of optical-fiber links.

2.5.3 Token Bus and Token Ring

Token bus and token ring networks are the implementation of a contention-free protocol called *token passing*. This protocol, which was developed by IBM, is based on the principle that only one station can transmit data through the network at a given time. The one-time right to transmit is provided by a *token*, a small 3-byte code packet which circulates through the link. The token circulates in a "free" mode or in a "busy" mode, as indicated by a single bit. Once a station A has picked up a free token it modifies it into the "busy" mode and appends it to a data frame, destined to station X. The frame circulates from station to station. The next station recognizes the "busy" token and reads the frame address. If the address destination is different, the frame is passed to the next station, and so on, until station X is reached. After being read by station X, the frame is returned by X to the originating station A, confirming good reception through a frame-status (FS) end-byte. Station A then politely resets the token to its "free" mode and sends the free 3-byte token to the next station (it is not permitted for station A to transmit again, immediately after this cycle. If it needs to transmit, the next station will then pick up the token and restart a cycle. If not, the token continues its circulation until it is found by a waiting station and/or by the monitoring station.

Token passing is a circular algorithm where all stations are given a deterministic and equitable opportunity to transmit in turn after a certain number of *token rotation times*. This is unlike in the CSMA/CD protocol where this opportunity is more or less random. It can also be implemented in bus or star network topologies, using round-trip or cyclic patterns, but the ring configuration remains the most common implementation.

At the beginning of a given frame, or embedded within a higher-level frame, the token can never be confused with information. This is because the first and last bytes (*start* and *end delimiters*) of the token contain recognizable "violation bits," such as in Manchester coding (Section 1.2). The central portion of the token is the *access control* (AC) byte. The AC byte has the format PPPTMRRR, where the PPP field defines priorities, T identifies the token status (T = 0 means free, T = 1 means busy with frame data under transmission). Bit M is used for monitoring purposes; the transmitting station sets it to 0, and a monitor station sets it to 1. If the monitor station receives a token with M set to 1, this indicates a full round-trip without transmission. A clean token is then re-transmitted by the monitoring station. The priority bits PPP can be used by a station having higher priority to "steal" the turn. The RRR fields permit a station to reserve a priority turn if a higher priority station has not taken over.

The field layout of a token-passing frame is shown in Figure 2.25. The 3-byte header is the "busy" token whose central byte (AC) was previously described. The end byte (FC) specifies whether the frame belongs to the MAC or the LLC layer. The next field contains the destination and source addresses. The payload field contains layer 1 (physical) and layer 2 (data link) information. Its length is variable, but the maximum allowable depends upon the bit rate, which determines the token-retention time for the network. For instance, it can be equivalently 4,000 bytes at 4 Mbit/s or 20,000 bytes at 16 Mbit/s. The next field concerns CRC error-correction. Then comes an end-delimiter (ED) field with information on the number of frames to follow, in the case of a sequence, and detected anomalies or errors. The frame closes with a frame-status (FS) byte, which, as previously indicated, acknowledges good receipt from the destination station.

The contents (payload) of the above frame can then be interpreted at OSI layer 2 by the MAC and the LLC sublayers. The field structure of the payload is made up of two MAC fields encapsulating a LLC frame. The LLC frame has four fields: DSAP, SSAP, control and data payload. The destination and source access points (DSAP, SSAP) provide a logical link corresponding to different application environments.

2.5.4 Fiber Distributed Data Interface (FDDI)

The next move is to implement a token ring network with optical fibers for the link. This is the *fiber distributed data interface* (FDDI), made to operate at 100 Mbit/s over an area deployment as large as 100 km. Such extended distances correspond

FIGURE 2.25   Token-passing frame structure, showing header (SD = start delimiter, AC = access control, FC = frame control), address, payload, error-correction (CRC), end-delimiter (ED) and frame-status (FS) fields, with corresponding byte sizes.

to *metropolitan area networks*, or MAN. If the fiber is multimode, the distance between stations in the ring should not exceed 2 km. If single-mode fiber is used, the distance can be up to 60 km. Note that FDDI can also be supported by twisted-pair wires, but the distance between stations should not exceed 100 m.

The FDDI network is based upon a *double ring topology*. Such an arrangement permits data to circulate in both clockwise and counterclockwise directions, as shown in Figure 2.26. For this reason, the stations or nodes situated on the ring path are called *double-attachment stations* (DAS). The DAS have four fiber ports corresponding to two incoming and two outgoing signals. To each DAS can correspond a number of *single-attachment stations* (SAS), which can be concentrated to the DAS through a multiport connector, or form local loops or buses linked to the DAS via bridges, as shown in the figure. More complex loop patterns can be formed with *double-attachment concentrators* (DAC), where SAS can be physically connected to only one of the two rings. The SAS play then the same role as DAS, but with only two fiber ports.

The main advantage of FDDI networks, as with all double-ring networks, is the ability to overcome cable breaks or failures by means of *self-healing ring protection*. The way self-healing operates is illustrated in Figure 2.27. Since the traffic is bidirectional, a physical connection failure between two given stations is immediately detected by both. The two stations then proceed to shunt the rings which results in the formation of a single, unidirectional ring. The price to pay is a loss of 50% of traffic capacity. But such a loss may not be significant if one of the

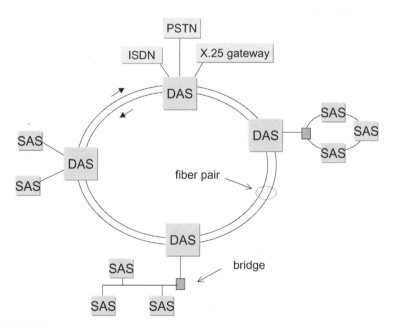

FIGURE 2.26 Token-passing, double optical ring network layout corresponding to FDDI, with access possibilities to PSTN, ISDN and X.25. SAS = single-attached stations, DAS = double-attached stations.

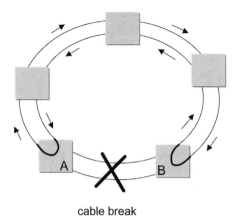

cable break

FIGURE 2.27 Principle of self-healing in double-ring networks. In the event a cable break is detected by two stations A and B, the stations immediately shunt the two rings so as to form a single, unidirectional ring.

rings is dedicated to *nonpriority traffic*, and can therefore be appropriated in case of failure. Note that self-healing ring protection is commonly used in all transoceanic cable systems.

The FDDI protocol operates at both OSI layer 1 (physical) and 2 (data link). The physical layer is divided into two sublayers, which concern two kinds of protocols:

Physical Media Dependent (PMD), defining the driving charactristics of the optical transducers.

Physical Layer Protocol (PHY), defining network synchronization and the encoding/decoding of the optical data.

With FDDI, layer 2 is only addressed through the lower medium access control (MAC) sublayer, which controls the token circulation. Finally, a *station management protocol* (SMT) controls the three-fold addressing of PMD, PHY and MAC sublayers, with the functions of initialization, fault detection and isolation/recovery.

The binary coding format for the optical signals is *nonreturn to zero inverted*, or NRZI (see Section 1.2 and Figure 1.10). The principle of NRZI is that light is turned on and off, according to the following rule: only "0" bits cause a change of signal polarity from ON to OFF or the reverse. One of the drawbacks of this coding format (like NRZ) is that long sequences of "1" bits resemble to an idle signal state having no clock tones or harmonics in its spectrum. In order to avoid loss of synchronization due to this absence of clock, a special coding was developed for FDDI, called *4B/5B*. The principle of 4B/5B is to code 4-bit words into 5-bit words, hence its name. Thus a byte (8 bits) corresponds to a two-symbols code word of 10 bits total. Table 2.3 shows the 4B/5B code correspondence for bit sequences 0000 to 1111. It is seen from the table that the initial sequence (1111)(1111) is coded into (11101)(11101), showing that a $4N$ bit sequence of "1" is converted into a code word containing a maximum of four successive "1".

TABLE 2.3   Correspondence between 4-bit words
(0000 to 1111) and 4B/5B code

Word	4B/5B	Word	4B/5B
0000	11110	1000	10010
0001	01001	1001	10011
0010	10100	1010	10110
0011	10101	1011	10111
0100	01010	1100	11010
0101	01011	1101	11011
0110	01110	1110	11100
0111	01111	1111	11101

Because 4-bit words are mapped into 5-bit words, the efficiency of this code is 80% (100 Mbit/s corresponding to 125 Mbit/s baud rate). However, there are $2^5 = 32$ possible symbols, which leaves $2^5 - 2^4 = 32 - 16 = 16$ symbols that can be attributed to network command and control symbols, as well as to frame delimiters. By convention, the code 11111 corresponds to the command "idle," and is placed at the beginning of each FDDI frame, both as a signaling flag and as a clock-synchronization code.[1]

In the process of data generation, transmission and reception, the three OSI sublayers interact as follows:

- At bottom sublayer 2 of transmitting station A, MAC sends to physical layer 1 the message word (1011)(0101)
- At layer 1, PHY encodes the message into the 4B/5B code-word (10111)(01011)
- Still at layer 1, PMD drives a light modulator to produce the optical NRZI signal (ON/OFF/OFF/OFF/OFF)(ON/ON/OFF/OFF/OFF)
- The signal is transmitted through the fiber
- At layer 1 of the receiving station B, PMD receives the NRZI signal and sends 5-bit code words to layer 2
- At layer 2, PHY de-codes the 5-bit words into 4-bit words
- Still at layer 2, MAC gets the original message.

As in token-passing networks, the token is made of a 3-byte ST/FC/ED (start-delimiter, frame control and end-delimiter). However, it is preceded by a *preamble*, which is a sequence of 11111 (see Footnote 1). As previously mentioned the preamble serves both for clock synchronization and flagging. Conventionally, the frames

---

[1]With the NRZI convention shown in Figure 1.10, the input bit sequence 11111 in fact corresponds to the idle NRZI sequence, while the input 00000 for which the NRZI signal polarity switches at each bit period, is the clock sequence of interest. It is just a matter of convention, i.e., whether the coding algorithm in the figure defines NRZI or its binary complement.

in FDDI are limited to 9,000 symbols (or 4,500 bytes). The field format with corresponding number of 4B/5B symbols (not bytes) is shown in Figure 2.28. The payload size has an upper bound determined by the size of the overhead information, the whole frame not exceeding 4,500 bytes.

As far as frame transmission is concerned, the main difference between the FDDI protocol and the basic token-passing protocol (see previous subsection) is the possibility for a node which has captured the free token to transmit several frames in sequence. The token rotation time thus increases linearly with the frame load.

2.5.5 Switched Multimegabit Digital Service (SMDS)

An alternate network version for MAN applications is based upon a double-bus ladder topology, as illustrated in Figure 2.29. Such a network type supports the *switched multimegabit digital service*, or SMDS, as it is called in North America. The European equivalent is referred to as CBDS, *for connectionless broadband data service*. SMDS/CBDS networks are designed to operate at 155 Mbit/s over 150 km distances. The two unidirectional buses are terminated by master and slave end stations. The network can also be arranged into a loop configuration with a single master station.

The SMDS protocol is called *distributed queue dual bus* or DQDB. It consists of using 125-μs synchronous frames generated by the end station. As shown in Figure 2.30, the frames contain a certain number of equal *time slots* of 57-byte length with a 48-byte data payload referred to as a *segment*. The 48-byte segment is of the same size as an ATM cell (see Chapter 3), which permits intrinsic compatibility with ATM-switched networks. The *frame header* contains the node's source and destination addresses. The *frame trailer* is used for node synchronization. Within each time slot, the *slot header* is made of a 2-byte delimiter field and a 2-byte control field concerning physical layer 1. The slot header is followed by a 1-byte *access control field* (ACF), which is detailed in the following text. The following *segment header* of 4-byte length contains a virtual channel identifier (VCI), similar to a logical channel number (LCN) for data link layer 2 addressing. Two modes of operation exist. In the asynchrounous case, referred to as *arbitrated operation*, the VCI field is set to 111 ... 1, meaning that all time slots are available to the node use. In the isosynchronous case, referred to as *prearbitrated operation*, the VCI field specifies which time slots are assigned to which node, in a partial or complete

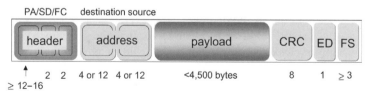

PA/SD/FC   destination source

| header | address | payload | CRC | ED | FS |

↑   2   2   4 or 12 4 or 12   <4,500 bytes   8   1   ≥3
≥ 12–16

FIGURE 2.28   Frame structure used in FDDI, showing header (PA = preamble, SD = start delimiter, FC = frame control), address, payload, error-correction (CRC), end-delimiter (ED) and frame-status (FS) fields, with corresponding 4B/5B symbol sizes.

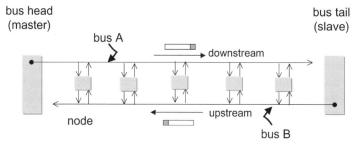

FIGURE 2.29 Double-bus layout of switched multimegabit digital service (SMDS) networks.

usage mode. With such pre-arbitration, a given node can be attributed guaranteed and uninterrupted traffic, unlike in the other mode of operation.

As seen in Figure 2.29, the frames propagate upstream (bus A) or downstream (bus B) until reaching the bus ends. Each node can communicate through either bus A or B, depending upon its relative position in the bus. Similarly to the Ethernet, the nodes can transmit (fill time slots with data) only at specific times, when all other nodes are not communicating. Each node is given a pair of counters, corresponding to buses A and B. The first bit of each time-slot ACF is "1" or "0" for time-slot "occupied" or "empty," respectively. The last four bits of the ACF consist in the *request counter*. In order to be able to transmit through bus A (for instance), the node makes a request by incrementing the ACF counter in bus B slots. Thus each node

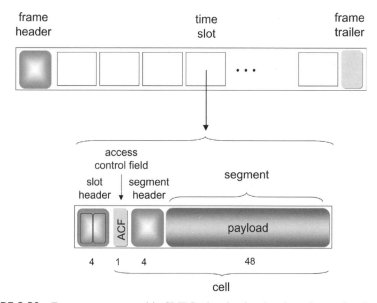

FIGURE 2.30 Frame structure used in SMDS, showing header, time slots and trailer. Each time slot is composed of a header and a cell. The cell is made up of an access control field (ACF) and a segment. The segment is made up of a header and a 48-byte payload.

is made aware of the number of stations downstream in bus A that are in the transmission-waiting mode. Conversely, each time an empty time slot ("0" in the first ACF bit) is detected in bus A slots, the associated request counter is decreased by one unit, with a minimum of 00. The value of the request counter in B thus determines the node position in the waiting queue. As this counter reaches 00, data can be loaded into the oncoming free time slot. The loading process is interrupted as soon as the request counter in B is increased by one unit, meaning that another node is calling for transmission. The same process applies for transmitting over bus B, while placing requests (or counter increments) on bus A frames. The bandwidth resources of the two buses are thus fully exploited, with decentralized allocation and control.

Note that the European CBDS version of the DQDB protocol enables variable-length payloads having up to 9,188 bytes length. Such an implementation of SMDS corresponds to a packet or *datagram* communication service. It can be viewed as a limited-range precursor of ATM networks (Chapter 3).

✓ EXERCISES

Section 2.1

2.1.1 Draw permanent connection diagrams corresponding to $N = 2$ to $N = 6$ subscribers. By induction, demonstrate the formula $N(N - 1)/2$ giving the number of required connections for any N subscribers.

2.1.2 Describe the sequence of events during a call from person A to person B within a local loop, corresponding to the following situations:

- B is absent (no answering machine)
- B is busy with another call
- Same as above, but the service offers the possibility of automatic call-back of B to A.

2.1.3 In reference to the analog repeater shown in Figure 2.3, calculate the net gain experienced by the signals leaking through the hybrids after one then N circulations though the loop. Assume $G = 11$ dB, an excess branch loss of $E = 4$ dB and a hybrid leakage loss of $F = 6$ dB.

2.1.4 The loss of optical fibers is 0.2 dB/km at 1.55 μm wavelength. How many amplifiers are required in an interoffice optical trunk of length 500 km where the signal loss would not exceed 15 dB between each repeater stage?

Section 2.3

2.3.1 What is the time duration of a 1-byte voice sample? Show it in two different ways.

2.3.2 How many framing and stuffing bits are included inside a T2 frame (refer to Figure 2.3b)?

Section 2.4

2.4.1 Using Table 1.2 in Chapter 1, write your own initials in ASCII code.

2.4.2 Calculate the byte size of a numerical photo having 640 × 480 pixels with 255 color-intensity levels for each of the red, blue and green tones. Calculate then the byte size of a

30-second movie clip based on this picture resolution, assuming 18 pictures per second. What is the bit rate required to transmit the video?

2.4.3 In X.25 networks, not more than 7 payload packets can be transmitted without receiving acknowledgment from the receiving terminal. Assume 1 Mbit/s, 128-byte packets travelling at the speed 3×10^8 m/s and a point-to-point distance of 600 km. How many packets approximately are in the line during transmission? How many complete transmission cycles (7 bytes plus acknowledgment received) can be made every second?

Section 2.5

2.5.1 Assume a network with four stations located on the four corners of a square with a 1-km side. Provide the minimum cable-length requirement for bus, ring, star and mesh network topologies and class them in order or length. (Advanced): There exists a bus-like connection configuration taking the shape of a letter H being squeezed sideways or a stretched letter X, looking like >-<. Find its minimum length and conclude as to what is the optimum topology for a 4-station network.

2.5.2 An SMDS network has a double bus of 75-km, node-to-node length. How many frames, time slots and cells are present at a given time in each of the buses? What is maximum number of time slots that could simultaneously be used by the nodes located at the bus extreme ends before transmission interruption? (Assume 155 Mbit/s rate and frame velocity of $c = 3 \times 10^8$ m/s.)

a MY VOCABULARY

Can you briefly explain any of these words or acronyms in the telephony and data network context?

ACF (SMDS)	Circuit (switching)	EBCDIC	Hypergroup
ADM	Collision	EIR	(frequency)
Amplifier	Contention	Encryption	Inbound (traffic)
Analog	CRC	Error correction	Instability
Application (layer)	Crosstalk	ESF	(repeater)
ASCII	CSMA/CD	Ethernet	Interleaving (bit)
B-ISDN	CTE	Exchange	Internet gateway
B-channel or B-line	D-channel or D-line	Extension	ISDN
Bidirectional	D4	FDDI	ITU
BRA	DAC	FDM	IXC
BRI	DAS	FIFO	ISP
Bridge	Data link (layer)	Flag (byte)	JPEG
Buffering	Datagram	Frame	Justification (bit)
Bus (topology)	DCE	Frame bit	LAN
Byte	Decibel	Frame relay	LCD
Call forwarding	Demultiplexing	Gain (amplifier)	LCN
CBDS	DTE	Gateway	LLC
CCBS	Digital	Group (frequency)	Logical channel
Cell	DMTF	GTE	Loss (transmission)
Central node	DMUX	H-channel	Loop
Central office	DQDB	HTTP	M12 ... M23
Centrex	DS0, DS1 ...	Hub (network)	(multiplexing)
CIR	Duplex	Hybrid	MAC

MAN
Mastergroup
  (frequency)
Mesh (topology)
Messaging
MPEG
Multiplexing
MUX
Network (layer)
NRZ
NRZI
NT1, NT2 (ISDN)
Nyquist (theorem)
OSI (model)
Outage
Outbound (traffic)
Overhead
PABX
Packet
Packet-switching
Path
Payload
PBX

PCM
PDH
PHY (FDDI)
Physical (layer)
Plesiosynchronous
PMD (FDDI)
POP
POTS
Preamble
Presentation (layer)
PRA
PRI
Protocol
PSTN
PVC
Queuing
Repeater
  (analog/optical)
Ring (topology)
RNR
Robbed-bit
  signaling
Router

Routing
RR
Sampling (rate)
SAS
SDH
Segment (SMDS)
Self-healing (ring)
Service
Session (layer)
Simplex
SMDS
SMT
SMTP
Star (topology)
STE (FDM)
STE (X.25)
Store-and-forward
Stuffing (bit)
Supergroup
  (frequency)
SVC
Switch
Synchronization

T1, T2 . . .
T-span
TDM
Token passing
Topology (network)
Transport (layer)
Tributary
Trunk or junction
Twisted-pair wire
Unidirectional
UTP
Voice mail
VC
VCI (SMDS)
WAN
X.25
X.28
X.75

CHAPTER 3

An Overview of Core-Network Transmission Protocols

This chapter describes the features of the top three main transmission protocols called SDH/SONET, ATM and TCP/IP. The issues of "mapping" IP into the former, including on the optical WDM layer are addressed. Special attention is also given to evolving Internet jargon.

The subsection closing the last section of this chapter, The *Internet and www jargon*, does not deal with the layer 2–4 protocols of TCP/IP and services, but with the top application layer where the Internet experience begins and ends for most users. We deemed it useful to describe as systematically as possible the set of concepts and neologisms associated with the use of the Internet from that top client layer (beginner or advanced), which can be seen as representing both a culture and an intricate technical jargon.

■ 3.1 SYNCHRONOUS DIGITAL HIERARCHY (SDH) AND SYNCHRONOUS OPTICAL NETWORK (SONET) PROTOCOLS

Synchronous digital hierarchy (SDH) in Europe and *synchronous optical network* (SONET) in North America and Japan are two near-equivalent protocols for voice and data transport in high-speed networks and point-to-point communication links. As the names indicate, SDH/SONET concerns *synchronous* traffic over

Wiley Survival Guide in Global Telecommunications: Signaling Principles, Network Protocols, and Wireless Systems, by E. Desurvire
ISBN 0-471-44608-4 © 2004 John Wiley & Sons, Inc.

optical network backbones. Both notions convey an idea of continuous data frame flow and higher signal bandwidth.

The SDH/SONET protocols operate at the bottom physical layer 1, as a mere transport service. This is in contrast to the higher-level (and asynchronous) types, namely ATM and TCP/IP, also described in this chapter. This does not preclude mutual compatibility. Indeed, Internet data packets can be first conveyed through ATM packets, then wrapped into SDH/SONET frames, and then all the reverse at some other network hub. Such successions of protocol conversions define new network layers' interfaces, inter penetration, integration and bandwidth demand evolution, as described in Chapter 2 of Desurvire, 2004. SDH/SONET is a general transport protocol which applies to electrical, optical, terrestrial radio and satellite network types.

Fundamentally, SDH/SONET is based upon the principle of *time-division multiplexing* (TDM). The protocol can be viewed as an elaborate version of the *plesiosynchronous digital hierarchy*, or PDH, previously described in Chapter 2. The TDM technique consists in combining a certain number N of *signal* tributaries at a given bit rate B into a *signal aggregate* at the higher bit rate $N \times B$. More specifically, multiplexing is made by byte-interleaving the tributaries after $1/N$ time compression, as illustrated in Figure 2.11 of the chapter. Applying the technique to voice channels (or circuits) at $B = 64$ kbit/s results into a hierarchy of TDM signals with increasing aggregate bit rates. As illustrated in Figure 2.12, the first TDM stage has a rate of $B' = 24 \times B$, the second of $B'' = 4 \times B'$, the third of $B''' = 7 \times B''$, and so on. At each PDH/TDM stage, a few extra bits are included for framing and stuffing purposes, which ensure remote synchronization. The standard PDH bit-rates, as applying to North America (named DSx,Tx), Europe (named Ex) and Japan (named Jx), are summarized in Table 2.2.

3.1.1 Limitations of Plesiosynchronous Digital Hierarchy

However successful in providing high bit-rate signals, PDH revealed itself to have certain limitations and even drawbacks.

The first drawback is the difficulty of extracting or inserting, at a given hierarchy level, subsets of tributaries at lower levels. This process, referred to as *add-drop multiplexing* (ADM) requires complete disassembling and reassembling of the payload. The corresponding demultiplexing and remultiplexing of the aggregate signal, in the example of T1 ADM from/to T3 frames, is illustrated in Figure 2.16. This is like having to unload and reload a truck for replacing a couple of boxes. Such a situation corresponds to a complete absence of *flexibility* in the way signals can be fed to or extracted from the aggregate traffic, and in the way network nodes can meet traffic evolution needs.

A second drawback of PDH is the proliferation of mutually incompatible hierarchy levels and subdivisions (see Table 2.2). The coexistence of these levels in networks requires intricate and costly conversion interfaces and makes uniform or smooth network development difficult. Furthermore, such a situation creates a lack of *interoperability* between networks from different local or national operators. Finally, we note from the aforementioned table that North-America/Japan and

Europe do not have the same PDH scales, which is an impediment for global network evolution.

A third drawback of PDH is the unwanted complexity associated with *synchronization*. Each hierarchy level has its own overhead of framing and stuffing bits and operates at different and independent synchronization cycles (hence the appellation *plesio*synchronous). This feature not only adds complexity but also increases the traffic overhead, which grows with the hierarchy level. A new and universal synchronization technique had to be found (namely introduced by frame *pointers*, see below).

A fourth drawback of PDH is that it does not allow either of the two important functions of *network performance monitoring* and *network operation management*. A new protocol was needed to provide such desirable functions, with the introduction of *overhead* bytes and dedicated frame fields. For instance, overhead bytes can indicate the physical point of frame assembly and disassembly, which provides useful information to both network monitoring and management. Another desirable feature is the possibility of delivering traffic *on demand*, on an as-needed basis, with upscalability potential. Put simply, operators could invest in the corresponding equipment for their inter-office trunking and use it at nominal capacity. The same equipment could also be reconfigured for higher capacities, as the demand increases, hence the *on-demand* capacity service concept.

3.1.2 SDH Framing Structure

In this subsection, we shall describe the specifics of the SDH protocol and progressive framing structure to the first assembly level (called STM-1). The higher-level frames (STM-*N*) are described in Section 3.1.3 after describing SONET framing up to the equivalent capacity level.

The SDH frames break down into *seven* subassembly-level categories, namely, in growing order of importance:

1. Containers (C),
2. Virtual containers (VC),
3. Tributary units (TU),
4. Tributary unit groups (TUG),
5. Administrative units (AU),
6. Administrative unit groups (AUG), and
7. Synchronous transport modules (STM).

The *container* (C) is the basic SDH payload frame. As in PDH, the frame rate is 8 kHz. The *C-4 container* is defined as having exactly 260×9 bytes $= 18,720$ bits. At 8-kHz cycles, the corresponding bit rate is 149.76 Mbit/s. Looking at Table 2.2 (Chapter 2), it is seen that C-4 can readily carry the European PDH payload of E4 (139.264 Mbit/s), corresponding to 2,048 voice circuits. One can also use smaller-size containers which are referred to (without straightforward numbering logic) as C-1 (C-11 or C-12), C-2 (C-21 or C-22), and C-3 (C-31 or C-32). Table 3.1

TABLE 3.1 Payload rates of SDH containers (in bits at 8-kHz frame rate) and compatibility with North-America/Japan (Tx, Jx) and European (Ex) PDH and ATM bit-rate standards

Container	Frame Size (bits)	Bit Rate	Compatibility
C-4	18,720	149.760 Mbit/s	E4 (139.264 Mbit/s) ATM (149.760 Mbit/s)
C-11	193	1.544 Mbit/s	T1/J1 (1.544 Mbit/s)
C-12	256	2.048 Mbit/s	E1 (2.048 Mbit/s) or T1/J1
C-21	789	6.312 Mbit/s	T2/J2 (6.312 Mbit/s)
C-22	1,056	8.448 Mbit/s	E2 (8.448 Mbit/s) or T2/J2
C-31	4,296	34.368 Mbit/s	E3 (34.368 Mbit/s) J3 (32.064 Mbit/s)
C-32	5,592	44.736 Mbit/s	T3 (44.736 Mbit/s) or E3 or J3
C-33	6,048	48.384 Mbit/s	ATM (48.384 Mbit/s)

shows the payload rates and compatibility with North-America/Japan and European standards, corresponding to each of the possible container groups or subgroups. It is seen that the frame sizes have been chosen to exactly match the PDH payload rates used in the two continental areas. Thus, European PDH levels are matched with the hierarchy C-12, C-22, C-31 and C-4 (for E1, E2, E3 and E4), and North-America/ Japan levels with the hierarchy C-11, C-21, C-31 and C-32 (for T1/J1, T2/J2, J3, and T3). Note that two container types, C-4 and C-33 can also carry ATM payloads.

The *virtual container* (VC) represents the first packaging stage for the previously-described containers. The one-to-one transformation between a container and a VC, referred to as *mapping*, consists in the addition of a 9-byte *path overhead*, or POH. The POH information, which is relayed between two assembly/disassembly points, serves the purposes of management and monitoring. Each container labeled C-x is thus transformed into a virtual container VC-x (e.g. C-4 becomes VC-4, etc.). See Figure 3.2 for illustration.

The next two stages concern *tributary units* (TU) and *tributary unit groups* (TUG). These two steps offer a variety of ways to multiplex VCs into groups of different sizes, which begins to be tricky. As Figure 3.1 illustrates, the TUs and the TUGs represent intermediate re-groupings that lead to different forms of VC-3 and VC-4. The most straightforward path is the one-to-one conversion of C-3 or C-4 into VC-3 or VC-4, as indicated by the dot–dashed lines in the figures. It is also possible to load the VCs with several TUGs. Such an operation is performed by *time-domain multiplexing* (or byte-column interleaving). For instance, VC-4 can receive three TUG-3, and VC-3 can receive seven TUG-2, as shown in the figure. In turn, the TUGs can be made by multiplexing TUGs of lower order (e.g., seven TUG-2 grouped into a single TUG-3), or more generally by a certain number of TUs (e.g., three TU-12 or four TU-11 grouped into a single TUG-2). Finally, the option exists to make up a TU-12 from either a VC-12 or a VC-11. As an illustration, consider the payload options for TUG-2. The finest grain (as shown in Figure 3.1) is given by four C-11, corresponding to the payload capacity

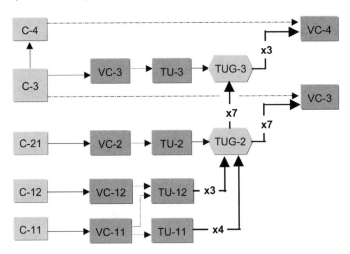

FIGURE 3.1 Assembly of virtual container (VC-*x*) frames from containers C-*x*, with different possibilities of their intermediate multiplexing (*x* times) into tributary units (TU-*x*) and tributary unit groups (TUG-*x*), as well as final TUG multiplexing into the VCs.

(Table 3.1) of $41.544\,\text{Mbit/s} = 6.176\,\text{Mbit/s}$ (four T1). Another option is three C-12, or $3 \times 2.048\,\text{Mbit/s} = 6.144\,\text{Mbit/s}$ (three E1). The final option is to fill TUG-2 with a single C-21 (6.312 Mbit/s). We leave it as an exercise (see the end of the chapter) to show that TUG-2 would not work with C-22, while TUG-3 works with all C3 container types.

The next stage of frame assembly concerns the *administrative unit* (AU) and its possible groupings (AUG). The transformation of a VC (namely, VC-3 or VC-4) into an AU (namely, AU3 or AU4) is made by adding a 9-byte *pointer field*. The pointer field indicates the position of the VC inside the AU. Figure 3.2 shows the different stages of assembly, for instance from C-4 to VC-4 (addition of 9-byte POH field) and then from VC-4 to AU4 (addition of 9-byte pointer field). Note that the frames are represented in *two dimensions*, with a convention of nine rows, and 260 byte-columns for the payload. The 9-byte POH field then occupies a single column, while the pointer field occupies a single 9-byte row at level 4. Conventionally, an AU4 directly converts into an AUG. The AU3s, on the other hand, are multiplexed by groups of three to form an AUG, as illustrated in Figure 3.3. In this case, the pointer field (row 4) contains three 3-byte pointers showing the location in the frame of the AU payloads, as well as their lengths. In addition, each pointer is used to compensate accidental phase variations of the payloads. Such variations are due to possible frequency mismatches between network nodes. The payloads can then be allowed to float within the allocated frame space, while the corresponding bit shifts (delay or advance) are recorded in the first two bytes of the corresponding pointer. This process is referred to as *negative* or *positive frequency justification*, corresponding to a phase advance or delay, respectively.

The making of the *synchronous transport module* (STM) requires the final adjunction of a *section overhead* (SOH) field. A *section* is an administrative network division between the regenerators and the multiplexers. The SOH is thus

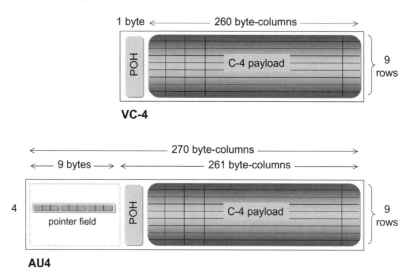

FIGURE 3.2 (Top) Assembly of container (here C-4) into a virtual container (here VC-4) into a 261-byte-columns × 9-rows matrix, showing inclusion of path-overhead (POH) field. (Bottom) Assembly of virtual container (here VC-4) into administrative unit (here AU4) into a 270-byte-columns × 9-rows matrix, showing inclusion of 9-byte pointer field at row 4.

split between a regenerator field (or RSOH) and a multiplexer field (or MSOH), which occupy the first three rows and the last five rows, respectively, as illustrated in Figure 3.4. The resulting frame is a 270 × 9-byte matrix, corresponding to *155.52 Mbit/s* (8-kHz frame rate). Such a bit rate represents the first granularity level of the SDH protocol, hence the frame name, STM-1. We note that the

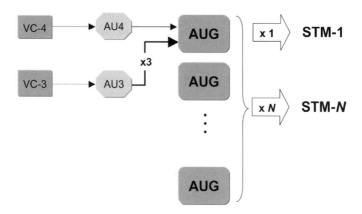

FIGURE 3.3 Assembly of administrative unit group (AUG) from a single AU4 or three AU3. After completion with specific fields (see next figure), the AUG frame is then converted into a STM-1 frame. Multiplexing *N* times AUGs results in an STM-*N* frame.

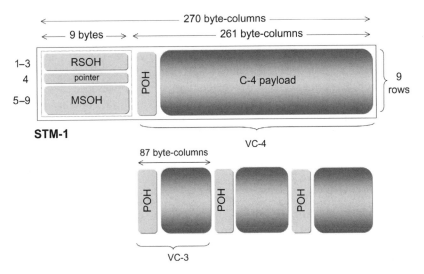

FIGURE 3.4 Completing a STM-1 frame (SDH) from a single AUG, showing inclusion of 8-row × 9-byte section-overhead fields for regenerator (RSOH, 3 first rows) and multiplex (MSOH, 5 last rows) network sections. The figures shows that the STM-1 frame can be made of a single VC-4 or three VC-3s.

corresponding payload is only 139.264 Mbit/s or 149.760 Mbit/s (E4 or ATM, see Table 3.1), which represents 89.5 to 96.2% payload efficiency. In fact the efficiency is somewhat lower, since E4 for instance carries 2,048 voice circuits at 64 kbit/s (Table 2.2), which represents 131.072 Mbit/s of voice payload. Thus the actual voice payload efficiency is 131.072/155.52 = 84.3%. But this is a small price to pay considering the advantages of SDH framing in terms of network monitoring and management, as compared to the equivalent PDH protocol (E4 efficiency being 131.072/139.264 = 94.2%).

The assembly of STM-*N* frames at higher SDH levels is detailed in Subsection 3.1.4.

3.1.3 SONET Framing Structure

The SONET (synchronous optical network) protocol represents the North America and Asian version of SDH. As its name indicates, SONET was developed for optical-cable rings, which provide high-bandwidth connection between network elements or exchanges (see subsection below). As with SDH, the function of SONET is to package various voice/data payloads, from T1 to T3, as an improvement of the former PDH multiplexing approach. But the terminology used in SONET is different. SONET frames are equivalently called either *optical circuits* (OC) or *synchronous transport systems* (STS). As we shall see, they look similar to STM frames, but they have different block lengths and detailed structure. Likewise, *virtual tributaries* (VT) play the same role in SONET as VCs in SDH. Occasionally, VTs and VCs do match at certain granularity levels, as will be shown.

The structure of the basic SONET frame, namely STS-1, is shown in Figure 3.5. The 90-column × 9-row frame is made of a 3 × 9-byte transport-overhead field, followed by a 87 × 9-byte *synchronous payload envelope* (SPE). The transport overhead has a section (SOH) and line (LOH) field of three and six rows, respectively. The SPE is made of a 1-byte column of path-overhead field and the payload. The payload contains two 1-byte fixed-stuff fields at columns 30 and 59. Thus, the actual payload consists in $(86 - 2) \times 9 = 756$ bytes. As in SDH, the frame rate is 8 kHz (8,000 frames per second), which corresponds to an effective payload bit rate of 756 bytes × 8 bits × 8 kHz = 48.384 Mbit/s. The full frame content (payload + overhead information) is 90 bytes × 9 columns = 810 bytes, giving a STS-1 frame bit rate of 810 bytes × 8 bits × 8 kHz = 51.84 Mbit/s. By definition, this is the first optical carrier (OC) bit rate, namely, *OC-1*. In case of phase mismatch, the SPE (POH + payload) can be sent though two consecutive STS-1 frames. The first frame is then filled up from column/row (K, N), then column/row $(4, N + 1)$, and so on, until there is no more space left in the SPE. The second STS-1 frame is then normally filled with the remnant bytes, i.e., from column/row $(4, 1)$ until the payload end is reached, i.e., to column/row $(N-1, K-1)$. Such an operation is called *mapping a floating SPE into STS-1*.

The noteworthy functions of the $3 \times 3 = 9$-byte SOH field are the following: frame synchronization (first two bytes being a fixed-pattern flag), frame specific identification (third byte), error monitoring (fourth byte), 64 kbit/s voice channel for maintenance staff (fifth byte, or 8 bits at 8 kHz = 64 kbit/s), and supervision between section-terminating equipment (last three bytes).

The noteworthy functions of the $3 \times 6 = 18$-byte LOH field can be described as follows. The first three bytes play a role conceptually similar to the 3-byte pointer in SDH, two bytes for addressing the SPE payload, and one byte for negative frequency justification. The following byte is for error location. Among the remaining bytes, nine are used for a 576 kbit/s data channel for supervision and one for a 64 kbit/s communication between section terminating equipment.

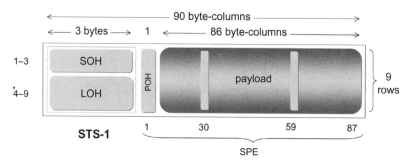

FIGURE 3.5 Framing structure of the STS-1 frame (SONET), showing transport overhead field (SOH and LOH), and the synchronous payload envelope (SPE). The SPE contains a path-overhead field (POH) and a payload field, which itself includes two fixed-stuff columns (numbers 30 and 59).

So far we have referred to the STS-1 payload as being contained inside the SPE. As with SDH with virtual containers (VC), the SPE can be made of different combinations of virtual tributaries (VT), as we shall describe now. The different VT types, called VT1.5, VT2, VT3 and VT6, are shown in Figure 3.6. The VT payloads are seen to range from 27 bytes (VT1.5) to 108 bytes (VT6). Each VT packet duration is $1/8\,\mathrm{kHz} = 125\,\mu s$. Thus, the payload bit rate of VT-6, for instance, is 108 bytes $\times$ 8 bits$/125\,\mu s = 6.912\,\mathrm{Mbit/s}$.

Having defined the VTs, we can now byte- or column-interleave them into specific groups, referred to as *SPE groups*. The standard way to make up such groups is as follows: four VT1.5, three VT2 and two VT3 (VT6 being its own group). We then note that all SPE groups have exactly 12 columns ($4 \times 3 = 3 \times 4 = 2 \times 6 = 12$). We also note that each group is made up of a single type of VT.

As described earlier, the SPE is made of 86 columns, including two fixed-stuff columns. The actual payload thus fits into 84 columns. Since each SPE group is 12 columns, we can fit $84/12 = 7$ SPE groups into the SPE. The various group arrangements from initial VT payloads are shown in Figure 3.7. Note that the seven SPE groups are byte- (or column-) interleaved together with the two fixed-stuff byte-columns.

From any possible VT arrangement into seven groups, the resulting payload is the same, namely $7 \times \mathrm{VT6}$, or $7 \times 2 \times \mathrm{VT3}$, or $7 \times 3 \times \mathrm{VT2}$, $7 \times 4 \times \mathrm{VT1.5}$. This corresponds to $7 \times 6.912\,\mathrm{Mbit/s} = 48.384\,\mathrm{Mbit/s}$ of actual payload, as was already shown. It is now possible to map the PDH voice circuits into the VTs and SPE groups, as summarized in Table 3.2. The table shows that STS-1 (or OC-1) makes it possible to carry from 28 T1/J1 up to 14 T2/J2 in individual SPE groups. With seven VT6, it is also possible to carry one T3/J3 or E3, but this high payload ($>30\,\mathrm{Mbit/s}$) must be broken into several SPE groups. The table only shows the frame arrangements made from SPE groups of the same type. In fact, SPE groups of different types can also be combined (up to seven per frame), which provides additional flexibility in PDH payloads. Since the complete STS-1/OC-1 frame has a total bit rate of 9×90 bytes $\times$ 8 bits $\times$ 8 kHz = 51.840 51.840 Mbit/s, the payload efficiency (at this hierarchy level) is $48.384/51.840 = 93.3\%$.

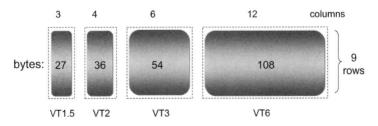

FIGURE 3.6 Different types of virtual tributaries used in SONET, with column/row structure, total byte contents and denomination (VT*n*).

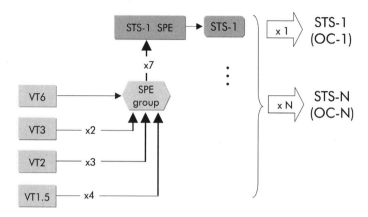

FIGURE 3.7 Completing a STS-1 frame (SONET) from seven SPE groups. Each SPE group is the combination of different possible types of virtual tributaries (VT*n*).

3.1.4 STM-*N* and STS-*N* Framing

Sections 3.1.1 and Sections 3.1.2, described the making of STM-1 (SDH) and STS-1 (SONET) frames. Both Figures 3.3 and 3.7 evoke the possibility of combining these frames into higher-order ones, as called STM-*N* and STS-*N*, respectively. The way this is done is shown in Figure 3.8. It is seen that in each case (SDH or SONET), the resulting *N*-frame strictly keeps the same structure as the original tributary frame (STM-1 or STS-1) but the number of columns in overhead and payload fields is multiplied by *N*. The byte interleaving is achieved though standard TDM technique, in either one or two stages. For instance, four STM-1 frames can be combined into

TABLE 3.2 **Payload rates of SONET virtual tributaries, compatibility with North-America/Japan (Tx, Jx) and European (Ex) PDH bit-rate standards, and corresponding STS-1/OC-1 payloads (seven SPE groups/frame)**

Virtual Tributary	Payload bit Rate (SPE Group)	SPE Group Compatibility	STS-1/OC-1 Payload
VT6	6.912 Mbit/s × 1	2 × T2/J2 (6.312 Mbit/s)	1 × T3 (44.736 Mbit/s) or E3/J3 or 14 × T2/J2 (6.312 Mbit/s)
VT3	3.456 Mbit/s × 2	2 × DS1-c (3.152 Mbit/s)	14 × DS1-c (3.152 Mbit/s)
VT2	2.304 Mbit/s × 3	3 × E1 (2.048 Mbit/s)	21 × E1 (2.048 Mbit/s)
VT1.5	1.728 Mbit/s × 4	4 × T1/J1 (1.544 Mbit/s)	28 × T1/J1 (1.544 Mbit/s)

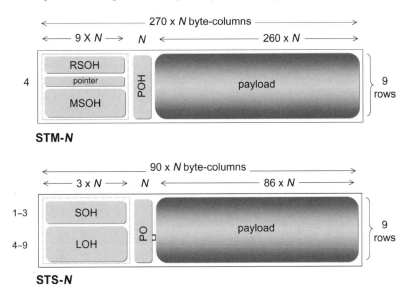

FIGURE 3.8 STM-*N* frames (top) or STS-*N* (bottom) frames resulting from the byte- or column-interleaving of *N* STS-1 or *N* STM-1 frames from SDH or SONET standards, respectively.

STM-4 using a 4 : 1 multiplexer. On the other hand, a STM-16 frame (for instance) can be realized by either one-stage 16 : 1 or two-stage 4 : 1 multiplexers.

The capacity resulting from the *N*-fold frame multiplexing is not the same for SDH and SONET, since they have different initial sizes. Remember that STM-1 has a 155 Mbit/s rate (actually 155.52 Mbit/s), while STS-1 has a 52 Mbit/s rate (actually 51.840 Mbit/s). But it is not by chance that STS-3 (3 × 51.840 Mbit/ s = 155.52 Mbit/s) has exactly the same bit rate as STM-1! This means that different multiplexing levels in SDH and SONET, and hence the payload capacities, happen to match when *N* takes certain integer values. The different frame bit rates and PDH/voice-circuit payload capacities are summarized in Table 3.3. It is seen that SONET capacities increase by multiples of three, while SDH ones increase by multiples of four. OC-3 matches STM-1 (155 Mbit/s), OC-12 matches STM-4 (622 Mbit/s) and so on. The progress in high-speed electronics through the '90s made it possible to move up the SDH/SONET capacities into the gigabit/s range. This led to the 2.5-Gbit/s (STM-16/OC-48) and 10-Gbit/s (STM-64/OC-192) standards currently in use in optical networks. At the time of writing, the 40-Gbit/s upgrade (STM-256/OC-768) is still confined to laboratory investigations, while 40-GHz electronic components are still under slow development. It should be noted that synchronization issues are far more complex at 40 Gbit/s, considering that the corresponding bit duration is 25 ps. Such a time frame is only one order of magnitude greater than the signal delay induced by a 1mm-long electrical wire. Clock frequency mismatches and random phase lags also make the implementation of 40-Gbit/s circuits substantially more difficult than previous 10-Gbit/s

TABLE 3.3 SONET (OC-*N*) and SDH (STM-*N*) frame capacities in bit rate, PDH channel payloads (North America/Asia and Europe) and corresponding number of 64 kbit/s circuits

SONET	SDH	Frame bit Rate		PDH Payload		Number of 64-kbit/s Circuits
		Approx.	Exact	North America/Asia	Europe	
OC-1	—	52 Mbit/s	51.840 Mbit/s	28 × T1/J1 or 1 × T3	21 × E1	672
						672
OC-3	**STM-1**	**155 Mbit/s**	155.52 Mbit/s	84 × T1/J1 or 3 × T3	63 × E1 or 1 × E4	2,016
						2,048
OC-9	—	466 Mbit/s	466.56 Mbit/s			6,048
OC-12	**STM-4**	**622 Mbit/s**	622.08 Mbit/s	336 × T1/J1 or 12 × T3	252 × E1 or 4 × E4	8,064
						8,192
OC-18	—	933 Mbit/s	933.12 Mbit/s			12,096
OC-24	—	1.25 Gbit/s	1.24416 Gbit/s			16,128
OC-36	—	1.8 Gbit/s	1.86624 Gbit/s			24,192
OC-48	**STM-16**	**2.5 Gbit/s**	2.48832 Gbit/s	1344 × T1/J1 or 48 × T3	1008 × E1 or 16 × E4	32,256
						32,768
OC-96	—	5 Gbit/s	4.97664 Gbit/s			64,512
OC-192	**STM-64**	**10 Gbit/s**	9.95328 Gbit/s	5376 × T1/J1 or 192 × T3	4032 × E1 or 64 × E4	129,024
						131,072
OC-256	—	13.2 Gbit/s	13.27104 Gbit/s			172,032
OC-768	**STM-256**	**40 Gbit/s**	39.81312 Gbit/s	21504 × T1/J1 or 768 × T3	16128 × E1 or 256 × E4	516,096
						524,288

The levels where both SONET and SDH standards do correspond are shown in bold. The associated bit rates are the ones used in optical communications (terrestrial and submarine).

technology. Fortunately enough, the need for the 40-Gbit/s upgrade has been alleviated for at least a decade since the advent of *wavelength-division multiplexing* (WDM). As described in detail in Chapter 2 of Desurvire, 2004; the technique of WDM consists in combining several optical channels in the same fiber. As a result, four wavelength channels at 10 Gbit/s, or $4 \times$ STM-64, provide strictly the same capacity as a single 40-Gbit/s (STM-256) channel. The channels are multiplexed and demultiplexed *optically*, which only requires a parallel implementation of 10-Gbit/s opto-electronics in the terminals. In contrast, the 40-Gbit/s SDH version requires 40-Gbit/s TDM circuitry for multiplexing and demultiplexing, and 40-Gbit/s opto-electronic receivers. Finally, WDM is not limited to any specific number of channels, which makes it possible (within the optical amplifier bandwidth) to achieve traffic aggregates as high as $320-640 \times$ STM-64 or $320-640 \times 10 \text{ Gbit/s} = 3.2-6.4 \text{ Tbit/s}$. A new generation of 40-Gbit/s-based SDH/SONET WDM systems yet represents a natural conceptual development of the current 2.5–10-Gbit/s version.

3.1.5 SONET/SDH Network Services

As mentioned earlier in this section, the SONET protocol was designed initially for optical networks based on ring topology. One added-value of this topology is the ability to implement *network protection*, as discussed in Chapter 2 (Figure 2.27). As in FDDI networks, the use of optical fibers provides extended bandwidth and transmission distance. We are no longer in the domain of local-area networks (LANs) of building-complex or campus sizes, but in that of metropolitan (MAN) and wide (WAN) area networks. According to their name, MANs can cover the areas of large towns and their suburbs, while WAN can be used to link such networks at a continental scale. In addition to the bandwidth/distance advantage (as required by MAN and WAN traffic), the SONET/SDH protocols provide intrinsic flexibility in voice and data transport. This is because SDH/SONET frames lend themselves to relatively simple assembly and disassembly stages, unlike PDH frames (see Chapter 2). Thus, it is possible to extract or insert traffic data at any network node, without breaking up the frame structure. Such an operation of retrieving/feeding tributaries is referred to as *add-drop multiplexing*, or ADM. In a given ring network, one must then chose the SDH/SONET framing level that best matches traffic needs. Figure 3.9 illustrates how SDH/SONET networks can be deployed as a means to connect different network types, with intermediary traffic levels. The interface between two SDH/SONET rings is the ADM, which for instance extracts one-quarter of the traffic from a given STM-N/OC-N frame. The other type of interface is the *digital cross-connect* (DXC), which links the SDH/SONET ring to a different network type, for example, public-switched telephone networks (PSTN) or LANs. Two SDH/SONET rings carrying different traffic types can also be linked by DXCs. A large PSTN can use a dedicated SDH/SONET ring to link a group of central offices.

As previously mentioned, SONET/SDH is a layer 1 protocol. Within this physical layer, the protocol has four levels of operation. The first is the *photonic layer*, where electrical data are converted into optical data via drivers, referred to as E/O

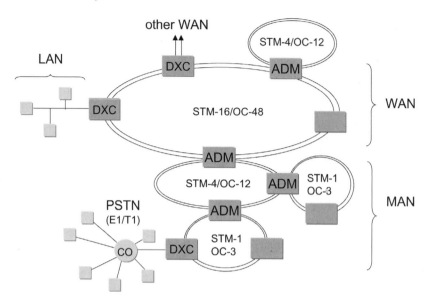

FIGURE 3.9 SDH/SONET ring topology, showing different subnetworks and traffic levels. The sub-network rings that handle similar traffic types are linked through add-drop multiplexers (ADM), while the others use digital cross-connects (DXC) as interfaces.

conversion, or the reverse referred to as O/E conversion (see Chapter 2 in Desurvire, 2004). The three other layers are activated by the different types of overheads (section, line and path) which have been described in the previous two subsections. The *section layer* concerns the monitoring of signal quality as it traverses line optical amplifiers. The *line layer* concerns synchronization and byte-to-byte multiplexing. Finally, the *path layer* manages the frame assembly/disassembly (towards higher/lower-order). It is responsible for the voice/data contents and integrity, from source to destination. It also monitors the overall operation of the *paths*, *lines*, *sections* and *segments* between two SDH/SONET nodes or network elements in the ring. Thus SDH/SONET rings provide the means to plug LANs, PSTNs and other network types into larger-scale, high-bandwidth MANs and WANs. It is customary to refer to the older network plant as the *legacy network*. This unique protocol (as implemented as either SDH or SONET, according to geography) makes it possible for a wide variety of legacy networks to exchange traffic, and also for the local PSTNs to stretch their capabilities with new voice and data services. As we have seen SDH and SONET are mutually compatible at specific rates such as 622 Mbit/s, 155 Mbit/s, 2.5 Gbit/s and 10Gbit/s. Therefore, these rates are used in submarine-cable systems which connect North America to Europe, or Europe to Asia.

As we have seen, SDH and SONET are very similar protocols, but they differ in several features. These concern frame synchronization, frame size and payload definitions (VC versus SPE, TU versus VT), bit rates, multiplexing/demultiplexing rules, overhead fields and related monitoring information, and error control.

These differences imply extra complexity in concepts, terminology and the need for dedicated SDH-SONET interfaces in both hardware and software. Yet the two standards do not have to coexist in a given continental area, and the interface issues are therefore only relevant at a global scale, such as in satellite and submarine-cable networks.

The luxury of the means to monitor, control and manage traffic down to the smallest granularity level of SDH/SONET provides a high level of network *reliability* and *flexibility*. Reliability involves frame integrity, signal quality, minimum guaranteed bit rate, degradation or failure identification, fault location, low outage probability, network protection and path restoration. On the other hand, flexibility concerns the possibility of on-demand traffic increase, traffic-path optimization, and capacity upgrade. A key advantage of SDH/SONET is that the network can be *remotely provisioned* (switching traffic and functions), *remotely tested* (trials and measurements), *remotely inventoried* (status of operation and use), *remotely customized* (local optimization) and *remotely reconfigured* (full optimization and framing/capacity upgrade).

The implementation of WDM (following that of optical amplifiers) in SDH/SONET rings represents the higher dimension for network reliability and flexibility. Indeed, the choice between multiple wavelengths for carrying traffic provides an additional means of protection/restoration, path and framing optimization, traffic aggregation and dimensioning, as described in Chapter 2 of Desurvire, 2004. The different signal wavelengths in the WDM network represent as many physical channels. The full channel capacity needs not to be activated at once, which offers the operator extra margin for future capacity upgrade and initial cost savings. The wavelength channels can also be leased, and in some cases independently owned or operated, by different network customers. This possibility of network resource sharing and cost saving allows new digital-voice/data services (and new operator entrants) to be deployed though a common physical platform and customer interface. Such services include for instance broadband-ISDN and ATM, high-definition TV (HDTV) and of course Internet service provider (ISP) applications. Note that SDH/SONET is not by itself a means to *increase* Internet speed, which is a question of end-user access bandwidth. Rather, it allows the conveyance of a high level of Internet traffic from the remote ISP all the way to the local PSTN. Network congestion is thus minimized, and hence the delay in receiving Internet data for the destination entity.

The implementations of ATM and TCP/IP on SDH/SONET are described in the two following sections, respectively.

◼ 3.2 ASYNCHRONOUS TRANSFER-MODE (ATM) PROTOCOL

The wide diversity of data formats and services, other than point-to-point voice communications, requires another type of protocol. This protocol would be fully adapted not only to such a format/service diversity but also to a burst-prone, random use, as characteristic of computer LANs and LAN-to-LAN communications, and more recently, of gigabit Ethernet (GigE) and IP traffic. This fundamental difference in data type and traffic defines two varieties of network use: *circuit-switched*, with relatively long and uninterrupted communication flows, and *packet-switched*, with relatively short and random-size communication flows. In Section 2.4, we described *packet-switched networks*, which are based upon the principle of switching and routing small data packets. Each packet contains its independent payload and destination, just like individual cars driving on a highway, each with their own random schedule. Upon reaching a highway crossing or exit, some of the cars split away and head towards a new intermediate destination, while new ones merge in.[1] The two network types also define a centralized, intelligent way of managing traffic (circuit), as opposed to a decentralized, uncontrollable or "dumb" way of using the network resource. Such a picture portraying the network *core* as intrinsically intelligent and the network *edge* or *access* as dumb, contrasting the big telephone service to the private computer LAN, is not valid anymore. The packet-protocol that could provide such fully integrated network intelligence is called the *asynchronous transfer mode*, or ATM. In this section, we shall first describe the ATM cell structure and then analyze its intrinsic functionality in multiuser, switched-network operation.

3.2.1 ATM Cell Structure

The basic structure of a data *packet* consists in an *overhead* field followed by a *payload* field. We can define the ATM *cell* as a packet of definite size, as opposed to variable size. The overhead field is then referred to as a the cell *header*. The standard cell structure is shown in Figure 3.10. It is seen that the header and the payload fields are made of 5 bytes and 48 bytes, respectively, amounting to a total of 53 bytes. The payload field has a small overhead of either one or two bytes, optionally. Thus, the overhead to actual-payload ratio is $6(7)/53 = 0.11(0.13)$, corresponding to about $11-13\%$ overhead or $87-89\%$

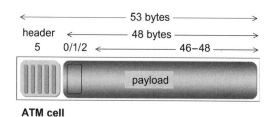

ATM cell

FIGURE 3.10 ATM cell structure, showing header (overhead) and payload fields sizes.

---

[1]To use the same simplistic comparison with SDH/SONET, one could conceive in this case of the railway system. Trains leave and arrive according to well-defined and regular schedules, up- or downloading passengers and freight at intermediate stations, always using wagon elements that are arranged in different payload types and sizes.

maximum payload efficiency. In ATM language, the partition of user data into payload chunks is called *segmentation*. Appending a header field to a given payload is called *encapsulation*.

The overall cell bit rate, as defined by the number of cell bits ($53 \times 8 = 424$ bits) divided by the cell duration can be 2 Mbit/s, 12 Mbit/s, 25 Mbit/s, 34 Mbit/s, 45 Mbit/s, 52 Mbit/s (OC-1), 155 Mbit/s (OC-3/STM-1), 622 Mbit/s (OC-12/STM-4), or 2.5 Gbit/s (OC-48/STM-6).

The 5-byte (40-bit) cell header is made up of six different control fields. During its transit through the ATM network elements, namely *switches* and *cross-connects*, some of the header contents are modified. First, we shall define the names and functions of the different header fields, which appear in the following order:

• **Generic Flow Control (GFC),** *4 bits*: this is an identifier only activated at the *user-network interface* (UNI), that lies between the customer's equipment (residential or business) and the ATM network. As the name suggests, its primary function is to regulate traffic at the multiplexer level, where cells coming from different customer equipment are being combined into a single flow. The GFC definition and actual use vary with suppliers. Past the multiplexing level, its space is then reallocated to extend the capability of the next field, called the VPI.

• **Virtual Path Identifier (VPI),** *8 bits*: a *virtual path* (VP) is assigned to groups of *virtual channels* (VC) that are multiplexed together. The path identifier corresponds to a given physical node-to-node segment; it can be changed by cross-connects as the cells transit through the nodes. In this process of VPI reassignment, the cells keep their destination address (VCI). Using the GFC, the VPI address field can be extended to 12 bits. This results in $2^{12} - 1 = 4,095$ possible VP identifiers.

• **Virtual Channel Identifier (VCI),** *16 bits*: a VC corresponds to a unidirectional and one-time user-destination address, as indexed by the VCI, to which the cell should be forwarded. The VCI is not modified during network transit. There are $2^{16} - 1 = 65,535$ possible VCIs. During the session, the same destination user can be granted more than one VCI, corresponding to simultaneous reception channels. Note that VCI numbers 0–15 and 16–31 are reserved for network management or idle/optional uses.

• **Payload Type Identifier (PTI),** *3 bits*: this field exerts several functions. There are $2^3 = 8$ possible values (from 000 to 111) to encode special information. For instance, 000 to 011 indicate absence or presence of network congestion from source to destination or the reverse; 100 and 101 indicate whether the cell is generated by an end-user or by a network segment, respectively. The two other codes have several possible uses, including for instance a sign that the cell is the last of a user series. The codes 110 and 111 are reserved for future functions use.

• **Cell Loss Priority (CLP),** *1 bit*: as the name indicates, the CLP means that the cell should (CLP = 1) or should not (CLP = 0) be discarded in case of switch/network congestion.

• **Header Error Control (HEC),** *8 bits*: this is an error-correction coding (ECC) field based on cyclic redundancy check (CRC) code. The CRC generating polynomial is $g(x) = x^8 + x^2 + x + 1$ (see Chapter 1, Section 1.6). Single errors can thus be detected and corrected within the cell header. Multiple errors can only be detected, not corrected. In this case (however much less likely), the cell is discarded.

We look next at the payload field of the ATM cell. As previously shown (Figure 3.10), the payload field is 48 bytes, including 0, 1 or 2 heading bytes of overhead. The true "information" field is thus 46 to 48-bytes long. The exceptional case of zero overhead (48 bytes payload) corresponds to AAL5 cell types (see Section 3.2.4 on AAL services). The first overhead byte has three subfields:

1. Convergence sublayer indicator (CSI), 1 bit: indicates whether a pointer is used. If yes, the second heading overhead byte is activated (see subsection below);

2. Sequence number (SN), 3 bits: indicates the possibility of missing cells or cells in the wrong position within a sequence;

3. Sequence number protection (SNP), 4 bits: used for error detection in CSI and SN fields.

The full cell payload (48 bytes) is scrambled through the generating polynomial $g(x) = 1 + x^{48}$ for error-correction purposes (see below in the subsection ATM adaptation layer).

There are up to seven different types of ATM cells, which can be identified as follows:

1. *Valid* cell: shows no header error or having a corrected error;

2. *Invalid* cell: had an uncorrectable header error;

3. *Idle* cell: inserted or extracted in order to adapt the cell rate (C) to the system transmission rate (C' < C), a function referred to as *cell stuffing* or *destuffing*;

4. *Assigned* cell: used by the ATM layer for specific network services;

5. *Unassigned* cell: also used by the ATM layer, but not assigned;

6. *Meta-signaling* cell: used to set-up or clear a switched VC connection (permanent VC connections do not need meta-signaling); see Section 3.2.2 for description of switched/permanent VC connections;

7. *Operation, administration and maintenance* (OAM) cell: as its name indicates.

3.2.2 Virtual Channels and Virtual Paths

As previously discussed (and in Section 2.4), the communication between two end-users in packet-switched networks is made possible by activating *virtual channels* (VC) or logical connections. Several independent VCs can be used at once by the communicating entities. As in telephone circuit connections, VCs are set up on demand (VCI attribution), then cleared off the network. This is referred to as the *switched virtual circuit* (SVC) mode of operation. Millions of users can access, at different times, one (or several) of the 65,535 VCs potentially available. In another mode of operation, referred to as *permanent virtual circuit (PVC)*, VCs are being permanently allocated between user terminals. Unlike telephone-circuit connections, VCs do not generally use a single, preassigned physical network link. Rather, a series of cells belonging to a VC transit the network through a physical path which depends upon the overall network use and node availability. The transit of VC cells is then determined by a succession of *virtual paths* (VPs). Each of the VPs is attributed at cross-connect/switch level, according to availability and traffic-optimization rules.

Figure 3.11 illustrates the hierarchy from VC to VP, from the customer-equipment (CEQ) to the network core. Cells belonging to the different VCs are first time-slot interleaved (a process referred to as *statistical multiplexing*) and are attributed a VPI to form a VP. The resulting cell stream is input to either a *cross-connect* or a *switch*. Both devices can handle cells that have been statistically multiplexed to form continuous VC and VP streams. Switching cells from one input physical port to one output port can be done in two ways. The function of the cross-connect is to move the stream from a given input port to a given output port, which is done on a *first-come, first-served* (FIFO) basis, with buffering being used to handle momentary congestions. This mere switching operation is usually made without changing the VPI contents. But the switch can also perform any VC rearrangement, resulting in new VP contents. Thus a given ATM cell (with unique VCI or destination) can hop between different VPs and physical links. From the figure, it may seem that at the switch (or cross-connect) output, *virtual path* and *physical connection* are the same notions. This is true if only one VP is attributed to the physical connection. With large switches, however, several VPs can be attributed to a single physical connection, as illustrated in Figure 3.12. The result is a stream of cells belonging to different VPs and VCs but sharing the same physical path. Put simply, a VC is

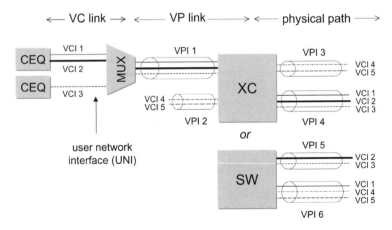

FIGURE 3.11 Virtual channels and virtual paths in ATM networks: virtual channels (VC) from different customer equipment (CEQ), as identified by VC identifier (VCI), are first time-domain multiplexed. The result is a virtual path (VP) as identified by a VP identifier (VPI). The cell flow is input to either a cross-connect (XC) or switch (SW). The XC assigns physical paths to VPs while keeping the integrity of the VC contents. The switch re-arranges the VCs into new VPs. VPs are attributed different output physical links.

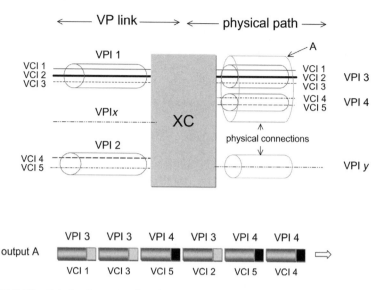

FIGURE 3.12 Relation between virtual channel (VC), virtual path (VP) and physical connection. In this example, two input VP (VPI 1 and VPI 2) are switched into a single output physical connection A containing two new VP (VPI 3 and VPI 4). The detail below shows the time-slot multiplexed cells in physical connection A, with different VP and VC identifications.

similar to a TDM channel, but with random time-slot assignment. A VP is analogous to a second-dimension for TDM channels, making it possible to manage a larger set of VCs as a single entity.

Note that the VPs are bidirectional. If we looked only at VC cells transiting from end-user A to end-user B and the others transiting in the other direction, the corresponding VPs would be identical and the network would look like permanently circuit-switched or connection-oriented, as illustrated in Figure 3.13. If we then look at another pair of end-users (C and D), the connection configuration appears different, and so on with the other communicating end-user pairs. This feature thus makes it possible for a multiplicity of end-users to share the limited number of physical connections between network nodes. Furthermore, the possibility of different virtual-path attribution and rearrangement at each node crossing intelligently optimizes the network throughp ut, so that node congestion is avoided or minimized.

3.2.3 ATM Protocol Reference Model (PRM)

According to the OSI model, the ATM protocol operates at both physical layer 1 (cell segmentation, encapsulation and transport) and data link layer 2 (virtual channel and path assignments). But the full operation and set of functions requires a more detailed layer description, which is referred to as the (ATM) *protocol reference model*, or PRM. The PRM identifies *four* operation layers, which can be described as follows, from top to bottom (Figure 3.14, left):

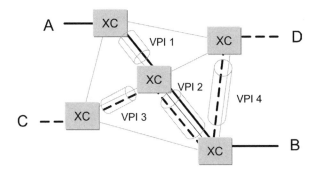

FIGURE 3.13 Physical bidirectional connection paths between network users, considering only user subsets (A,B) (C,D), and intermediate virtual paths (VPI *n*).

- *Upper-layer service* is concerned with unspecified information to be communicated through the ATM network. This information can be of user type (namely voice, data and video), of control type (setting up or clearing VC connections), and of management type (monitoring, configuring and signaling of network elements).

- *ATM adaptation layer* (AAL) has its responsibilities decomposed into two sublayers:

  (1) *segmentation and reassembly* (SAR): generation of 48-byte payload chunks without headers from input data, or the reverse process (reconstruction of the initial data from cell payloads after header discarding).

  (2) *convergence sublayer* (CS): distinguishes four types of service classes, which are referred to as "class A" to "class D" services; see Section 3.2.4 for a detailed description.

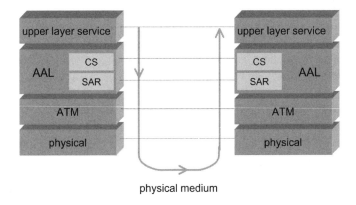

FIGURE 3.14 The different ATM-network layers according to the ATM protocol reference model (PRM), with functional activation from network origin (left) to destination (right), through the physical medium. Dotted horizontal lines show imaginary, peer-layer to peer-layer communications.

- *ATM layer* is given five main responsibilities:

  1. Statistical cell multiplexing (forming a continuous cell stream from random inputs) and demultiplexing,

  2. creation of cell header and encapsulation,

  3. managing cell transport (VP and VC switching, assignment of virtual channel (VCI) and virtual path (VPI) identification),

  4. identification of the payload type (user cell or network cell and types thereof),

  5. cell-flow control through GEC header field at UNI level (insertion of idle/ unassigned cells for ensuring continuous cell stream (although thought of as a physical-layer function).

- *Physical layer* is concerned with cell delineation, synchronization, bit-error control (HEC header field), payload scrambling/de-scrambling, variable-rate data modulation (emission) and recovery (detection), E/O and O/E conversions for optical-fiber links, and overall, physical-medium connectivity and transport. For ATM over SONET/SDH, the function of payload mapping (column/row byte-alignment into frames) is also performed at this level.

The process of cell generation, transport, and reception from end to end can be simply visualized as traveling through the stack of successive protocol layers, first from top to bottom, and then from bottom to top (Figure 3.14). The upper layer service first passes on the user data to the AAL; the AAL segments it into 48-octet payload chunks and passes them to the ATM layer; the ATM layer creates the header and encapsulates the chunks into ATM cells; the cells are then sent to the physical layer. After the cell has been physically transmitted through the ATM network, this process is repeated in reverse, leading to the recovery of user data at the upper-layer service level.

The ATM protocol reference model (PRM) provides a view more detailed than that of the open-system interconnection (OSI) model which is reviewed in Section 2.4. From the OSI reference, ATM is a layer 2 or data link protocol, in contrast to SDH/SONET (layer 1, physical) and TCP/IP (layer 3, network). In short, layer 3 provides additional features such as static/dynamic path optimization and elaborate routing functions, which are not available in ATM.

3.2.4 Adaptation Layer (AAL) Service Types

In Section 3.2.3, we mentioned that the AAL layer, which is immediately above the ATM layer, provides four types of service classes, referred to as A to D by the International Telecommunications Union (ITU-T). These services differ from each other according to three criteria: source-to-destination timing relation, bit rate and connection mode, as shown in Table 3.4. For voice and video services, a timing relation is required, because of the need for real-time interactivity. The communication channel is also used for a certain period of uninterrupted time, which justifies a *connection-oriented* mode of operation. In voice/video applications, the bit rate

TABLE 3.4 ITU-T service classes in the ATM adaptation layer (AAL), showing requirements for source-to-destination timing relation, bit rate, and connection mode, with application examples

AAL Service	Class A (AAL-1)	Class B (AAL-2)	Class C (AAL-5, AAL-3/4)	Class D (AAL-5, AAL-3/4)
Timing relation	required	required	not required	not required
Bit rate	Constant	variable	variable	variable
Connection mode	connection-oriented	connection-oriented	connection-oriented	connectionless
Application examples	CBR voice/video	VBR voice/video	VBR voice/video, X.25, frame relay	IP, SMDS

(CBR = constant bit rate, VBR = variable bit rate)

can be either *constant* (CBR) or *variable* (VBR). Other possible services concern data transfers; the data can represent large or small files, corresponding to connection-oriented or connectionless modes. The transfers may or may not require a timing relation. For instance, Internet data transfer (TCP/IP on ATM, see Section 3.3) and switched-multimegabit data service (SMDS, Section 2.5) would require no timing relation and a connectionless/VBR operation mode. The different possible combinations of these three criteria define the A–D service classes and applications, as shown in Table 3.4.

The segmentation and encapsulation of ATM cell payloads depend upon the service class. The corresponding cells are named AAL1, AAL2, AAL3/4 and AAL5, and have the following characteristics:

AAL1: class A service (see Table 3.4), adapted to CBR voice/video and SONET/SDH over ATM. Also referred to as "circuit emulation," similar to a hard-wired circuit connection. The payload header, which is 1-byte long (see Figure 3.10) is used for controlling data segmentation and reassembly (SAR sublayer of ATM layer). The first bit of this header is a CS indicator, the next three act as a sequence counter and the last four bits provide error-correction on the header byte.

AAL2: class B service with VBR, adapted voice and video signals which present silences (voice) or idle/motionless modes (video). Several voice channels represented by mini-cells can be multiplexed and embodied into a single ATM cell. The service is also adapted to compressed-video payloads. Reassembly at SAR sublayer requires, within the payload field (47 bytes), an information-type (IT) prefix field and a length-indicator (LI) postfix field, followed by a CRC field for error correction.

AAL3/4: class C (connection-oriented) or class D (connectionless) VBR service, adapted to variable-length packets with size up to 64 kbytes.

Packet segmentation and reassembly at SAR sublayer requires a 2-byte payload header and a 2-byte payload trailer, leading to 44-byte actual information payload. The last cell in the sequence (up to complete payload filling) has a 4-byte payload-field trailer.

AAL5: class C (connection-oriented) or class D (connectionless) VBR service, also adapted to variable-length packets with size up to 65.5 kbytes. The last cell in the sequence has a 1-byte trailer with LI and CRC fields. Unlike in the previous AAL3/4 format, there is no protection against cell missequencing. The difference with AAL3/4 is that there is less payload overhead, which makes it more practical and efficient. This feature is illustrated by the fact that AAL5 is often referred to as the "simple and easy adaptation layer," or SEAL.

As Table 3.4 illustrates, typical examples of class C and class D payload types are X.25 or frame-relay for the first (connection-oriented) and TCP/IP or SMDS for the second (connectionless).

3.2.5 ATM Network Connection Types and Service Classes

Different types of *network connection* are possible. The ITU-T distinguishes five connection types:

1. Point-to-point (bidirectional),
2. Point-to-multipoint (unidirectional),
3. Multipoint-to-point (unidirectional),
4. Multipoint-to-multipoint (bidirectional),
5. Point-to-multipoint (bidirectional).

The first connection type corresponds to the classic use of telephony, interactive video and data transfer. The second is adapted to *multicasting*, i.e., the one-way communication of voice, video or data to a collection of users, such as in TV *broadcasting*, for instance. The third corresponds to the reverse operation where a collection of users is in a one-way communication with a single point. For instance, this single point could receive video signals/clips and data recorded from different locations for global network monitoring or surveillance purposes, as in national operator headquarters. The fourth connection type corresponds to a full implementation of multimedia services, where a collection of users can interact together in real time. A straightforward application is audio- and video-conferencing with the involvement of multiple geographical sites. The last connection type represents a centralized version of the previous service, where bidirectional communication or interactivity is possible only with a central or command station. Clearly, all network connection types have their own requirements in terms of *quality of service* (QoS), a set of guaranteed performance parameters which are defined further in the following text.

We described AAL *services classes* in Section 3.2.4 as they have been defined by ITU-T. These classes differ from each other by criteria such as bit rate (constant or variable), connectivity (connection-oriented or connectionless) and timing relation (required or not required). A somewhat different way to define ATM network services, as followed by the *ATM Forum* standardization body, is based upon the concepts of *real time* (RT) and *nonreal time* (NRT) network use. Table 3.5 lists five possible service classes according not only to RT (1–2) and NRT (3–5) uses but other types of QoS parameters as well. These can be described as follows:

1. Real-time constant bit rate (CBR): with *guaranteed peak cell rate* (PCR), and *cell-delay variation tolerance* (CDVT); adapted to voice and video applications;

2. Real-time, variable bit rate (RT-VBR): same as (1) with PCR and CDVT, but bit-rate is adapted according to real-time needs with guaranteed *sustained cell rate* (SCR) and *maximum (cell) bursts size* (MBS); adapted to compressed voice and video signals;

3. Nonreal-time, variable bit rate (NRT-VBR): same as (2) without time-relation constraint; adapted to random or burst-prone traffic, such as with LAN-to-LAN communications and Internet;

4. Nonreal-time unspecified bit rate (UBR): same as (1) with PCR and CDVT, but without time-relation, loss, delay and bandwidth constraint; adapted to usual computer communications applications, frame relay and TCP/IP;

5. Nonreal-time available bit rate (ABR): same as (4) but with guaranteed *minimum cell rate* (MCR), which reserves network resource for minimum cell loss and bandwidth use; furthermore, feedback is provided to ensure fair access to the network, while bandwidth is provided on a best-effort, availability basis; adapted to computer communications for data file transfers, e-mail, overnight maintenance, and other nonpriority traffic uses.

Other QoS indicators used in the above service classes are the *peak-to-peak cell delay variation* (CDV), the *maximum* and *mean cell transfer delay* (max-CTD,

TABLE 3.5 ATM Forum service class definitions and QoS descriptors with guaranteed performance in cell loss, cell delay and connection bandwidth

Service Class		Descriptors	Loss	Delay	Bandwidth	Feedback
CBR	RT	PCR, CDVT	yes	yes	yes	no
RT-VBR		PCR, CDVT, SCR, MBS	yes	yes	yes	No
NRT-VBR	NRT	PCR, CDVT, SCR, MBS	yes	yes	yes	no
UBR		PCR, CDVT	no	no	no	no
ABR		PCR, CDVT, MCR	yes	no	yes	yes

(CBR = constant bit rate, VBR = variable bit rate, UBR = unspecified bit rate, ABR = available bit rate, RT = real time, NRT = nonreal time, PCR = peak cell rate, SCR = sustained cell rate, MCR = minimum cell rate, CDV = cell-delay variation tolerance, MBS = maximum burst size)

mean-CTD) and the *cell loss ratio* (CLR). The ATM Forum has also defined a variety of ATM services which can be briefly described as follows:

Cell relay service (CRS): the most commonly used ATM service, with minimum QoS requirement;

Circuit emulation service (CES): same as CRS with variable bandwidth and payload sizes, in connection-oriented mode;

Voice and telephony over ATM (VTOA): features yet to be defined, in view of class A service availability;

Frame-relay bearer service (FRBS): provides internetworking connectivity between frame-relay and ATM networks (frame/cell conversion), as well as a bridging function between frame-relay networks;

LAN emulation (LANE): provides a seamless bridge between two LANs of similar protocol (e.g. Ethernet, token ring)

Multiprotocol over ATM (MPOA): provides the equivalent of a routing function (OSI layer 3) between LANs, but more overhead-intensive than TCP/IP.

3.2.6 Mapping Protocols Over ATM and the Reverse

The process of encoding data from a given protocol (A) into a new protocol (B) is called *protocol mapping* (of A into/over B). By definition, ATM can be used to carry traffic of different type and bandwidth requirements, through the principle of payload segmentation into single-size cells having variable/scalable bit rates. Since LANs, MANs and WANs carry different types of protocols or "clients" with different bandwidth/service needs, mapping them over ATM provides extensive possibilities for network interconnectivity and interoperability (e.g. LAN to LAN, LAN to MAN, MAN to WAN...). As seen in the previous subsection, ATM can bridge LANs of similar protocols, for instance frame-relay (FRBS) and X.25 as a class C service. It can carry voice channels (*SDH/SONET over ATM*, with DS1-DS3/E1-E4 payloads of PDH), as a class A service. This mode of operation is referred to as *pure ATM transport* for SDH/SONET traffic, not to be confused with the reverse operation (see below). It can also carry TCP/IP packets (*IP over ATM*), as a class D service, which is further described in Section 3.3.

The other way around is the transport/mapping of ATM cells over other protocols such as wireless transport (WATM), digital subscriber line (DSL), and SDH/SONET, for instance. The implementation of WATM opens a variety of low-bandwidth mobile services, as further described in Chapter 4. Implementation of *ATM over DSL* advantageously offers the flexible-bandwidth capabilities of ATM in classic twisted-pair wire infrastructures such as in use in public-switched telephone networks (PSTN), as discussed in Chapter 1 of Desurvire, 2004. Here, we shall only briefly describe the features of ATM over SDH/SONET.

Mapping ATM over SDH/SONET can be done in two modes of operation: *embedded* and *hybrid*. The first mode is fully transparent to the SDH/SONET layer, meaning that ATM data are carried like any payload type, without specific network function and traffic management other than provided for SDH/SONET. ATM cells are only encapsulated "as is" into VC-4 containers (see Section 3.1

and Exercises at the end of the chapter). In the hybrid mode, both SDH/SONET and ATM traffic types are transported. Contrary to in the previous case, the network elements can discriminate between the two and switch the payloads accordingly. This requires that the switches be able to manage both types of traffic. The ATM Forum defines different ways to map ATM payload cells into SONET frames with specific byte-alignment rules (fixed or floating). These are referred to as STS-3c and STS-12c, corresponding to OC-3 (155 Mbit/s) and OC12 (622 Mbit/s) payload types. In the SDH case, the ATM cells are mapped row by row in the different container types (C-11 to C-4).

Chapter 2 in Desurvire, 2004 provides further description of the issues of network evolution and integration, which take into account the progressive penetration of SONET/SDH, ATM and IP in different sublayers (referred to as access, metro, edge and core layers).

■ 3.3 TRANSMISSION CONTROL (TCP) AND INTERNET (IP) PROTOCOLS

As its name suggests, the *Internet* represents a means to link together networks of nonuniform types. The Internet is a *connectionless packet service,* which is based upon a two-component protocol: the *transmission-control protocol* (TCP) and the *Internet protocol* (IP). Their combination, TCP/IP, is also referred to as the *Internet Protocol suite.* By extension, the name Internet is commonly used to (incorrectly) designate both the associated protocol and its higher-layer service applications. More generally, one conceives of *the Internet* as being a global and seamless network, also called the *World Wide Web* (WWW). More accurately, the Internet represents a set of interconnected networks, which communicate through the TCP/IP protocol. A set of networks connected by *IP routers* form an internet. In this section, we shall review and detail the different concepts involved in all the above, from the TCP/IP suite to Internet applications and services.

A first important observation is that TCP and IP have separate layer attributions. The IP functions operate at *network layer 3* of the OSI model (see Figure 2.18). On the other hand, the TCP functions operate at a higher level which is similar to *transport layer 4* of the OSI model. To provide a simplified view, IP is in charge of *packet routing* between the network nodes (hence the justification of *IP router),* while TCP is concerned with the *end-to-end user connectivity.* This is the reason why IP is often called a "layer 3" switching protocol, in contrast to ATM, which is a "layer 2" switching protocol. A main difference between the two switching approaches (IP and ATM) can be described as follows. The ATM packets, called *cells,* have a small/fixed size and share virtual channels/paths through the network; the flow of ATM cells is continuous. On the other hand, the IP packets, called *datagrams,* have long/variable sizes and are traveling independently of each other; they appear at random times, singly or by bursts. Such a fundamental difference between ATM and IP has a major impact on switching definition and algorithmic complexity. It has caused a "layer 2 versus layer 3" debate, which reflects preferences on how to effectively operate a network with its own pros and cons (cost, reliability, quality of service, scalability, management). Such a debate is now irrelevant due to the fact that IP may be viewed as one of the best ATM clients, and also because the layer 4 TCP now ensures comparable quality of service. As we shall see, there is more in the Internet than a mere "best effort" packet communication.

3.3.1 The TCP/IP Suite and Application Layers Stack

Just like ATM, which was described in the previous section, the Internet can be conceived of as made from a stack of intermediate protocols, each having its own sublayers and functions, as illustrated in Figure 3.15. This picture represents the current Internet-protocol standard, which has evolved over the years from a primitive ocean containing no less than 40 different protocols. The main surviving ones of interest are called:

- IP (ARP, RARP, RIP, OSPF, ICMP, IGMP) and TCP (UDP);
- FTP, HTTP, SMTP, SNMP, and TelNet.

As seen from Figure 3.15, the first group correspond to the TCP/IP layers, while the second group represent applications that come on the top of the stack. We shall focus first on the TCP/IP level. The functions of the two protocols can be briefly described as follows:

- Internet protocol (IP) layer:
  (a) provides "best-effort" packet routing through the network,

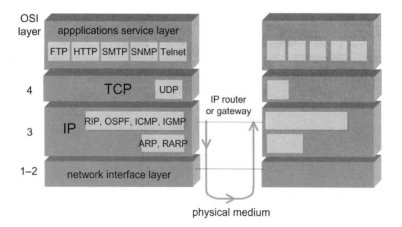

FIGURE 3.15 Protocol stack corresponding to TCP/IP suite, with respective sublayer fields, equivalent OSI-model layers and path followed by the data between two end-users (see text for layers description).

(b) provides connection-less interconnectivity between different subnetworks (e.g., X.25, FDDI, Ethernet, ATM),

(c) controls congestion events, and

(d) specific IP applications include:

ARP (*address resolution protocol*),

RARP (*reverse address resolution protocol*),

RIP (*routing information protocol*),

OSPF (*open shortest-path first protocol*),

ICMP (*Internet control message protocol*) makes it possible to exchange supervisory information between network nodes and error messages,

IGMP (*Internet group management protocol*) makes it possible to manage multicast communications between groups of hosts.

The applications ARP, RARP, RIP, OSPF, ICMP and IGMP are further described in Section 3.3.6 entitled *IP-layer functions.*

- Transmission control protocol (TCP) layer:

(a) establishes reliable *connections* between terminal users (initiate, accept, send/receive data, clear and break);

(b) ensures message integrity and error control;

(c) recovers from network failure;

(d) UDP (user datagram protocol) is a simpler but less reliable alternative to TCP.

The TCP and UDP functions are described in detail in a later subsection entitled "TCP and IP datagram/packet structures (IPv4/IPv6)."

We focus next on the *application services* layer. The applications can be briefly described as follows:

- FTP (*file transfer protocol*): enables one to transfer pre-formatted files (e.g., text)
- HTTP (*hypertext transfer protocol*): makes it possible to download web pages in HTML format
- SMTP (*simple mail transfer protocol*): program to exchange e-mail messages and notes
- SNMP (*simple network management protocol*): program managing non-uniform networks
- Telnet: terminal emulator for connecting to a distant host computer.

The applications FTP, HTTP, SMTP, SNMP and Telnet are further described in Section 3.3.7 entitled "*Applications service-layer functions.*" But, we shall first consider the issues of IP network *connectivity*, *addressing* and *routing* as well as the *header structure* of TCP/IP datagrams.

3.3.2 The Internet and Internet Connectivities

At the very bottom of the stack, the *network interface layer* (Figure 3.15) provides, as its name implies, data formatting and coding (data link layer 2 in OSI model) and physical connectivity through hardware (physical layer 1 in OSI model). The network types being thus connected through this interface can be anything from Ethernet, X.25, token-ring, FDDI, ATM, or SDH/SONET. As mentioned previously, the key element which enables the interconnection of these different networks is the *IP-router*, also called IP *gateway*. The IP gateway is able to achieve any data transformation from one network to the other, thus ensuring full end-to-end connectivity. The communicating entities which are attached to a given network are called *hosts*. Figure 3.16 shows how an *internet* can be formed out of a variety of networks and their associated hosts. In this internet, hosts are identified by an *IP address*. As shown in Section 3.3.3, the IP address consists of a *network address* and a *host address* within this network. On the Internet, hosts act either as *clients* or as a *provider of services*. A client is an entity which requests a given service. The provider, hosted somewhere in the Internet, handles this request through an application called a *server*, which in reply provides the requested service. The internet thus establishes a set of *client/server relations* between a variety of hosts. It must be clear now that what is called "the Internet" represents the global version of the previous picture. Networks A, B or C shown in Figure 3.16 can be anything from LANs, MANs, PSTNs and ... smaller internets.

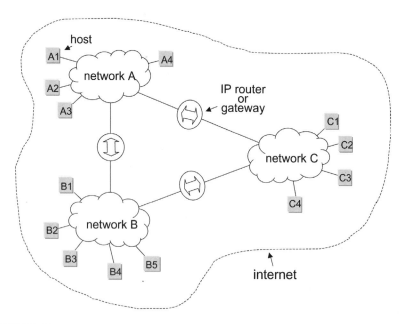

FIGURE 3.16 Making up an internet from different network types, showing protocol layers involved in the network hosts and IP routers. The labels A1, A2,..., C4 represent IP addresses, which identify each host within the internet.

The envelope which encapsulates this interconnected environment has become the Internet.

3.3.3 IP Addressing Format

The IP address, which identifies both subnetworks and hosts, is made of 4 bytes or 32 bits. This definition corresponds to a fourth (and current) version of the IP addressing protocol, which is referred to as *IPv4*. In this IPv4 address, the first two bytes and the last two bytes represent the (sub)network and the host addresses, respectively. They are also called *netid* and *hostid*. It is customary to write IP addresses in decimal with dot separators, i.e., W.X.Y.Z. Thus the number

128.8.15.113

or, in binary representation:

10000000 00001000 0001111 01110001

corresponds to the network identified by 128.8 (netid) and to the host 15.113 (host ID) within this network. The only authorities which can provide the network part of the IP addresses are called *regional internet registries*, or RIR. There are three RIRs: the American registry for Internet numbers (ARIN) for the Americas and sub-Saharan Africa; the Réseaux IP Européens (RIPE) for Europe, Middle East, and parts of Africa; and the Asia-Pacific Network Information Centre (APNIC). There are five classes of IP addresses, as summarized in Table 3.6:

- *Class A* is identified by "0" in the first IP address bit. The netid is given 7 bits, corresponding to $2^7 - 2 = 126$ networks, and the hostid is given 24 bits, corresponding to $2^{24} - 2 = 16.777214$ million hosts. The total thus represents 126×16.77 million $= 2.11$ billion addresses. The subtraction of 2 in this calculation is due to the two reserved addresses where all bits are either zero or one.

- *Class B* is identified by "10" in the first IP address bit. The netid is 14 bits ($2^{14} - 2 = 16,382$ networks) and the host ID is 16 bits ($2^{16} - 2 = 65,534$ hosts). The total thus represents $16,3282 \times 65,534 = 1.07$ billion addresses.

TABLE 3.6   Classes of IP addresses

	byte 1		byte 2	byte 3	byte 4
Class A	0	x x x x x x x	x x x x x x x x	x x x x x x x x	x x x x x x x x
Class B	10	x x x x x x	x x x x x x x x	x x x x x x x x	x x x x x x x x
Class C	110	x x x x x	x x x x x x x x	x x x x x x x x	x x x x x x x x
Class D	1110	x x x x	x x x x x x x x	x x x x x x x x	x x x x x x x x
Class E	11110	x x x	x x x x x x x x	x x x x x x x x	x x x x x x x x

Showing class-identifier field (bold), network address field (netid, shaded in gray) and host address field (hostid) sizes

- *Class C* is identified by "110" in the first IP address bit. The netid is 21 bits ($2^{21}-2 = 2.09$ million networks) and the host ID is 8 bits ($2^8-2 = 254$ hosts). The total thus represents 2.09 million $\times$ 254 = 0.530 billion addresses.

- *Class D* is identified by "1110" in the first IP address bit. The remaining address field corresponds to a common address shared by multiple hosts, which makes possible *Internet multicasting* (sending IP datagrams to multi-point hosts)

- *Class E* is identified by "11110" in the first IP address bit. The remaining address field is reserved for future use.

The arrangement into the first three classes A–C permits IP addresses to be attributed to internets of various sizes and configurations, from the ones with small numbers of networks but with a virtually unlimited numbers of hosts (Class A), to the ones with large numbers of networks with relatively small numbers of hosts (Class C). The total number of addresses available through Classes A–C is 2.11 billion + 1.07 billion + 0.53 billion = 3.71 billion, or about 4 billion.

A certain number of specific addresses have been reserved for different network uses. An address with all netid and hostid bits set to zero (0.0.0.0) mean "this host"; it can be used at system startup but is never a valid destination address. Addresses with all netid bits set to zero (0.0.Y.Z) mean "host [Y.Z] on this net"; the responding hosts use the valid network address, which is then used by the original sender. Addresses with all hostid bits set to one (W.X.255.255) or with both netid and hostid bits set to one (255.255.255.255) are used for local broadcast; these are never used as a valid source address. Finally, the specific Class A address 127.X.Y.Z (127 = 00111111) is used for designating the *loopback network*, which processes data in local system interfaces. Table 3.7 shows the conventional range of IP addresses (minimum to maximum values) that can be assigned to any Internet subnetwork and host. The first Class A networks to be registered by the RIR/ARIN were granted permanent addresses such as 9.X.Y.Z (IBM), 10.X.Y.Z (ARPANET), and 12.X.Y.Z (AT&T), for instance.

The rapid growth of internet hosts has created the need to break up the networks into a set of subnetworks, which are referred to as *subnets*. The subnet structure is introduced by the network administrator, who assigns two different address fields within the netid. Only hosts belonging to that network (identified by this two-field netid) recognize the subnet structure. For Class B networks, a possible subnetting rule (a) consists in using the first byte as a subnet ID and the second byte as a host ID for this subnet. Considering that the first byte has only 6 free bits (see Table 3.6), this provides up to $2^6-2 = 62$ subnets, each having up to $2^8-2 = 254$ subnet hosts. Another possible rule (b) consists in attributing the first 12 bits to the subnet ID ($2^{12}-2 = 4{,}094$ subnets) and the remaining 4 bits to the hosts ($2^4-2 = 14$ subnet hosts). Clearly, rule (a) is adapted to networks having a limited number of subnets, each having many subnet hosts, while rule (b) is adapted to networks with a large number of subnets, each having a limited number of hosts. Note that some network elements can have more than one IP

TABLE 3.7   Conventional ranges of IP addresses by classes, from minimum to maximum values, as expressed in decimal (top) and binary (bottom) forms

	min	max
Class A	1.1.0.0 00000001 00000001 00000000 00000000	126.0.0.0 01111110 00000000 00000000 00000000
Class B	128.0.0.0 10000000 00000000 00000000 00000000	191.255.0.0 10111111 11111111 00000000 00000000
Class C	192.0.1.0 11000001 00000000 00000001 00000000	223.255.255.0 11011111 11111111 11111111 00000000
Class D	224.0.0.0 11100000 00000000 00000000 00000000	239.255.255.255 11101111 11111111 11111111 11111111
Class E	240.0.0.0 11110000 00000000 00000000 00000000	247.255.255.255 11110111 11111111 11111111 11111111

For clarity, the class-identifier field (heading bits in bold shown in Table 3.6) is underlined.

address, which makes it possible to connect different network types. This is the case of *multiple-home* hosts and IP routers (see below).

The RIR policies for address attribution (in terms of classes and regions) have been designed to avoid exhaustion in the different internet classes. These policies can be briefly summarized as follows:

Class A: netids from 64 to 127 are indefinitely reserved for future needs; other IDs attributed on an individual case basis;

Class B: netids assigned only to organizations able to prove a need of over 32 subnets and 4,096 subnet hosts: by default, Class C netids are being attributed;

Class C: the netids are attributed region by region, the first byte ranging from 192 to 207 (208–223 being reserved, see Table 3.7 for Class C). The regional attribution is made according to the following rules (Z = hostid):

Multiregional:	192.0.0.Z	to	193.255.255.Z
Europe:	194.0.0.Z	to	195.255.255.Z
Others:	196.0.0.Z	to	197.255.255.Z
North America:	198.0.0.Z	to	199.255.255.Z
Central/South America:	200.0.0.Z	to	201.255.255.Z
Pacific Rim:	202.0.0.Z	to	203.255.255.Z
Others 1:	204.0.0.Z	to	205.255.255.Z
Others 2:	206.0.0.Z	to	207.255.255.Z

Any corporation can develop its own *private internet*, in which case the network is called an *intranet*. The corporation might also extend its intranet to a set of external networks (e.g., belonging to preferred customers and partners). In this case, one refers to the whole system as an *extranet*. While the intranet/extranet IP addresses are not bound to any RIR attribution, they should not be used in the outside world of the Internet. In private internets, the IP addresses are not globally unique. The allowed addresses are of the form:

10.0.0.0 (single Class A networks),

172.16.0.0 to 172.31.0.0 (16 possibilities for Class B networks),

192.168.0.0 to 192.168.256.0 (256 possibilities for Class C networks).

In spite of all these provisions to face a growing number of internet hosts, the problem of eventual *IP address exhaustion* is still looming in the near-future. This is not only because the number of computer stations will always increase, but also because of the appearance of new types of "unmanned" web-activated receivers (home TVs, "smart appliances," pacemakers, surveillance, guidance, machinery, etc.). As previously seen, the 32-bit addressing protocol of IPv4 makes it possible to allocate nearly 4 billion possible addresses, which is over the current number of Internet users: 110 million in Y2000, 200 millions in Y2001, but ... 1.8 billion projected for 2005. For this reason the standardization body, called *Internet engineering task force* (IETF), has been mandated to study an upgraded addressing system, which is called *IPv6*.

A first requirement for IPv6 is that it could be used as the new IP addressing system for the next 30 years. Other requirements for IPv6 are to have a means for practical transition/migration from IPv4 and to be able to coexist with it, to reduce the number of network classes in order to simplify routing tables, while introducing new service classes that allow data type identification. In addition, advanced features like authentication and encryption should be included. To avoid confusion, the IPv6 datagrams are called *packets*. Network elements (hosts and routers) are called *nodes*. The details of IPv4 datagrams and IPv6 packet structures are provided in Section 3.3.4.

3.3.4 Datagram Routing

We shall now describe how datagram routing is actually performed through the network. Figure 3.17 shows how an internet (or intranet) can be assembled from a set of corporate LANs and IP routers, with outside connections to two other internets (here ARPANET and AT&T, for instance). Note that within each local network, routers and multiple-home hosts have different IP addresses. Each router conveys the IP message to the network and host it is destined for. For instance, if LANCELOT wants to send a message to some host in the AT&T internet, the corresponding datagrams will go first through MELUSINE (acting as a router) and then TALYESIN (IP router).

Each host and router in the network possesses its own *routing table*. Such a table lists all possible routes to the other IP hosts and routers in the network. The routes can be direct (host in the local network), indirect (host in different local network) or by default (if route not identified from previous cases, a default router is assigned). As illustrative examples, the routing tables of host ARTHUR and router TALYESIN

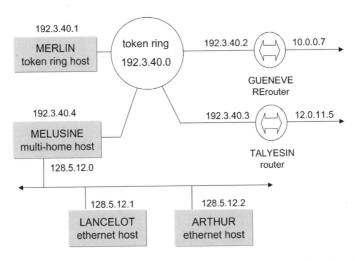

FIGURE 3.17   Example of an internet comprising a token-ring network and an Ethernet, with IP routers connecting to two other internets. The names of the different network elements (hosts or routers), and their corresponding IP addresses within each network domain, are also shown.

of Figure 3.17 are shown in Table 3.8. Upon receiving a datagram, each router follows the logical sequence of tests:

Destination address matches the next local network? YES: deliver directly, NO: go to next step;

Is there an indirect route? YES: deliver to router as indicated; NO: go to next step;

Is there a default route? YES: deliver to default router as indicated; NO: go to next step;

Send error message ("network unavailable") through ICMP.

Since intranet hosts can be accessed from the outside via the Internet, special measures must be taken to protect them against unauthorized use and different types of attack. Specific network elements placed next to the intranet gateways, called *proxy servers*, were developed for this purpose (Figure 3.18). Proxy servers are also able to detect computer *viruses* which could be embedded as executable files, and therefore deny the datagram entry (see more on viruses in Section 3.3.11). Conversely, proxies can also prevent users within the corporate intranet from accessing IP addresses of a nonprofessional or forbidden character. Thus, the proxy monitors, controls, authorizes, or denies datagram transfers from the outside and to the outside, which creates different types of blocking functions, or *firewalls*, as illustrated in Figure 3.18. Internet firewalls are of four types:

Packet filtering: Refuses the entry or exit of packets corresponding to unauthorized service or format (e.g., restricted-access URLs within intranets, nonprofessional or content-sensitive Internet Web sites, maximum e-mail

TABLE 3.8   Routing tables for host ARTHUR and router TALYESIN in Figure 3.17

	To Reach Address	Route To
ARTHUR	128.5.Y.Z	deliver directly
	192.3.40.Z	128.5.12.0
	10.X.Y.Z	128.5.12.0
	12.X.Y.Z	128.5.12.0
TALYESIN	128.5.Y.Z	192.3.40.4
	192.3.40.Z	192.3.40.0
	10.X.Y.Z	192.3.40.0
	12.X.Y.Z	deliver directly

The first column shows the IP address of the destination network, while the second shows the IP address of the next or neighboring router.

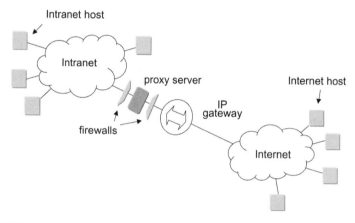

FIGURE 3.18   Intranet and its connection to the Internet though an IP gateway. A proxy server, surrounded by network firewalls, checks addresses and contents, and grants or denies permission of access/destination for input/output datagrams.

upload/download-sizes), which protects against various forms of intranet/ Internet abuse and attacks;*

Dual-homed Gateway: Concerns types of "screened hosts" which are assigned two internet addresses; two-way IP traffic is not only filtered but also analyzed by the proxy which blocks traffic in case of unauthorized use or packet type;

Screened host: acts as a traffic filter gateway for both packets and applications; the application gateway is referred to as the *bastion host*, to which all incoming untrusted traffic is routed prior to reaching the host;

---

*Firewalls equipped with *deep packet inspection*, check entire packet headers and payloads at wire speed. One distinguishes *static* from *dynamic*, or *stateful*, packet inspection. The latter checks packet streams in set-up connections at *OSI layer 4*, which prevents undesirable packets from traveling the protocol stack up to the application layer.

Screened subnet: Same as previous, except that all traffic (incoming and outgoing) is forced to pass through the bastion host, with packet filtering on each side, representing three consecutive levels of security.

Because the Internet/intranet proxy servers are able to check out not only the source/destination addresses but also the content of the IP traffic, they have full and complete access to the information that can be exchanged between the communicating hosts (e.g., e-mail and file attachments). This conveys the idea that IP networks are hardly secure for *privacy use*. This is true for internet and more so for corporate intranet communications. With the Internet, the only solution for securing privacy and confidentiality (but with various security and legality requirement levels), is *encryption* (see Chapter 3 in Desurvire, 2004).

Because of the different types of networks traversed, it can happen that the IP datagram size exceeds that permitted in a given section. In this case, the datagram must be fragmented into smaller pieces. These pieces travel independently through the network to be finally collected at the end destination for complete datagram reassembly. This process of fragmentation and re-assembly is described in more detail in Section 3.3.5, when considering the IP header.

Internet addressing (W.X.Y.Z) represents both a universal and local identification system for network hosts and node elements, at IP-layer level. It should not be confused with electronic-mail addressing (username@domain) and Web-site locators (http://www.domain). These address types correspond to specific Internet applications handled by SMTP and HTTP, respectively, at the applications service layer. Both are described in separate following subsections.

3.3.5 TCP and IP Datagram/Packet Structures (IPv4/IPv6)

After the description of SONET/SDH and ATM protocols Sections 3.1 and 3.2, the reader must now be familiar with framing/packet structures and their encapsulation into byte-fields and allocated functions. The TCP/IP datagram (or packet for IPv6), is yet another protocol, which is only a bit more rich and complex than the preceding ones. Note that the datagram is not a frame (as in SONET/SDH), but a packet, meaning that, as in ATM, the different byte-fields follow each other in an uninterrupted sequence. It is only for visual convenience that these fields are represented in rows and columns.

The IP datagram has a maximum length of about 65 kbytes. As shown in Figure 3.19, it is made of a *IP header* followed by a *TCP segment* or *body*. The IP header is at least 20 bytes long, with the possibility of adding another 4 bytes of optional/padding fields. The TCP body includes a TCP header, which like the IP header is 20–24 bytes-long, followed by a *data body*.

We focus first on the *IP header*, the structure of which is detailed in Figure 3.20, arranged for visual convenience into five to six rows of 4 bytes. The different fields, lengths and functions of the IP header can be described as follows:

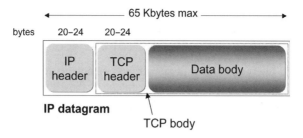

IP datagram

TCP body

FIGURE 3.19 IP datagram structure, showing IP header and TCP body, with respective byte sizes. The TCP body includes a TCP header and data body of payload.

Version, 4 bits: Identifies the IP standard, for example version 4 or IPv4;

Internet header length (IHL), also called header length (HLEN), 4 bits: Defines the IP header length as being either 20 bytes or 24 bytes, exactly, and measured by 32-bit (4-byte) units (thus HL = number of 4-byte blocks, i.e., 5 or 6);

Service type, 8 bits: Describes the type of service in terms of delay, speed and reliability; the field is decomposed into a 3-bit precedence indication (0 = normal, 111 = network control level), followed by single D, T, R bits (D = 1 means low delay, T = 1 means high throughput, R = 1 means high reliability), and two unused bits;

Identification, 16 bits: Permits the reconstruction, at receiver level, of fragments of the same datagram sharing this identification (see *datagram fragmentation);*

Total length, 16 bits: corresponds to the total IP datagram length in bytes, including the IP header; since the field is 16 bits long, the maximum allowed number of bytes is exactly $2^{16}-1 = 65,535$ or 65.535 kbytes;

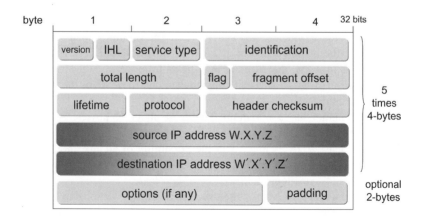

FIGURE 3.20 Byte sequence forming IP header (IPv4), as shown by rows for visual convenience. See text for description of the different byte fields.

Flag, 3 bits: The first bit indicates whether datagram fragmentation is allowed (0 = don't fragment), the second whether or not the fragment is the last of a series (0 = more fragments, 1 = last fragment), the third is unused; comparison of the number of fragments received with the total length field number makes it possible to verify that no fragment has been lost in the reconstruction;

Fragment offset, 13 bits: Indicates the offset of the fragment with respect to the original datagram, starting from zero and as expressed in 8-byte multiples; an all-zero fragment offset field (in addition to a zero flag field) indicates an unfragmented datagram;

Lifetime, 8 bits: Indicates in seconds how long the datagram or fragment should live in the network (namely, up to $2^8 - 1 = 255$ seconds, corresponding to about 4 minutes); at each node crossing, the lifetime is decreased by one unit or second (datagram or fragment being destroyed for zero);

Protocol, 8 bits: Defines the protocol to be used for the higher layer. The values (indicated here in decimal) and their protocol correspondence can be for instance: 6 = TCP and 17 = UDP for transport, 1 = ICMP for network control and supervision;

Header checksum, 16 bits: Controls the integrity of the IP header's value at each node or router crossing, in a way similar to CRC or FCS error control; the checksum must match the datagram contents, otherwise the datagram is discarded.

Source and destination IP addresses, 4 bytes each: These are the binary IP addresses (W.X.Y.Z in decimal) described in the previous subsection, referring to the source (sender of the datagram) and the destination (ultimate intended recipient of the datagram); the two addresses never change during transit through the network nodes; through the routing table, the sending host knows to which neighboring node/router to send the datagram, but this information is not in the IP header; likewise, the IP router knows, from its own routing table and upon reading the destination address, to which neighboring node it must forward this datagram; the routing process is repeated until the final destination is reached;

Options, maximum of 4 bytes: An optional, variable-length field making it possible to introduce network testing or control functions; these include: recording of route and node traversed, pre-specification of route, security/handling restrictions, time-stamping (record of time when datagram was handled by given router); each option occupies a single byte (bit 1 = flag, bits 2–3 = definition of two possible option classes [11 and 11 unused], bits 4–8 = option number in the class selected); option classes and numbers so far define up to 8 possible options;

Padding: Used to complete the previous option row up to a length of exactly 32 bits.

Datagrams can be *fragmented* to allow their transport/handling through different physical networks having limited frame size or length. Networks are characterized by a *maximum transmission unit* (MTU), as expressed in bytes. For instance, Ethernet and FDDI networks have MTUs limited to 1,500 and 4,470 bytes, respectively. For the Internet, the requirement is a minimum MTU of 68 bytes (noting however that fragmentation is technically possible at even smaller scales). This allows for a maximum TCP/IP header size of 60 bytes and a fragmented data payload of 8 bytes (allowing for a reserve considering that IPv4 headers have a maximum header field of 48 bytes). The operation of fragmentation must be implemented when a router receives a datagram from a large-MTU network to be forwarded to a small-MTU network. Since fragmentation must be indexed by multiples of eight bytes (fragment-offset field, above), the fragment size is chosen accordingly. In short the fragment payloads must be multiples of eight bytes, except the last one (e.g., 24 or 64 bytes yield only one fragment, 25 or 125 bytes two fragments, etc.). The fragments are assembled by concatenating the same header as the original (except for flag, fragment-offset and route-option fields), with the fragmented data body. The fragmented datagrams then travel through the network on their own independent path, all the way to their ultimate destination, as illustrated by Figure 3.21. Fragments can be refragmented should they encounter a network with smaller MTU. Should any of the fragments be lost in this process, it will not be possible to reassemble the original datagram and its other surviving pieces are therefore discarded. *Datagram reassembly* is always made under time constraint. Received fragments are buffered in the order they arrive, waiting for outstanding ones. When the timer expires, or if errors are detected in any of the received fragments, reassembly is aborted and the whole datagram is discarded.

This description concerned the IP protocol and datagram fragmentation/ routing. Figure 3.15 shows the list of different control functions operating at the IP layer, which have the names ARP, RARP, RIP, OSPF, ICMP and IGMP. These functions are described in detail in the next subsection. At this stage, we shall only state that the most important feature of IP is that datagrams are sent from origin to destination without the establishment of any connection. This is referred to as a *connectionless* transmission mode. It is clear that datagrams (and any of their fragments) can be lost at some intermediate network node, or datagram re-assembly may fail because of delays incurred by some of the fragments. This feature shows that IP only represents a *best-effort*, connectionless transmission. It is essentially an unreliable means for packet delivery. The protocol TCP (or alternatively, UDP, see below) has been developed at the network layer immediately above IP in order to compensate for these deficiencies. We must now describe the features of the *TCP header*.

The TCP header structure is detailed in Figure 3.22, arranged for visual convenience into five rows of 4 bytes. One extra 4-byte row (or more) can be included for options. The different fields, lengths and functions of the TCP header can be described as follows:

> *Source* and *destination ports*, 2 bytes each: While IP establishes a route between
> source and destination hosts, as identified by their IP addresses, TCP

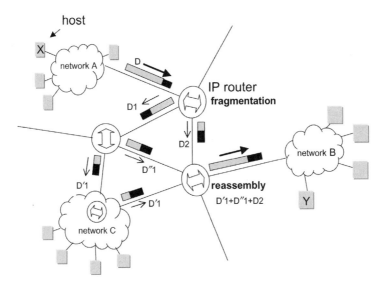

FIGURE 3.21 Principle of IP datagram fragmentation. Host X of network A is sending a datagram D to host Y of network B. The datagram D and its payload is first fragmented into D1 and D2 in order to comply with the maximum allowed frame sizes (MTU) of neighboring network routers. Likewise, the datagram fragment D1 and its payload is fragmented into D'1 and D''1. Finally, the original datagram is reassembled by the router controlling the destination network B, which is done under a specific time constraint. Each fragment header is a replica of the original datagram header, shown here in black. Among other features, the fragment headers keep the information on their subpayload position with respect to the original one.

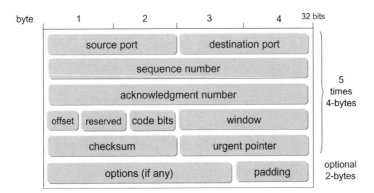

FIGURE 3.22 Byte sequence forming TCP header (IPv6), as shown by rows for visual convenience. See text for description of the different byte fields.

establishes an application or program connection between these two hosts, as identified by standard software *ports* in the hosts. These ports represent the ultimate source/destination within the communicating machines. The combination of TCP ports and IP addresses is referred to as a *socket*. An IP address with a TCP port number is called a host *endpoint*.

Sequence number, 4 bytes: Number of the first data byte in a transmitted TCP segment (see *segment* definition below); used by receiver host to rearrange segments if they arrive out of order or as duplicates; segments that are lost are retransmitted;

Acknowledgment number, 4 bytes: Signals the position of the next byte in the window (see *window* definition below), as expected by the receiver; also indicates how many bytes have been successfully received and acknowledged;

Offset, 4 bits: Specifies in 32-bit multiples the full length of the TCP header, including options and padding (locates the actual start of data/payload field);

Reserved, 6 bits: Unused;

Code bits, 6 bits: Six individual flags indicating, when set to "1", the purpose of the TCP message, for example "reset" (call for resetting connection), "synchronize" (to request and synchronize a new connection), "acknowledge" (to acknowledge the connection request) and "finish" (no more data to send);

Window, 2 bytes: Gives the number of bytes that can be received and buffered before acknowledgment (see below for description of *window*);

Checksum, 2 bytes: Controls the integrity of the TCP header's value at each node or router crossing, in a way similar to CRC or FCS error control;

Urgent pointer, 2 bytes: When set to nonzero, indicates that payload corresponds to urgent data.

Options: Optional, variable-length field making it possible to introduce network control or monitoring functions; for instance, these include: setting maximum allowable segment length (as indicated by receiver at connection initiation), time-stamping (record of time when segment was handled), information from receiver to sender on successful segment transmission.

Padding: Used to complete the last option row up to a length of exactly 32 bits.

As previously stated, TCP establishes a working connection between two endpoints or host machines. While the hosts are identified by their IP addresses, the TCP ports identify what application/software type is concerned in the connection. Within a given local host, several connections can be independently established via the same TCP port, but concerning different remote hosts having different IP addresses. TCP thus makes it possible to establish *virtual connections* between a simultaneous set of users. An example is provided by file transfer and e-mail: in a local server host, the TCP ports with numbers 21 and 25 (corresponding to FTP and SMTP, respectively) can be made accessible to any distant client host, which enables the server to send datafiles and messages to them, and the reverse.

The connection's initiation, agreement (of what software/port to use), control and termination between any set of endpoints is handled by TCP, in a full duplex mode. While IP defines a connectionless communication path for the datagrams, TCP controls the data flow, in the form of *segments*. Each byte in the original data stream is assigned a *sequence number*. The data are blocked into individual segments of N bytes which are assigned the sequence number of the first segment's byte. Thus consecutive segments of 500 bytes have sequence numbers: $1, 501, 1001$, etc. The full segment is made of the TCP header followed by the data block. TCP segments can carry different types of messages, such as acknowledgments, requests to open/close/reset a TCP connection, or mere payload data.

The segment flow is controlled in such a way as to allow the receiving host to buffer no more bytes than it can process. Segments are made of a series of bytes to be sent and acknowledged. Bytes are acknowledged one after another, but several bytes can be sent before the corresponding acknowledgments are received by the source. The flow control is ensured by a *sliding-window* technique. This sliding window is defined by two pointers within the byte sequence, whose positions shift in time, as illustrated in Figure 3.23, for two different times. The window size is defined by the number of bytes that can be sent without acknowledgment. The left and right pointers show the first and last bytes to be sent; the intermediate pointer shows the location of the next byte to be sent without delay. The window is allowed to move to another position when all bytes of a segment have been sent and acknowledged. Each acknowledgment from the receiving host indicates how many bytes were received and how many can still be sent. The window size is thus allowed to change according to the last instruction. An increased window size means that more bytes can be received. A reduced window size (down to zero) means a request to decrease (or stop) the byte transmission. This principle of variable window size makes it possible to maximize the data flow,

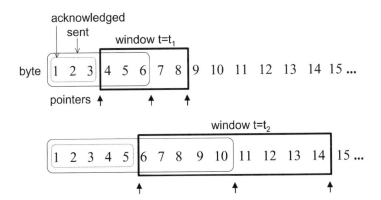

FIGURE 3.23 Principle of sliding window for TCP data flow control at two different times $t = t_1$ and $t = t_2$ with pointers showing first byte sent in the window, first byte ready to be sent and last byte of window to be sent. The window slides ($t = t_2$) when all bytes have been sent and acknowledged, with the possibility of changing size.

without congesting the receiving-host buffer. It also makes it possible for two machines having different speeds and memory sizes (e.g., microcomputer and mainframe computer) to communicate together and adapt their send/receive flows accordingly. In fact, each endpoint maintains two sliding windows: one for the bytes to be sent, one for the bytes to be received and acknowledged. Therefore, a total of four windows (two per endpoint) is used during a given TCP connection.

As mentioned in a previous subsection, UDP (*user datagram protocol*) is a simpler alternative to TCP, but does not offer the same level of reliability. As opposed to TCP, it operates in connectionless mode, meaning that no connection is made between the endpoints for message transmission. The 4-byte UDP header only consists of the source and destination ports (1 byte each), a message length (indicating full length with data, 1 byte) and a checksum field (1 byte). This header and following data field is called a *UDP datagram*. Like TCP, UDP uses IP for best-effort datagram delivery, but without the features of data flow control, message acknowledgment and message reordering. For this reason, UDP can be well adapted to reliable local networks, but not to internets.

In Section 3.3.3 dedicated to the IP addressing format, it was mentioned that the current standard corresponds to a "version 4," also referred to as IPv4. Overlooking an experimental version 5 leads to IPv6, which corresponds to the next generation of IP networks. With IPv6, the IP datagrams, which are called packets, have a different format. The rationale for IPv6 is a dramatically extended addressing space (see below), believed to be sufficient to meet the demand for the next 30 years. New service features are also introduced with IPv6, such as (but not limited to) recognition of data/payload type, efficient network routing, encryption and authentication. One should note that in IPv6, the term *node* is used for both routers and hosts. The routers can dynamically determine the *maximum transmission unit* (MTU) corresponding to the networks to be traversed up to the destination. This enables the router to avoid or minimize packet *fragmentation* and optimize the packet path. The minimum requirement for IPv6 implementation is a MTU of 1,280 bytes.

The IPv6 header format is shown in Figure 3.24, with field arranged by rows for visual convenience. Without options, the header is 40 bytes long, twice that of IPv4. The first two rows (8 bytes) are for control. The source and destination addresses are 16 bytes (128 bits) each, which is four times as much as in IPv4 addresses, described further below.

We first focus on the control field, which present the following subfields:

Version, 4 bits: Indicates that packet is IPv6;

Traffic class, 1 byte: Indicates class of traffic (see below);

Flow label, 20 bits: Used to introduce a *flow* relation between packets, for instance to be used in real-time transmission service between source/ destination; enables router to forward packets efficiently with reduced header checks;

Payload length, 2 bytes: Indicates length of packet in byte units, excluding header; set to zero if length exceeds 65,635 bytes, corresponding to "*jumbo payload*" (true length shown in option header);

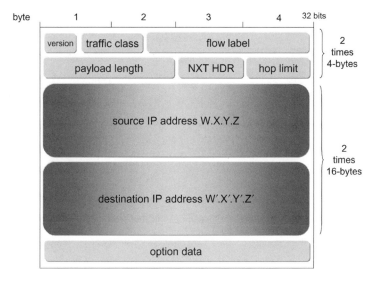

FIGURE 3.24 Byte sequence forming IP header, as shown by rows for visual convenience. See text for description of the different byte fields.

Next header, 1 byte: Indicates type of header to be found next to IPv6 header (eg. *option, routing, fragment, security, authentication, extension,* or *hop-by-hop options* headers, see description below);

Hop limit, 1 byte: Same as the *lifetime* field in IPv4 header (see above), but measured in number of router (node) hops rather than in seconds; decreased by one unit at each hop traversed until reaching zero, upon which the packet is automatically discarded.

The IPv6 header is not limited to the previous 40-byte field. Indeed, optional *extension headers* can be included, as illustrated in Figure 3.25. In this case, the top next-header field is set to zero. As the figure shows, the extension headers have a next-header field and a length field (1 byte each). The values (X, Y, Z ...) set in the next-header fields indicate the header option type. For instance:

The value "43" is used for a *routing* header, which enables the source to pre-define the path, for instance to ensure higher reliability or security. The predefined path is described in the extended-header data field through a series of IPv6 node addresses, corresponding to all nodes involved from source to destination. In this case, the routing nodes forward the packet without further analyzing its contents.

The value "44" indicates a *fragment* header. The extended-header data field indicates the fragment offset (number of bytes relative to the start of the original message where the fragment begins, in 8-byte units) and the fragment identifier (to allow packet reassembly).

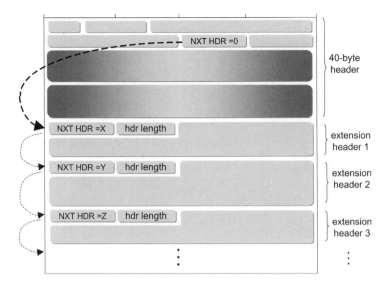

FIGURE 3.25 Principle of extension headers. The top next-header field (nxt hdr) is set to zero, indicating extension header immediately following destination IP address. Extension headers label types (X, Y, Z . . .) and total-lengh fields.

The values "50" indicates that the payload immediately following encapsulates *security* data, which is encrypted. This option is referred to *encapsulating security payload* or ESP. The payload is decrypted at destination after an *authentication* test. The extension-header data field contains a 4-byte *security parameter index* (SPI) which indicates the security protocol chosen in conjunction with specific destination IP addresses. The SPI is followed by a sequence number and authentication data.

The value "51" indicates an *authentication* header (AH). Failure of authentication results in packet discarding. This ensures protection against *denial of service* (DoS) attacks where nodes (routers and hosts) are flooded by "bogus" packets.

The value "0" means a *hop-by-hop options* header. This last feature makes it possible to introduce options to be examined at each router or destination node. The data field of the (hop-by-hop) options header is organized according to the so-called *type-length-value* (TLV) option format. The type portion indicates to the node how to handle the packet. The length and value fields specify the optional length and choice. A noteworthy type option is the *jumbo packet length*, which is used if the length exceeds that allowed by the 16-bit top-header field, i.e., $2^{16} - 1 = 65,535$ bytes. The allowable jumbo packet length is coded on 32 bits, making possible sizes up to $2^{32} - 1 \approx 4.3 \times 10^9$ bytes (!).

The number of addresses that can be handled by IPv6 is huge. Indeed, coding addresses with 128 bits corresponds to as many as $2^{128} - 1 \approx 3.4 \times 10^{38}$ possibilities. This compares with IPv4, which is based on 32-bit address coding, or $2^{32} - 1 \approx 4.2 \times 10^9$ possibilities, or approximately 100 billion billion billion times fewer. As seen in a previous subsection, the IPv4 addresses are expressed in decimal form. With IPv6, they are expressed in hexadecimal, in the form of 8 numbers separated by colons. Here are two possible examples of IPv6 addresses:

0801:0000:0000:0000:AAAF:0000:DD1E:5A6C

A8EF:123B:0000:0000:0000:3C3B:0025:47E2

By convention, leading zeros can be omitted; all-zero fields can be written either with a single zero and consecutive zero-fields as double colons, i.e., for the previous examples:

801::AAAF:0:DD1E:5A6C

A8EF:123B::3C3B:0025:47E2

Note that the double-colon abbreviation can only be used once in the address. Other consecutive all-zero fields present in the address must be replaced by a single 0.

As previously seen, the third or second field of the IPv4 or IPv6 headers defines a *service type* or a *class of traffic*, respectively. Both service types or traffic classes correspond to a preassigned *quality of service* (QoS). As with ATM, the Internet QoS is primarily concerned with issues of connection bandwidth, packet delay and packet loss. In TCP/IP, nonreal-time applications (Web and e-mail) are fairly tolerant of transmission delays but not of packet loss, while it is the opposite for real-time applications (voice, multimedia . . .). The first level of QoS defined by the IP layer is *best-effort* delivery. In the layer immediately above, TCP provides a QoS of end-to-end connection reliability and message integrity (recovery of packet loss, error control). Packets are processed on a *first-come, first-served* basis, without any control of delay, or predefined bandwidth allocation. Thus, the QoS provided by TCP/IP altogether only corresponds to a minimum level of network access/use and traffic quality. Controling packet delay and bandwidth within a given internet connection represents a higher QoS level. Such level can be reached by new service classes referred to as either *integrated services* or *differentiated services*. With *integrated services*, the stream of datagrams relating to a specific sender-to-receiver connection can be distinguished from the rest of the internet traffic, referred to as a *flow*. Such a (data) flow is given routing priority with minimum packet delay, resulting in guaranteed end-to-end connection bandwidth and network-resource reservation. The *traffic class* (not to be confused with the data flow) is defined by the source/destination IP addresses and corresponding ports identification (defining higher-level application). Packets belonging to the same traffic class are given identical treatment and priority by the routers. It is clear that such conditions are best suited to real-time voice/multimedia applications, which otherwise work poorly. *Differentiated services* offer the same features as integrated services, but are applicable to a set of multiple network users, each having different QoS prerequisites. Such a differentiation makes it possible to optimize the use of the network resource, while minimizing flow-control overhead and congestion for hosts and routers.

3.3.6 IP-Layer Functions

In Section 3.3.1 we listed a certain number of IP-layer functions, referred to as ARP, RIP, OSPF, ICMP and IGMP. These functions are also shown by their labels in the OSI diagram of Figure 3.15. Here, we shall provide further details on what they are and do.

As its name indicates, the *address resolution protocol* (ARP) provides the means to convert the symbolic IP addresses (X.Y.Z.W) into 48-bit physical/hardware addresses where hosts or routers can be found. This one-to-one conversion, or mapping, is referred to as *address resolution*. Such a resolution can be done statically or dynamically. The static operation is effected by consulting a table, also called an *ARP cache* or *address-resolution cache*. The cache memory provides a *binding* between the two address types. The dynamic operation corresponds to the situation where the binding cannot be found in the cache. In this case, a message called an *ARP request* is broadcast by the source through the network. The destination machine which recognizes its own IP address then sends back an *ARP reply*, directly and only to the source. The source's ARP cache is then updated for subsequent and future use. Since the cache memory has a finite size, only the most recently acquired addresses (or bindings) are stored. The ARP request broadcast from host A also includes its own address binding. This makes it possible for all network hosts to update their cache, in particular the destination host B, which is more likely to correspond with host A in the future. The data corresponding to ARP requests/replies are conveyed through specific ARP packets, to be handled at the OSI's physical layer.

The *reverse address-resolution protocol* (RARP) makes it possible to achieve the opposite operation to ARP, corresponding to the case where a host's physical/hardware address is known, but its IP address is not. This is for instance the situation of a "diskless" machine which stores data on a remote server. The RARP requests and replies are handled the same way as in ARP. However, the network must have at least one server which keeps an update of the network's IP-address mapping.

The *routing information protocol* (RIP) is one of the standard means for network routers to exchange information concerning paths and routing topology and update their routing tables accordingly. Every 30 seconds, RIP broadcasts this information in the form of packets that contain a list of *vector-distance pairs*. Each of these pairs, (a,I), indicates an IP network address (a = X.Y.Z.W) and the integer number of hops (I) that it takes (or costs) to reach this particular network from the source router. Such vector-distance information makes it possible, from source to destination, to identify a variety of possible routing paths (routes) and their associated number-of-hops cost. Just minimizing the hop cost is generally not sufficient to optimize the route, since the subnetworks traversed may differ in technology performance or speed. Also, several equal-cost possibilities may be offered, making the route selection ambiguous. Finally, router failures might occur, which requires rapid updating for the routing tables to minimize packet loss. These are issues handled by RIP. If the number of possible hops is large, the update might take a long time to propagate and diffuse through the network. Such a situation may lead to relatively slow or nonconverging (inconsistent) routing reconfigurations, called *loops*. In order to ensure minimum convergence or reduce risks of routing loops, RIP is designed to handle a maximum of 16 hops.

Considering the rapidly-growing size of internets, even considering 16-hop limits, the total information represented by vector-distance lists can be enormous. Broadcasting of all this information, even through the local network defined by the 16-hop limit can also represent sizeable bandwidth waste. As an alternative to the vector-distance algorithm, the *shortest-path first* protocol (SPF) makes it possible to avoid this excess-information dilemma. Following SPF, routers periodically interrogate all their neighboring routers on their link status (available [up] or not available [down]). Thus the first available routing path [current list of (up) links] can be identified and communicated as valid routing tables. With the *open SPF* (OSPF) protocol, routers are grouped into independent subnetworks, called *areas*. In these areas, the routing information and periodic updates are kept on a local and autonomous basis. This makes it possible to define *virtual network topologies* which abstract away from unimportant and minute local routing details.

The *internet control message protocol* (ICMP) is a means of reporting to the source any events of packet loss due to routing failures. By itself, this reporting is not a way to correct the failure, or even to faithfully account for all possible failure occurrences. For fragmented datagrams, ICMP messages always (and only) concern failures relative to the first segment. ICMP messages are encapsulated into IP datagrams (this not making ICMP a higher-level protocol). The ICMP data inform the source about the failure type, including for instance: unreachable IP-address destination (e.g., machine disconnected or unable to process, or network access denied by firewall), intermediate router congestion or fragmentation/size issues, exceeded datagram lifetime (detection of routing loops), and incorrect parameter setting in IP header. The ICMP can also be constructively used to tell a source host whether or not a given destination host is reachable and furthermore, how long it takes to connect to it. This information is obtained through a sonar-like process with "ping" (sending echo request) and the following "echo" reception. Such an operation is called to "ping a host."

The *internet group management protocol* (IGMP) is associated with IP-address multicasting, or dynamic, multipoint delivery of IP datagrams. As seen in a previous subsection, IP multicasting uses *class D* addresses. In decimal notation, these range from 224.0.0.0 to 239.255.255.255 (see exercise at end). Like ICMP, the IGMP messages are carried by IP datagrams (yet remaining an IP-level protocol). When a new host wants to join or leave the multicast, it sends a corresponding IGMP message (membership declaration) which reaches all multicasting routers involved. The multicasting routers maintain the information of host participation for their subnetworks through periodic polling (representing 1 request per minute, at most). Only a single acknowledgment from a given subnetwork host is sufficient to keep the broadcast to this subnetwork; the lack of any acknowledgment after repeated polls terminates the broadcast in that specific subnetwork. Thus IGMP enables the dynamic management of IP multicasting with router congestion avoidance.

3.3.7 Applications Service-Layer Functions

In this subsection, we briefly review some of the *applications service-layer functions*, called FTP, HTTP, SMTP, SNMP and Telnet, which come on top of the TCP layer, as shown in Fig 3.15. We also mention other related functions such as TFTP and NFS/AFS.

Access to data files from a remote server can be gained in two different ways. Either a copy of the file is transferred from the server to the requesting client (file copy transfer), or the client is given access to read/write the file directly in the remote-server machine (on-line shared file access). As its name indicates, the *file transfer protocol* (FTP) corresponds to the first approach. It is the basic protocol used for exchanging data files between two distant machines, namely a server and a set of clients, each one being able to send and request/receive files. Such FTP files can be text, articles, books, images, sounds, videos, multimedia, course work or anything else. Although FTP is designed to be used by programs, it is also an interactive interface, i.e., to be controlled by humans for different options. An FTP session is initiated by the client who must then provide a succession of commands. The first command (ftp://host/domain ...) *connects* the client to the remote host. The second command logs in the client (specification of the client's username ID and password, for authentication). In some cases, the connection can be automatically authorized (referred to as *anonymous* FTP, which uses *user name = anonymous* and *password = guest*). The client then selects a directory, checks which files are available and specifies the transfer mode. By default, the transfer mode is a byte stream of which the client has specified the data type (ASCII, EBCDIC, see Chapter 1, or raw 8-bit/1-byte packets). The files are then copied and the client disconnects from the remote server. It is not necessary to know the programming language to use FTP. Indeed, the FTP servers can be reached via a web browser (see the following text on HTML) using a uniform resource locator (URL) such as *<ftp://host/domain>*. The directories are displayed and files can be downloaded by clicking on the page's *hyperlinks* (see below). There are also *search engines* (see last subsection on www jargon), which help to find and retrieve FTP files (a corresponding URL being <http://www.ftpfind.com>, see HTTP, below).

A much simpler version of the preceeding, which eliminates complex client/ server interactive features, is provided by the *trivial file-transfer protocol* (TFTP). Consistently, TFTP is designed to run on top of UDP (see description in above subsection), unlike FTP which runs on TCP. The TFTP data are sent through 512-byte blocks. The blocks include a 2-byte header (called *opcode*) that specifies one of five block functions: "read request," "write request," "data," "acknowledgment" or "error." Note that TFTP can also be used to remotely initiate (or "boot") diskless machines. As mentioned at the beginning, the other way to access files is to read them directly in the remote server. The client accesses and handles the data as if they were located in its own machine. The protocols used for this emulation are the *network file server* (NFS) and the *Andrew file system* (AFS), the latter enabling multiple-server operation.

Some documents, such as Internet web pages, are written in the *hypertext mark-up language* (HTML) format, which is described in a following subsection. The transfer of HTML files is controlled by the *hypertext transfer protocol* (HTTP), which runs over TCP. The software initiating the HTTP session, called a *transaction,* is a *web browser* (see Section 3.3.11). The browser manages the HTTP transaction with the server through four basic steps: opening the connection (generally using TCP default port 80), sending a request to the server, getting the server's response and closing the connection. While HTTP does not keep any record of successive connections, the Web browser can memorize some information on the client's requests in the form of *cookies*. The HTTP transaction can be made through the intermediary of *gateways* and *proxy servers*. Apart from network security purposes, the main feature of gateways and proxies is their ability to store generic information in a temporary *cache memory*. This "cache" then allows rapid downloading of HTTP files containing information from earlier transactions (e.g., generic Webpage titles, pictures and logos), which allows faster IP traffic and reduces network congestion. The browser calls the server host though the command *http://host/domain . . .* , which is referred to as a *uniform resource locator* (URL). If the primary host to be called is the *World Wide Web*, the URL is called a *www address* (see the following subsection on Web-site addressing). The HTTP message contains a header specifying the message type ("request," "response," etc.), followed by the message body (e.g., HTML file). The secured version of HTTP, which uses encryption, is referred to as S-HTTP or HTTPS.

Next to Web-site browsing, the other most popular Internet application (or service) is *electronic mail*, or *e-mail* for short. Exchanging e-mail messages is just another form of data transfer between client and server, which in this case is handled over the TCP layer through the *simple mail transfer protocol* (SMTP). The mail is usually delivered to a local SMTP server. This server holds the mail and delivers/copies it on request from the end recipient. The particularity of mail delivery is that in order to send his message, the client does not need to know of the destination host's availability or readiness to receive it. The mail-handling process relies upon a background technique called *spooling*. The message data are temporarily stored in a "mail spool area", with a time/origin/destination label. The local mail server then identifies the IP address of the destination SMTP host and attempts to make a TCP connection with it. Once the connection is established, the mail is copied (or transferred) to this destination host, for remote spooling. Upon client/server acknowledgment of successful transfer, and only in this case, the mail is removed from the local server spool. Possible repeated connection failures are also time-recorded, which results in corresponding error messages to the originator. The reception of an e-mail message is usually notified by the recipient's mail server through a message such as a *"You have mail"* icon. In large corporations, employees keep their mailboxes constantly open, so that incoming messages can be responded to as they arrive, in order of priority. Personal or individual e-mail addresses take the form of <name@domain> or <alias@domain>. These mailbox addresses are recognized at SMTP level, and therefore should not be confused with the IP addresses, which are used at IP-layer level (see more on *e-mail addressing* in Section 3.3.8). The content of mail messages is usually coded in 7-bit ASCII format (see Chapter 1). As well-known by all e-mail users, messages can include *file attachments*, which are generally non-ASCII. Such attachments are then encoded into ASCII through a protocol called *multimedia/multipurpose Internet mail extension* (MIME). The MIME attachments have a header specifying the contents type. The type is then recognized by the recipient host and the attachment is (usually) displayed in the received message under the form of hypertext icons. Clicking on the icon calls for the corresponding application, which can concern text, images or video, making the file run or display. Because such attachments can possibly contain *computer viruses* or other hazardous self-executable programs (see last section on *www jargon*) they are normally preceded by a warning message from the mail server. Corporate intranets systematically check out all attachments for such contents, recognizing viruses or hidden executables and blocking them. In order to avoid intranet or internet congestion, mail servers can also deny the transfer of oversized attachments (e.g., ≥ 1 Mbyte) and also compress them for saving spool memory and achieving more rapid IP transfer. Actually, the popularity of e-mail is such that all types of files, even of very large sizes, are now more commonly transferred via SMTP under the form of e-mail attachments rather than directly via FTP. Any Internet/intranet user can indeed resort to this most convenient way to exchange all kinds of files, requiring only a few seconds of initial training. See more on various e-mail use and issues in Section 3.3.11.

As another service running over TCP/IP, the protocol *Telnet* makes it possible for a set of client machines to execute programs on a remote server, playing the role of *virtual terminals*. The *network virtual terminal* (NVT) consists in a virtual display (or "printer") with a keyboard. The keystrokes are passed to the remote host as if the keyboard were directly attached to it. NVT messages are sent in 7-bit ASCII data in the form of 8-byte packets (recalling that ASCII includes nonalphanumeric keyboard characters which control text and page, e.g., "enter," "carriage return," "tab," "erase character," etc.). A similar mode of operation of Telnet is called *terminal emulation*. It is based upon EBCDIC data and, unlike NVT, has graphic capabilities. These Telnet applications make it possible for clients to connect to on-line databases, library catalogs and other real-time chat services (see Section 3.3.11). Another service provided by Telnet is remote program/job execution. This is made possible by the *remote execution command* (REXEC) and the *remote shell* (RSH) protocols of Telnet. REXEC automatically performs login to the distant server with password, while RSH does not require any authentication. Finally, Telnet provides the possibility of creating, within a set of cooperating hosts, a *distributed computing environment* (DCE) with shared files, databases and resources. This is also referred to as *remote interactive computing*. Applications are remotely executed under the form of *threads*, representing blocks of program codes. *Multithreading* consists in concurrently executing different blocks of the same program (e.g., program procedures/subroutines). The DCE is securely operated under a service which performs authentication, grants authorizations and privileges with different levels of authority, manages groups, enforces network policies and keeps a registry of user accounts.

To complete our description of main service-layer applications running over TCP/IP, we finally consider the *simple network management protocol* (SNMP). The principle of SNMP is to manage the internet by examining and controlling a set of hosts such as IP routers, gateways and participating client terminals. Each of these hosts contains a *management agent*, or *SNMP agent/entity*, with which the managing client (manager) can interrogate and communicate. Several managers from different network locations can perform this function on a given network host, while only one of them is given the authority to change the host's information/configuration or control it (e.g., routing table modification). Within SNMP, the *management information base* (MIB) is a companion standard which defines the types of data, or *object variables*, which need to be checked. Each object (IP, TCP, UDP, ICMP) has its own set of MIB variables. Examples of variables are the router statistics of input/output traffic, the status and number/addresses of its network interfaces, its routing table, its datagram loss rate, its number of TCP or UDP connections, or its number of ICMP error messages, for instance. All this data is compactly coded through the *abstract syntax notation 1* (ASN.1). The ASN.1 syntax (pronounced A-S-N "dot" one) points without ambiguity to the type of MIB variable, among other useful network-management objects. The syntax is both acronymic (for human reading use) and numeric (for machine use). The ASN.1 name representation consists of a set of groups organized under a tree structure. The top group is the standard organization (*ISO = 1*, *ITU = 2*, joint ISO/ITU = 3), under which is the organization (e.g., *org = 3*), then the department (e.g., U.S. Department of Defense, *dod = 6*), the network (e.g., *internet = 1*), and the final subtree: directory (*dir = 1*), management (*mgmt = 2*), experimental (*exp = 3*) and private (*priv = 4*). This series of group numbers form the ASN.1 name, for instance:

iso.org.dod.internet.mgmt, or equivalently, 1.3.6.1.2

Within the management subtree, MIB data are grouped under *mib = 1*. The numbers immediately following correspond to the MIB variables, which distinguish IP, TCP, UDP or ICMP object/group types. For instance, the number of IP datagrams received, called "IpInReceives," is identified by "3" under the object/group type *ip = 4*, thus forming the full ASN.1 name

*iso.org.dod.internet.mgmt.mib.***ip.ipinreceives** or 1.3.6.1.2.1.**4.3**

The variable *IpInReceives* is one out of 42 possibilities under the IP group. In contrast, the object/groups ICMP, TCP and UDP have 26, 19 and 7 variables, respectively. The network interface group (*if = 2*) includes 23 variables, such as *IfNumber* (number of interface attachments), *IfType* (type of interface), *IfMtu* (interface maximum allowed datagram size or MTU), *IfINErrors* (number of inbound datagrams having errors), to quote only a few illustrative examples. The MIB data are then collected by the host's managing agent (SNMP entity) and transferred to the *network managing station* (NMS) under the form of SNMP messages. Likewise, the NMS can send SNMP requests to the host's managing agent. Network management and associated protocols is another field in itself, which is beyond the scope of this book.

3.3.8 E-mail Addressing

As seen in the previous subsection, e-mail addresses are handled by SMTP. As is well-known to millions of users, e-mail addresses take the form of ⟨name@domain⟩

or sometimes ⟨alias@domain⟩. The domain following the @ sign refers to the mail server or "post office." The associated conventions widely vary according to the internet service providers (ISP) and any other local intranets. For instances typical ISP addresses are of the form ⟨xxx@yahoo.com⟩, ⟨xxx@aol.com⟩, ⟨xxx@hotmail. com⟩ and ⟨xxx@wanadoo.fr⟩. In international corporations, the domain can be complemented by either a .com field (e.g. ⟨xxx@alcatel.com⟩) or .country field (e.g., ⟨xxx@alcatel.cn⟩ for China). Educational, governmental or military institutions can have e-mail addresses terminated by .edu, .gov or .mil (e.g., ⟨xxx@stanford.edu⟩, ⟨xxx@senate.gov⟩, ⟨xxx@pentagon.mil⟩). As concerns the recipient's name standing before the @symbol, a most straightforward nomenclature is *firstname.lastname* (e.g., ⟨emmanuel.desurvire@xxx⟩ or, for saving space and generating more user possibilities, ⟨e.des@xxx⟩, ⟨edes@xxx⟩ ⟨ed@xxx⟩, ⟨emmanuel4@xxx⟩, etc.). Such a system makes it very easy to memorize (possibly reverse-guess) e-mail addresses on a *universal*, *world-wide* and *quasi-permanent* basis (an illustrative example, not to be used lightly: ⟨president@mycompany.com⟩). In addition, most ISPs allow up to five e-mail addresses to be used by a single subscriber, making it possible to distinguish family members with different nicknames or *aliases* (⟨alias1@xxx⟩, ⟨alias2@xxx⟩, etc.) which relate to the unique dominant subscriber address ⟨dominant@xxx⟩. *Alias e-mail addresses* are also used in big corporations by working groups or teams of registered/authorized users to receive messages collectively. A growing number of individuals possess both private and corporate e-mail addresses. While on travel, employees can check their mailbox using a secure *remote access server* (RAS), which only requires a corporate login/password and a modem connection to the PSTN. E-mail services also extend to wireless networks, namely mobile terminals or cellular telephones (see Chapter 4). E-messages and their attachments can thus be forwarded anywhere to anyone in the world, provided the destinee has a registered or known e-mail address. See more on e-mail use, issues and *netiquette* in Section 3.3.11.

3.3.9 Web-Site Addressing

Internet web sites and other Internet files are identified by their *uniform resource locator* (URL). The URL is nothing but the address of a file (or set of files relating to a directory) stored in a host computer. Entering a URL into a browser activates the process whereby this file is downloaded to the client's computer, making the corresponding Web pages or files/directories appear in the client's monitor. URLs are alphanumeric, case-insensitive (except for filenames), and have the syntax:

> *protocol://host/domain*
>
> *protocol://host/domain/directory*
>
> *protocol://host/domain/(directory/path/)filename*
>
> *protocol://host/domain/(directory/path/filename)/page/...*

The most used protocol for websites is HTTP. Internet text files or catalogs/directories can be also be reached through the protocols FTP and Telnet, for instance. In these three cases, the URLs take the form:

http://host/domain/...

ftp://host/domain/...

telnet://host/domain/...

The most popular Internet host is *www* (World Wide Web). Intranets of large companies "x" have their hosts designated by names such as *xww* (e.g., *aww* for Alcatel).

The domain is usually formed in two sections, i.e., ⟨secondary.top-level⟩. The top-level domain (some being called *root domains*) indicates a purpose or category, for instance:

.*com* for commercial enterprise,

.*edu* for education,

.*gov* for government,

.*mil* for military,

.*org* for nonprofit organizations or institutions,

.*net* for network centers or service provider, and

.*int* for international organizations.

The top-level domain list is still expanding. New names can be approved by the *ICANN* (Internet corporation) or *IANA* (Internet Assigned Numbers Authority), for assigned names and numbers), such as .*biz*, .*pro*, .*info* and .*coop*, for instance.

The *country domain* is implicit for the U.S. sites, notwithstanding that there exists a nomenclature for the different states in the U.S. (e.g., .*ca.us* or .*nj.us* for California or New Jersey, respectively). Other countries are indicated by a two-letter *Internet country code*: examples are .*fr* (France), .*uk* (United Kingdom), .*jp* (Japan), or .*au* (Australia), the list ranging from .*ae* (Arab Emirates) to .*zw* (Zimbabwe). Top-level domains outside the United States can also be formed with country codes, for instance .*co.uk* (for "commercial" in the UK corresponding to .*com* in the United States) and .*ac.uk* (for "academic" in the United Kingdom, corresponding to .edu in the United States). Note that since the Internet was developed, the .*com* and other top-level root domains are no longer restricted to only U.S. hosts.

The above URL addressing is generally easy to memorize or directly guess without needing an Internet search by keywords (e.g., http://www.wiley.com). However, the actual Web-site address used by the network is not alphanumeric, but binary. Therefore, the URL is a high-level symbolic name which must be converted into an IP address. Such a conversion is made through a software called *Internet domain name system*, or *DNS*. The DNS is handled by a *(domain) name server*, which is able to effect the two-way mapping between URL and IP address (or by default, delegate the task to a higher-authority domain name server). Such a URL/IP-address conversion process is transparent to the client and is referred to as *domain name resolution*.

The main programming language for defining text, graphics and other contents of Web page documents is the *hypertext markup language* (HTML). It describes *tags*, which are basic blocks such as page background (color/pattern/image), text (font/style/size/color), titles (headers), paragraphs (justification/length), lists (bullets), tables (row/columns) and pictures (icons, logos, photos . . .). One of the key features of HTML, which explains the success of internet Web sites, is to support *hypertext* commands, which are embedded links to other HTML pages. The user just clicks on a text portion or an icon or image to see another HTML page appear in his machine's screen (see Section 3.3.11 on *www jargon*). Several programmed functions can be embedded in HTML files and called from there, for instance:

> CGI (*common gateway interface*) scripts: Short program realizing interactive functions from a web page, for instance filling out and submitting a form, opening an e-mail session ("contact us" hyperlink), displaying the number of page visits or "hits," etc.

> *Java applets:* Short programs realized in Java language, which enhance Web page functionality and dynamics, e.g., pop-up menus, real-time calendars/clocks/stocks-values, news-flashes, calculators, button/mouse interactions, welcoming homepage sounds/videos, moving images in banners and various other page fields, etc.

The software which runs over TCP and enables a client to access HTML files (i.e., download Web pages) from a distant server is defined by the *hypertext transfer protocol* (HTTP), as described in Section 3.3.9. See more on *Web page browsing* in Section 3.3.11 describing *www jargon*.

3.3.10 Mapping IP Over ATM, SDH/SONET, and WDM

The TCP/IP protocol was designed to work in a connectionless mode, unlike other protocols such as SDH/SONET or ATM in connection-oriented service mode. We recall that IP is a "best effort" packet delivery, while TCP ensures a secure end-to-end connection at the OSI level immediately above. But nothing forbids such a secure connectivity being provided by other connection-oriented protocols, such as SDH/SONET or ATM, meaning that IP payloads can also be transported by SDH/SONET frames or ATM cells. In Section 3.2 we saw how ATM cells can be transported through SDH/SONET frames. More generally, the process of *encapsulating protocol X data* (e.g., ATM) *into protocol Y data* (e.g., SDH) is called *mapping X over/onto Y*, and is nicknamed for short "*X over Y*" (e.g. "ATM over SDH"). Considering here internet data, the options *IP over ATM*, *IP over SDH* or *IP over WDM*, define many possible classes of service where TCP/IP data ("client payload") are being transported/managed within ATM, SDH or WDM networks. In this subsection, we review these different mapping possibilities and consider as well the transport of *voice over IP*, a service referred to as *VoIP*.

The rationale of *mapping IP over ATM* is to exploit the advantages of ATM networks and related AAL service classes, with QoS such as bit-rate and timing/delay control, and other contractual network use (see Section 3.2 and below with WDM). IP over ATM could be initially conceived as the mere operation of using the same addressing for IP and ATM source/destination hosts. But recalling that ATM switches operate at OSI layer 2 (data link), and IP routers at OSI layer 3 (network), this would require complex redesigning, namely to integrate routing into ATM switching. It makes more sense to run IP as a higher-level protocol on top of ATM, referred to as the *overlay model*. In this model, there is no with no logical coupling between IP and ATM addresses. A first approach, called *classical IP over ATM* consists in segmenting the ATM network into many sub-IP networks, between which ATM achieves end-to-end connections. The sub-IP networks, also *logical IP subnetworks* (LIS) are then connected to each other through IP routers, as illustrated in Figure 3.26. Within a given LIS, IP hosts communicate through local ATM addresses which requires a specific *address resolution protocol* (ATMARP). The ATMARP server contains tables with all ⟨IPaddress-ATMaddress⟩ pairs. If the IP address is not recognized by the ATMARP server, as corresponding to a host outside the LIS, the packet is forwarded to the IP router. The packet is then forwarded to the next LIS, and the process continues until the local ATMARP server recognizes the destination ATM host. Inter-LIS communication through such a hop-by-hop forwarding is time consuming.

To avoid traffic congestion, the *next-hop routing protocol* (NHRP) was developed. In short, NHRP makes it possible to identify the ATM address of a destination host located outside a given LIS, therefore creating a shortcut. The IP data are then transferred throughout as ATM packets, by-passing IP-routers. A second approach for IP over ATM is called *LAN emulation* (LANE), which we mentioned in Section 3.2. The LANE protocol bridges small local-area networks such as Ethernet or token-ring, providing seamless connections. Thus IP can run over ATM through the combination of LANE and NHRP, which like the classical approach, by-passes

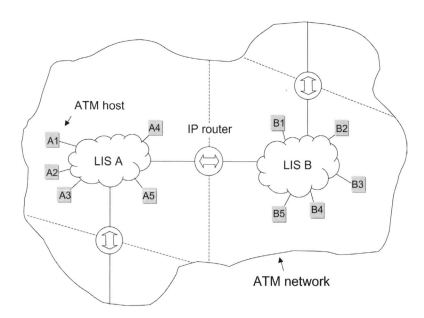

FIGURE 3.26 Principle of partitioning ATM network into independent logical sub-IP networks (LIS) for servicing IP over ATM.

IP routers. Encapsulation of IP into ATM cells is made in three steps. First, an 8-byte trailer field for *logical-link control* (LLC) and other functions, is appended to the IP payload-field (up to 65,535 bytes) to form an ATM *protocol data unit* (PDU). The ATM-PDU is then handled by the ATM *adaptation layer* AAL5 (see Section 3.2), which adds an 8-byte overhead (field length, CRC) to form an AAL-PDU. Padding is used to make the AAL-PDU size an integral multiple of 48-bytes ATM payload. Then ATM cells can be formed by segmenting the AAL-PDU into many 48-byte payloads with 5-byte ATM headers, resulting in the 53-byte ATM cell (Figure 3.10).

The above describes how IP can be mapped into ATM. Since ATM can be mapped into SDH/SONET, we can have *IP over ATM over SDH/SONET*. This operation is the same IP over ATM mapping as outlined previously, followed by ATM over SDH/SONET mapping, as previously described in Section 3.2.

Direct mapping of *IP over SDH/SONET* makes sense at first in long-haul and high bit-rate transport, for instance at continental (intercity) or transoceanic (intercontinental) scales. One thus loses the advantages of a connectionless network (such as the Internet) in order to exploit in full those of point-to point networks offering higher reliability (low error-rate), greater bandwidth (bit rate, number of available channels) and service availability (uncongested routes with global traffic dimensioning and global network protection). In fact, a substantial fraction of the long-haul traffic (terrestrial and submarine) concerns internet data, i.e., IP over SDH/SONET, which is progressively taking over the traditional voice payload. The above concerns point-to-point, long-distance transport, such as in submarine-cable systems. In terrestrial optical networks, however, IP over SDH/SONET can be analyzed from a different and more complex perspective, namely through the concept of "*IP over WDM*," which we describe further below. Whether for point-to-point long-distance or network transport, mapping of IP over SDH/SONET consists in the same three-stage encapsulation process. First, the IP datagram (IP header + data body) is encapsulated into a *point-to-point protocol* (PPP) packet, which provides error control and link initialization features. The resulting PPP packet is then transformed into a frame using the *high-level data-link control* (HDLC) protocol, which provides frame delineation and size definition. The end result, the IP/PPP/HDLC frame is then encapsulated into either a SDH *container* or a SONET *synchronous payload envelope* (SPE), whose frame formats are detailed in Section 3.1.

The concept of *IP over WDM* is short for a multi-layered service definition. WDM networks offer high-bandwidth transport (optical fibers) and switching characteristics (wavelength channels or tributaries) at the OSI physical layer. Therefore, the wavelength channels/tributaries, can transport any type of traffic such as FDDI, (Gigabit) Ethernet, ATM or SDH/SONET. But nothing forbids IP datagrams being directly fed to the WDM layer, without encapsulation into these protocols, which represents the "overlay model" of IP over WDM. The different multilayer approaches for IP over WDM are summarized by the diagram shown in Figure 3.27. The approach of *IP over ATM over SDH/SONET over WDM* represents the most diversified network service. The WDM optical backbone provides the highest possible bandwidth for IP traffic, while using a classical SDH transport network with adequately managed STM-N/OC-M bit-rate hierarchy.

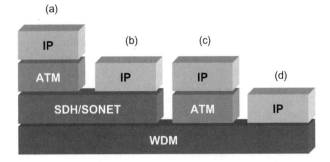

FIGURE 3.27 Different network layering approaches corresponding to IP over WDM: (a) IP over ATM over SDH/SONET over WDM, (b) IP over SDH/SONET over WDM, (c) IP over ATM over WDM, and (d) IP over WDM.

The ATM layer immediately above provides connection-oriented transport, with the three advantages of: (1) QoS management; (2) dynamic or variable traffic granularity, through switched or permanent virtual channels (SVC or PVC; see Section 3.2); and (3) optimized traffic management through the creation of virtual paths (VP; see Section 3.2). It is clear however that such a layered network is highly complex to manage, let alone the major issues of network evolution and investment/operating costs. Also, the multiple protocol stack introduces subtantial *cumulative overhead*, corresponding to inefficient use of bandwidth (see below). A simpler architecture, is *IP over ATM over WDM*. This approach consists in directly feeding the WDM optical network with IP-over-ATM cells. Beside the reduced overhead and simpler implementation, the advantage of this architecture is that, ATM being asynchronous, there are fewer timing constraints between network nodes and terminals. The other simplifying alternative is IP over SDH/SONET over WDM, which consists in feeding the WDM network with IP-over-SDH/SONET frames. As previously discussed, this approach is advantageous for high-bit-rate, long-haul transport such as in submarine cable or continental/terrestrial links. Also, the overhead is reduced to a minimum. It can be shown (see EURES-COM, 1999) that for a 350-byte IP datagram, and for instance using SDH, the total overhead is only 3%. Compare this with IP over ATM over SDH, which yields 19% overhead, and IP over ATM over SDH which yields 22% overhead. Finally, it is possible to feed the WDM optical network directly with IP datagrams, the true form of IP over WDM. As a matter of fact, this approach is not as simple at it may look. This is because a certain number of control features (otherwise provided by ATM and/or SDH-SONET) are absent and must be reintroduced by some other means. For instance, the traffic is not protected against failures such as fiber-cable breaks, power transients, misconnections, timing jitter and other impairments inherent to optical transport. To combat impairments due to optical noise accumulation and fiber nonlinearities, it is possible to implement *forward error correction* (FEC), as described in Chapter 1. However, FEC introduces overhead, especially in long-haul and high-capacity (terabit/s) transmission. Regardless of signal quality issues, it is clear that new types of network protection and restoration must be implemented in the WDM layer. It should be noted that network protection and restoration is also performed at the IP layer, through OSPF and RIP which control IP-routing tables, as described in a previous subsection on IP functions. Finally, it is worth mentioning that the IP layer can control the wavelength path through the *multiprotocol lambda switching* (MPλS), a variety of *multiprotocol label switching* (MPLS). See Desurvire, 2004 for a comprehensive treatment of this subject.

In Chapter 2, an extensive description is made of the various techniques used to convert analog voice signals into digital signals. The resulting voice traffic is shown to be handled through a fairly complex byte-to-byte and frame-to-frame multiplexing process referred to as *plesiosynchronous digital* (PDH) and *synchronous-digital* (SDH) *hierarchies*. The possibility of transporting voice directly over IP, called VoIP, is emerging as a new class of Internet service, different from that provided by the traditional public-switched telephone networks (PSTN). Since IP is a connectionless packet protocol, is clear that voice quality can be seriously affected in the absence of differentiation and prioritization protocols for the corresponding traffic. The same remark applies to video data, which together with voice require secure connections with minimal delay and packet loss. Interestingly, VoIP was first used in countries having under-developed PSTNs, using computer microphones/speakerphones as a way (though of very poor quality) to make cheap phone calls. One-way voice messages, with resolution quality up to compact-disk-grade, can also be carried by e-mail (SMTP/IP), although the size of the file attachments may limit duration or require fast-Internet lines. As the technology of VoIP will mature through differentiated traffic, the age-old telephone conversation will be progressively substituted by video and multimedia communications, at both private and business levels. One may expect VoIP to just represent "free bee," incurring no special billing within the advanced multimedia services. This may become true at least when the revenues of voice services will fade in comparison of those from data services, a perspective still far ahead in the future.

3.3.11 The Internet and www Jargon

Against any expectations from telecom professionals, the Internet has within less than a decade become extremely popular. Its level of popularity does not even compare with those at the beginnings of mobile telephony, home computers, compact-disk music, or color television. This is because a very large number of users already had telephone lines and computers, and Internet access only requires a small and relatively inexpensive connection box called a *modem*. Such a modem (for MOdulator/DEModulator) makes it possible to convert the computer's digital data into telephone's analog signals, and the reverse. A second requirement is to have software able to recognize what could be coming from the Internet. But it was made available for free. The fact that the Internet is so easily and cheaply accessible through standard telephone lines only represents a first level of explanation. If all that the Internet had to offer was dull screens showing text and data files (as was the case in its very beginnings), it would not have been so attractive. But new applications were developed to generate nice, colorful and animated page displays

(HTTP), with the possibility of sending/receiving *electronic mail* (SMTP) including file attachments and many other functions (see the following text about *e-mail*). With such tools and services, professional and private people rapidly discovered that the Internet represents an unlimited resource for information search, gathering and sharing, a channel for interactive dynamics and group discussions and overall, a collective means for freedom of expression. Shared by about 200 million users, the *internauts*, the "Net" (or the "World-Wide Web," www) has produced a culture of its own. Accordingly, an Internet jargon rapidly developed, happily mixing software application, service or utilities concepts, new acronyms, fashionable tools and a wealth of neologisms (i.e., words made up to describe new concepts). Even a superficial account of this cultural side of the Internet is beyond this book's scope. However, it is worth briefly explaining some of the Internet/www jargon, being aware that, considering its rate of growth, it will remain hopelessly incomplete. Other definitions which may not appear in the following have been described earlier in this section. While the reader of this book might be wholly familiar with most of this jargon, it remains an interesting and challenging exercise to try to formalize it or even explain it to peers, such as parents or children.

Here it goes. The Net experience begins when a computer is connected to the Internet via a modem (the user says "I am connected"). The computer is then *on-line*, and gets directly to the *homepage* of some Internet service provider, from which different options are offered. These may include, for instance, consulting and responding to electronic messages/mail (or *e-mail*), looking at news, checking the local weather and being exposed to pop-up commercials. Options can be selected through *menu buttons, hypertext links (hyperlinks)* or *icons. Mouse-clicking* on any of these activates a command that opens up a new *page* appearing inside a *window* on the screen. This is already inside the Internet environment, but only at a *gateway* level. From there, it is possible to go exploring the rest of the Internet world, which is done by selecting one of the *Web browsers.* The most popular browsers are *Netscape* and *Internet Explorer*, but others are *Mosaic, Macweb* and *Netcruiser*, for instance. The browser converts the HTML files into *web pages* that display text, images (fixed or animated), video and sounds. It can be conceived as a client to remote servers/ hosts where *Web sites* are *hosted* (meaning both functionally and physically in geographical location). Web sites consist of a set of Web pages grouped together via hypertext links. In turn, the websites and pages can be linked to other websites (usually through a page/button called "(useful) links").

A Web site is accessed through its *address*, or URL for *uniform resource locator*. The URL is to be typed into the browser's "go (to)" field or is hidden within a hyperlink. It is of the form *http://www.websitename.domain*, as described in Section 3.3.9. Such a nomenclature makes it easy to memorize, if not directly guess, the URL for Web site access (e.g., "Wiley" naturally yields www.wiley.com). The operation of *browsing the Web* is also called *navigating* or *Web surfing*. The exploratory nature of this navigation or surfing, explains the neologism "internaut," copied after "astronaut/cosmonaut." All browsers are equipped with *search engines* (popular ones being *Altavista, Ask Jeeves, E-Spotting, Excite, Google, LookSmart, Lycos, Netscape Search, Nomad, Overture, Voila!* and *Yahoo!*). These are programs that, based upon certain searching algorithms, find from millions of registered Web sites a selection that may contain predefined *keywords*. Search engines keep their databases updated by use of *spiders, crawlers*, or *know-bots (also called robots)*. These are programs that run through the entire www, in constant search for new Web pages and report them to the search engine. More sophisticated versions of search engines are of the *meta-search* (or *multithreaded*) and *subject-specific* types. The first type (e.g., Metacrawler, Ixquick, SurfWax, Dogpile and Profusion) make it possible to simultaneously activate different search tools, each one having its own set of databases. The second type, which does not target the entire www database, can provide rapid convergence through a subject-oriented searching approach (see their directories such as Beaucoup, Searchengines and Searchengine-colosuss). Internet or Web *searching*, not to be confused with internet/Web *surfing*, is the action of submitting one or a few keywords to the search engine, in a dedicated "find" entry field. The result of this search is a more or less extensive Web site index with titles, captions and hyperkinks. Some keywords can give rise to lists indexing thousands of items, some none at all, depending upon their scope, i.e., broad or specific. It is worth also mentioning *Lynx*, a dedicated text browser that only reads text files in any www page, without images or other features, allowing one to gain considerable downloading time. *Internet search* is an art that requires both intuition and practice. To keep the search in focus and hit the target faster, it is possible to use *boolean logic* (AND, OR, and NOT) to associate keywords in a meaningful way. Selecting the language opens other dimensions. Exploring the different URLs coming out from the search may reveal lots of interesting websites, which in turn may contain other useful links. It is wise to include them in a personal list of *bookmarks* or *favorites*, because otherwise one may not be able to re-discover or relocate them. It is a good idea to save the result of a search session as html files, for future reference or investigation. These files may contain tens to hundreds of URLs which one may be short of time and patience to check out immediately. Exploiting all this information and making sense of it is an art which many internauts tend to overlook, confusing serendipity with hastiness.

Internet search, browsing and surfing gets one to and into websites. The action of getting there and hopping through the associated pages is called *visiting*. Visiting a Web site is usually a quick operation, which rarely requires careful attention and long access time. The site visit is guided through a hierarchy of fields, titles, menus, hyperlinks and icons. Menus can be extended by use of small windows called *pop-up menus*. The pop-up menu is a list from which a set of keywords can be selected (e.g., country sites, business departments, product families . . .). Figure 3.28 shows an example of a homepage of a large-size company Web site. The duration of the website visit closely depends upon the site's purposes and contents. Some sites almost reduce to a single homepage, with a side-menu of links to pages still *under construction*. Generally, Web sites include a *presentation* (statement of purpose) and *frequently asked questions* (FAQ, pronounced F-A-Q)[1] pages, which help the user to rapidly understand their use. Most of the millions

---

[1] Nicely translated into French as "Foire Aux Questions," illustrating how popular English acronyms may shape new concepts in other languages.

browser toolbar   Web-site URL   Web-site homepage

book-
marks
access

menu
field

hyper-
links

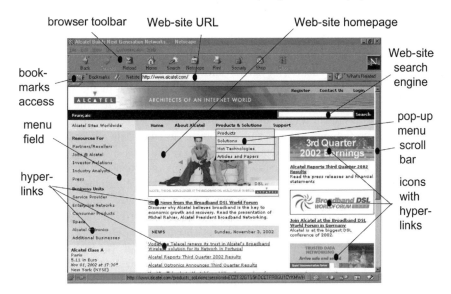

Web-site
search
engine

pop-up
menu

scroll
bar

icons
with
hyper-
links

FIGURE 3.28 Web-site homepage example for a large-size company, showing commands, menus, hyperlinks and many other typical features (with permission of Alcatel, ©2002)

of registered web sites only consist in a set of pages more or less artistically presented and more or less frequently updated, which deserve a hasty visit or possibly a bookmark. But other Web sites do offer a very useful resource: for instance, the possibility of reading *on-line magazines (e-zines)*; of searching past articles in journal databases; of making hotel/flights reservations (*e-booking*); or of consulting legal documents from government agencies or filing tax. Large-size documents or files can generally be *downloaded* on the client's hard drive, for printing, storing or future study. There are many types of *download*, usually available for free, which concern text, photos, music, video clips and generic/trial software (*shareware/freeware*). The action of *on-line shopping*, or *e-shopping*, filling up an *e-cart* and *checking out* using one's credit card number or *consumer ID* can be quite rapid, just like rushing through a big department store with an exact idea of what to get.[2] Some websites or pages can be accessed only by means of a user *login* (identification) and *password* (for *authentication*), meaning that some kind of subscription or permission of access is required to benefit from the contents, such as professional-society journals. All the above actions and perceptions resort to virtual reality. For the client's computer, visiting a website is nothing but being connected to a distant server and downloading Web-page files in HTML format. The client and server remain obedient to the many commands activated by

---

[2]Choosing and sending gifts to relatives at different postal addresses including "gift-wrapping" messages is also possible, but this takes practice and patience. This observation illustrates that nothing intelligent or meaningful is within only a "three clicks" reach.

mouse-clicking hyperlinks, entering keywords, selecting bookmarks and toggling between browser windows. In this multiple page-downloading process, all generic titles, pictures, logos and other page frames are temporarily stored in the client's *cache memory*. The *cache* (for short) allows a rapid reloading of previously visited pages (at the expense of computer memory space). In the process, the browser is allowed (unless the option is deactivated) to store what are called *cookies*. Cookies are short code-messages that a distant server can use to interpret the client's profile/preferences and customize its web site accordingly. The browser capability can be extended by *plug-ins*. Plug-ins are programs called upon whenever the browser identifies certain forms of coded sounds, pictures or video. As the name indicates, they can be "plugged" at once into the client's computer, and are usually downloadable as freeware or shareware.

Of all Internet services, *electronic mail, e-mail* (or *courriel* in Canada) is one of the most phenomenal impacts, whether for private or professional use. E-mail has almost replaced the telephone because it is a more expedient way to communicate without disturbing the recipient and does not require immediate response or interactivity. Likewise, e-mail has almost replaced *post mail* (regular paper mail, also known as *snail mail*), since text and picture documents can be included as electronic-file *attachments*. There is no day at an office that does not begin with *checking e-mail messages, opening and saving attachments* and responding to the higher-priority or more recent ones. The process of checking and responding to one's e-mail is no guarantee that the list of unread messages will decrease, sometimes it is the contrary. In large companies, a reason for e-mail proliferation is the practice of *copying messages* to a subset of colleagues (team, partners, hierarchy...). The commands *reply to all* (sends the reply to all persons listed) and *reply with history* (includes the previous message texts in the reply) can generate non-converging e-mail exchanges. Another practice is *forwarding*, which consists in sending a received e-mail to another addressee or group (with a *for your information,* or FYI tag), usually without concern for confidentiality. By default, forwarded e-mail may include the entire history of previous messages and file attachments. When uncontrolled, such forwarding floods the Internet and the companies' intranets with an excess of irrelevant communications or data. Depending upon the contents, forwarding can threaten the originator's status in terms of privacy and professional or private relations. Because e-mail, including any file attachment, can be forwarded again and again by single clicks to third parties, outside secure intranets, it is in no way a secure means of communication. Two major plagues also caused by e-mail forwarding are *hoaxes*, and *viruses*. The hoax takes the form of a general alert to be immediately relayed to your friends and colleagues, playing on some emotional reaction. The alert is false news or made-up stories with no other purpose then to saturate the Internet and mischievously manipulate people for months and sometimes years. Many dedicated websites describe the incredible variety and longevity of these hoaxes (use "hoax" as a keyword to find and bookmark them). Typical hoaxes concern attractive ways to make money fast, receive products for free in exchange for e-mail forwarding, or false alerts for *computer viruses*. Most declared viruses do not exist, which hides the reality of the real and active ones. Real computer viruses are far more poisonous than hoaxes, since they are purposefully designed to alter, to manipulate or even to irremediably destroy computer resources and data. Viruses are short executable programs which can be activated without one's knowledge or intention, for instance by opening e-mail attachments such as innocuous pictures or *e-gifts*. The virus then inserts its own message commands into the client/victim computer system, provoking different types of damage. Viruses can be designed to have a specific activation date or to be progressively operated without the host being the least aware (computer *worms*). They can install a permanent program that directly exploits the host computer resource (connectivity, memory) to use the Internet resource for their own purpose. These are referred to as *backdoors* and *Trojan horses*. By means of e-mail forwarding with the effect of a geometric progression, viruses can propagate and multiply in a matter of hours throughout the entire WWW. It can take several days to become aware of a *virus attack*, to publicize *virus alerts* and to develop and install screening countermeasures (*virus shield*). Internet connections offer many opportunities for spying and tracking private information, for purposes of advertisement, profiling and other intrusive monitoring. Programs that are designed for this activity are referred to as *spyware*. Spyware generates *Web bugs* (Web site images monitoring host information), computer *worms* (data packets placed in the host), *spying cookies* (host-user profiling), and *scripts* (short programs activated without the user's awareness). Intranets are protected against virus attacks and other backdoors by powerful proxy servers and firewalls. Personal computers can be equipped with *antivirus* and *spy-screening* software, most of which is available as freeware or shareware. Such a protection remains safe however only as long as it is updated to include all the latest virus/spy versions.[3] Several dedicated Web sites can be consulted to learn about the variety and danger of viruses and spyware (use any of the above keywords for search).

Since e-mail is a new form of rapid and informal communication, a set of rules had to be defined for proper use and good conduct. These rules represent the *netiquette* (for network étiquette). According to netiquette, short e-mail messages which do not start by "Dear XXX and finish by "Best regards" are wholly acceptable. However, the most polite and formal e-mail communication format can be a regular letter, as included in a separate electronic-file attachment, possibly with a prescanned official *e-signature*. E-mail messages should exclude capitals words that generally sound like shouting (PLEASE, DO NOT, IMMEDIATELY, etc. . . .). They should also exclude ambiguous phrasing which could have the potential of being emotionally misinterpreted (irony, joke, over-familiarity). To circumvent such a possibility, *smileys* (or *emoticons*) have been developed. Smileys consist in an alphabet of alphanumeric symbols, which more or less exactly communicate the mood or emotion (sympathy, humor, irony, sadness . . .) associated with a message sentence or statement. Although the list of smileys is rich and unbounded, only very few are commonly used, such as :-) or :-(for happiness or sadness, and ;-) for joke or semi-irony, for instance. The fact that smileys had

---

[3]Such a protection is to be completed with a cheap PC firewall, which is practically indispensable with today's fast-Internet (DSL) "always-on" connection services.

to be developed to complement the expression of writing or the language itself illustrates how different e-mail messaging is compared to ordinary communication by traditional voice or hand-written letters. The netiquette also calls for the use of reasonably short messages (and small attachment sizes), as well as a rapid response turnaround. A temporary acknowledgment, followed later by a fully developed reply, is considered most polite, however rarely observed. Acknowledgment messages (such as "thank you!") can also be sent using only the message title field, including at the end a "(no text)" mention. This represents a most effective and polite way to end an acknowledgment loop. In many corporations, a golden rule is to systematically *copy* to all persons whose names (or function) may be mentioned in an original message. It is however impolite to systematically copy the line hierarchy of the destinee, to whom such a prerogative belongs. In most corporations, an essential netiquette behavior is to send ahead of time, to your boss and other colleagues, a copy of the slide presentation you will make in a forthcoming meeting, even should this be done late at night the day before! This way, all meeting participants will be informed and have the opportunity to get acquainted/ prepared or to introduce constructive comments/corrections prior to your presentation. Receiving the slide presentation by e-mail either before or after a meeting is generally expected from participants and customers. In big corporations, the practice of systematically copying messages to other colleagues can also lead to a form of *e-mail congestion*. The result is that copied messages may systematically be overlooked or discarded, especially if the person copied to is high in the hierarchy. It is wrong to expect that all copied mail will be read and acted upon. It is also wrong to expect that nobody will read it, which is a true paradox of this copy/forward communication culture.

Observation of the above netiquette rules, most of them being unwritten, makes it possible to avoid saturating e-mail servers and adressees with masses of *junk mail*. Note that junk mail can also come from regular *e-advertisement*, whether stemming from mutually agreed *e-mail subscription* or foist upon users through targeted campaigns. The action of flooding the Internet with e-mail messages, whether by automation or self-multiplication is called *spamming*, which is a form of *net abuse*.

Other important components of the emerging Internet culture are *bulletin board systems (BBS*, also called *bboards), newsgroups, discussion forums (fora)*, and *Internet chat*. All these are variants of open and public discussions, whether in delayed time (forum) or real time (chat). It is customary to *register* to a newsgroup, using one (or several) *pseudonyms* for anonymity, and enter a personal *password* for authentication. *Usenet* is a worldwide, distributed network which offers thousands of newsgroups and forums classed by interest keywords (as classed into a set of *hierarchies*). The BBS offers a set of "walls" where people can *post* messages, usually in the form of short statements, in order to initiate or continue a discussion on a defined subject. Contrary to popular belief, the subjects being covered go far beyond politics and societal issues (e.g., disabled and health-related support groups). It is possible to *crosspost* a single message onto several forums in a given BBS. The *post* is then commented on and sometimes *rated* by the newsgroup, giving rise to unbounded chains of replies. Here also, a specific netiquette must be observed in the postings, otherwise it may lead to the risk of adverse group reactions. The group negatively reacts to what is called a *flame*, or *flaming*, i.e., the action of posting immature, provoking, or offending messages. *Spamming* is also used to qualify the negative behavior and action of *excessively cross-posting* or *multiposting* (ECP/EMP) messages that are irrelevant to the discussions or of commercial nature. Newsgroups can be *moderated* by a leader, who filters, edits and posts the received messages, in order to avoid the above effects and raise the level and quality of the discussion. Newsgroups can also be left free-running, or *unmoderated*. While owned by no specific entity and distributed worldwide, the previously mentioned Usenet relies upon some minimum local sponsorship, administration and supervision; however, newsgroups can *propagate* all the way through the entire network with relative absence of control. The unmoderated formula works well in professional-community Web sites, for instance, where people can post their reactions under recent Web site articles. Such posts can be rated on a scale of 1 to 5 by an editing board, for the sake of interest and pertinence, which may guide the serious participants' interest. These communities can also organize *webinars* (a form of *e-seminar* or *e-convention/conference*) which are special on-line events where registered participants can attend *slide presentations* and chat through live multicast/broadcast, voice/video applications (voice being serviced through either ISDN or VoIP). The real-time version of a newsgroup or discussion group, as established between a set of connected users, is called *Internet chat*. The discussion takes place in a virtual *chat room*, which participants enter and leave at will. Internet chat can also take place in parallel with e-mail, when a group of correspondents (relatives, friends) are automatically made aware of *who is on-line*, making it possible to exchange short messages (including voice) in parallel to regular e-mail and Internet browsing. This and Internet chat are two approaches for *instant messaging*. Examples of corresponding service providers are Internet *ICQ* ("I seek you"), *America's Online Instant Messenger (AIM)*, and *Internet relay chat* (IRC). These new means for multi-level, real-time communications can understandably disorientate previous generations only accustomed to two-party telephone. But the practice is expanding rapidly, as proven by the widespread use in large corporations of mixing audio-conferencing (ISDN and PBX), in connection with web-based multi-media tools. The combination of these different services and tools makes it possible to create *virtual teams* operating around the clock and over the time zones.

As with Internet browsing and searching, making sense of discussion forums, participating in newsgroups, internet chats, or attending public/corporate webinars is an art of learning and experience, ranging from the "*newbie*" stage to the confirmed stage. It is not necessarily an impersonal and lonely adventure, since *cybercafés* and other public training centers have been opened in most towns. The Internet is also making its way into schools, offering new teaching/learning potentials, as well as access to huge educational databases and resources. The penetration process is far slower in countries under development, as local telephone networks may not be equally developed all the way through the metropolitan, suburban and geographically remote areas. Yet the combined development in these areas of submarine-cable, satellite, terrestrial-microwave and public-telephone networks, accompanied with cheaper terminal appliances might rapidly bridge this ultimate gap and open Internet services to practically *anyone* on earth.

There is no doubt that the above Internet services, from e-mail to newsgroups and databases, offer new opportunities for the public sharing of ideas, opinions and beliefs, representing an extended development of "democracy," "right of speech" and "group expression." Predictably, such a phenomenal development comes together with its own shortcomings and potentials for abuse and attack. The concept of *net abuse* refers to any unethical, mischievous or ill-intended use of network resources. A long and nonexhaustive list of such abuses includes the already mentioned *hoaxes, virus attacks, excessive cross-/multiposting* (ECP/EMP), *e-mail bombing* which is followed by the client/server *denial of service* or DoS, automatic *bounce-and-forward mailings,* newsgroup disruption (*spamming*), post erasures or *"silencing," Web-site attacks, search-engine flooding*, intrusion and takeover of memory/node resources in corporate and public hosts (*theft of service* or ToS), installing *nested backdoors* in private computers, breaking in to restricted-access government networks and banks. This list is meant to quote only a few examples of net abuse, which range from unhealthy/immature games to straight criminality. It can be the matter of website/firewall *cracking* by people defining themselves as smart *hackers,* in a group of otherwise inoffensive, non-conformist and eccentric *cyberpunks.* Hacking is usually an attempt to get public prestige by proving telecom network weaknesses (e.g. making it into the PSTN and PBXs for free, referred to as *phreaking*), to nefarious criminal activities. These can consist in generating and using "valid" credit card numbers (*carding*), downloading proprietary/sensitive information from protected networks, and implanting home-made viruses. Such an activity should not be confused with *code-breaking* or *code-cracking* in cryptography or encryption. Depending upon its conditions of use and purpose, code-breaking can be a service to a nation (wartime intelligence) or just a mere mathematical contest to get international recognition and prizes (game of "cipher punks." See detailed description of network communications and security issues, including cryptography protocols in Desurvire, 2004.

Finally, it is needless to recall that the Internet helped to develop classical forms of organized crime at the international level, from trafficking of illegal materials to terrorist networking and related propaganda. It should be noted that counter-measures against such networks are developing at a rapid pace, making possible better observation and dismantling. Indeed, if the Internet is a new communication tool for new forms of crime, it is also the place where these activities can be monitored and tracked down. The only limit to such monitoring is the capability of identifying delinquent contents out of millions of Web sites and billions of electronic messages. By use of remote proxy servers ("remailers"), messages can be "anonymized," making it difficult or possibly impractical to identify the originator's IP address. The use of advanced cryptography, as allowed in some countries, makes the investigation far more difficult. Notwithstanding the rights of using cryptography for privacy and any honest purposes, it remains true that *internauts* are accountable for their network use, from visiting websites, to posting messages, and sending e-mails or documents. At any moment of use, either source or destination computers (or both) can be identified from their IP address, of which ISPs offer no confidentiality guarantee. See Desurvire, 2004 for a more in-depth treatment of these privacy/security concepts.

This description of Internet/www jargon and reality is more than enough to illustrate the high complexity of the *Internet environment*, with its highly beneficial resources on one hand and its fragility against attacks or abuse, on the other. In this report, there is nothing new compared to the early development of past mail and telephone services, except for the unbounded globality offered by Internet technology and reach. As before, the human/economical benefits surpass by and large the deviations and threats, to shape a society with higher information, education, entertainment, and wisdom.

✓ EXERCISES

Section 3.1
3.1.1 Show that after the conventions of Figure 3.1, (1) the tributary unit group TUG-2 cannot be loaded by a container of C-22 type, and (2) the tributary unit group TUG-3 can be loaded by any of the C-3 container types.

3.1.2 A town of 10,000 homes (95%) and businesses (5%) must be serviced with a new SONET telephone system. Assume that homes are equipped with a single telephone line, while businesses have two lines. Also assume that the daily average traffic pattern is 5 incoming/outcoming calls for homes and 50 for businesses. What is the required OC-n frame capability for handling the full average traffic? Same question if a margin of 50% must be provided to handle network congestion.

Section 3.2
3.2.1 How many ATM cells can be embedded into a STM-1 frame, using VC-4 containers?

Section 3.3
3.3.1 An IP datagram with a data payload of 2,800 bytes must traverse two networks (A, B) which have MTUs of 1,488 bytes (A), and 648 bytes (B), respectively. Describe how the datagram must be fragmented from source to destination, keeping the number of fragments minimum and assuming a header size of 48 bytes. How many fragments are received at destination, with what payload sizes?

3.3.2 If all IPv6 addresses were assigned to a square meter of the Earth's surface, how many addresses would each square meter contain? Same question with a square micron.

3.3.3 Convert the Ipv6 address 801::AAAF:0:DD1E:5A6C to binary format.

3.3.4 Show that *class D* addresses for IP multicast applications range from 224.0.0.0 to 239.255.255.255.

a MY VOCABULARY

Can you briefly explain each of these words or acronyms in the core-network protocol context?

AALn (ATM)
Abuse (Internet)
Access (network)
Add-drop
 multiplexing
ADM
Administrative unit
 (group)
AFS (internet)
AH (IP)
AIM (web)
Alert (virus)
Alias (e-mail)
Anti-virus (web)
Attachment (e-mail)
Attack (virus, web)
Authentication
ALL (ATM)
Applet (Java)
ARP (TCP/IP)
ASN.1 (TCP/IP)
ATM
ATM Forum
ATMARP
 (IP over ATM)
Attachment
AU (SDH)
AUG (SDH)
Aggregate
Backdoor (virus)
Bastion host (IP)
Bboard (web)
Bit
Bookmarks
Boolean (logic)
Broadcast TV
Browser (web)
Bulletin board
Byte
Cache (memory)
Carding
CBR
CDV (ATM)
CDVT (ATM)
Cell (ATM)
CES (ATM)
CGI (script)
Chat (Internet)
Circuit-switched
 (network)

Client
CLP (ATM)
CLR (ATM)
Connection-
 oriented
Connectionless
Container
Cookie(s)
Copy (e-mail)
Core (network)
Courriel
Cracking
Crawler (Web)
CRC
Cross-connect
Cross-posting
CRS (ATM)
CS (ATM)
CSI (ATM)
CTD,
 max/min (ATM)
Cybercafé
Cyberpunk
Datagram
Destuffing (cell)
DCE (Telnet)
Digital
 cross-connect
Discussion group
Distance-vector
 (algorithm)
DNS (Web)
DoS (Internet)
Download
 (Internet)
DSL
DXC
e-booking
e-mail
e-shopping
e-zine
E/O conversion
ECP
Edge (network)
Emoticon
EMP
Encapsulation
ESP (IP)
Extension
 header (IP)

Extranet
FAQ
Favorites
FEC
FIFO
Firewall
Flame (Web)
Flaming (Web)
Flexibility
Forum (Internet)
Forward (e-mail)
Frame
Fragmentation
 (datagram)
FRBS (ATM)
Frequency
 justification
FTP (TCP/IP)
FYI
Gateway (IP)
GFC (ATM)
GigE
Hacker
HDTV
Header
HDLC
 (Internet/SDH)
HEC (ATM)
Hoax
Homepage
Hop-by-hop
 options (IP)
Host, hosting
Hostid (IP)
HTML
HTTP (TCP/IP)
HTTPS (TCP/JP)
Hyperlink
Hypertext (link)
ICANN (Web)
ICMP (TCP/IP)
Icon
ICQ (Web)
Internaut
Interoperability
Intranet
IP
IP address
IPv4
IPv6

IRC (Web)
ISP
ITU-T
Jumbo payload (IP)
Junk mail
Keyword (Web)
Know-bot (Web)
LAN
LANE (ATM)
LIS (IP over ATM)
Login
Lynx (web)
MAN
Mapping
Management
 (network)
Management
 (agent)
MBS (ATM)
Menu (Web)
Meta-search engine
Meta-signaling
 (ATM)
MIB (TCP/IP)
MIME (Internet)
Modem
Moderation (web)
Monitoring
 (network)
MPLS
MPλS
MPOA (ATM)
MTU
Multicasting
Multiplexing
Multiposting
Multithreading
MSOH
Net abuse
Netid (IP)
Netiquette
Newbie
Newsgroup
NFS (internet)
NHRP
 (IP over ATM)
NMS (internet)
NRT
NVT (Telnet)
OAM

OC-N (SONET)
O/E conversion
On-demand (traffic)
On-line
OSI (model)
OSPF (TCP/IP)
Overhead
Packet
Packet-switched
(network)
Page (Web)
Password
Path overhead
Payload
PCR (ATM)
PDH
PDU (ATM)
Phreaking (PSTN)
Physical connection
Plug-in
POH
Pointer (frame)
Pop-up menu
Port (SW)
Posting (Internet)
PPP (Internet/SDH)
PRM (ATM)
Protocol mapping
Proxy (server)
Pseudonym
(Internet)
PSTN
PTI (ATM)
QoS

RARP (IP)
RAS (internet)
Reassembly
(datagram)
Reliability
(network)
Reply (e-mail)
Resolution
(address)
REXEC (Telnet)
RIP (TCP/IP)
RIR (IP)
Robot (Web)
Router (IP)
Routing table (IP)
RSH (Telnet)
RSOH
RT
SAR (ATM)
SCR (ATM)
Script (Web)
SDH
SEAL (ATM)
Search engine
Searching (Web)
Segmentation
Server
Smileys
SMTP (TCP/IP)
SN (ATM)
Snail mail
SNMP (TCP/IP)
SNP (ATM)
Socket

SOH
SONET
Spamming
SPE (SONET)
SPE group
SPF (IP)
SPI (IP)
Spider (Web)
Spooling
Spyware (Web)
Statistical
multiplexing
STM-N (SDH)
STS-N (SONET)
Stuffing (cell)
Subnet
Surfing (Web)
Switch
Tag (HTML)
TCP
TCP/IP
TDM
Telnet
TFTP (Internet)
Thread
Time
stamping (IP)
TLV (IP)
ToS (Internet)
Tributary
Tributary unit
(group)
Trojan horse
(virus)

TU (SDH)
TUG (SDH)
UDP (TCP/IP)
UNI
Upscalability
URL
Usenet (Web)
VBR
VC (SDH)
VCI (ATM)
Virus (computer)
VPI (ATM)
VT (SONET)
VTOA (ATM)
Virtual circuit
Virtual container
Virtual path
Virus (Internet)
Visiting
(Web site)
VoIP
WAN
WATM
WDM
Web (the)
Web bugs
Webinar
Web site
Worm (virus)
www

CHAPTER 4

Wireless Communications

In this chapter, we first review the basic physics of radio signals. Then we describe mobile communications, satellite-based communications and fixed-wireless communications, including the case of free-space optical carriers.

■ 4.1 BASIC PHYSICS OF RADIO-WAVE SIGNALS

In this first section, we shall describe the fundamental properties of *radio waves*. We first take a close look at how radio waves are generated by antennas. Different types of antennas are reviewed, including those using dish reflectors and antenna arrays, which introduces the concept of *radiation pattern* and *antenna directivity*. We describe then the decomposition of the spectrum into *radio wavebands* and their ranges of application. We analyze in detail the propagation of radio waves through the atmosphere and their reception by antennas, which introduces the concept of *power budget*. Different physical effects such as *attenuation by rain* and reflection/refraction onto the higher atmospheric layers (*ionosphere*) are described. We consider then parasitic effects associated with *multipath interference*, and their impact on reception by mobile antennas. We conclude this introduction with the analysis of noise and the concepts of *signal-* or *carrier-to-noise ratio*.

4.1.1 Generation of Electromagnetic Waves

To understand how *electromagnetic (EM) waves* can be generated and radiated in space through antennas and recaptured at a distance through other antennas, one needs some familiarity with elementary "electrostatics." From the simple concepts of electrostatics, it is possible, without mathematical tools, to get a sense of what EM-waves are made of and how they can be generated through an antenna, which we describe in Section 4.1.3.

Wiley Survival Guide in Global Telecommunications: Signaling Principles, Network Protocols, and Wireless Systems, by E. Desurvire
ISBN 0-471-44608-4 © 2004 John Wiley & Sons, Inc.

What is known as *electric current* flowing through a metallic wire corresponds, on an atomic scale, to the movement of elementary electric charges, the *electrons*. The movement of electrons is produced by an electric field which originates from the *difference of potential* existing between the two wire ends, as expressed in units of *volts* (V). Here, we shall not be concerned with defining "potential." We shall just consider that an *electric field* exists whenever two points in space have different potentials. Being a vector, the electric field has strength (graphically represented by vector length) and a direction in space which points from the higher to the lower potential. Figure 4.1 illustrates a static electric field generated by two such potentials. It is seen that the field varies in strength and direction according to the local position in space. The curves corresponding to constant field strengths and to which the local field is tangential are referred to as *field lines*. When the field lines are all parallel, the field is said to be *uniform*. As the figure shows, the electric field is approximately uniform in the vicinity of the potential axis.

Consider next a cylindrical region characterized by a uniform electrostatic field, E_{ext}, as generated between two external potentials at points A and B. Let us suddenly introduce a metallic rod inside this region, the rod being oriented parallel to the direction A-B, as shown in Figure 4.2. Because of the electric field, the electrons in the rod experience a force F, called the *Coulomb force*. The Coulomb force is proportional to the field strength and the electron charge, q, i.e., $F = qE_{\text{ext}} = -|q|E_{\text{ext}}$ (the minus sign comes from the fact that the electron charge is negative). The Coulomb force thus causes electrons to move opposite to the field direction towards point A, leaving many positive charges in the other rod extremity, B, as shown in the figure. This charge separation in space, induced by the applied electric field, is referred to as an *electrostatic dipole*. As the charges separate from each other and build up at the two rod extremities, a new electric field E_{dip} is created, pointing from B to A (see Figure 4.2). The dipole field, which is opposite in direction to the external field, E_{ext}, grows in strength as the number of charges build up at each

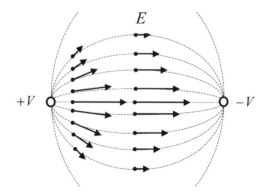

FIGURE 4.1 Generation of a static electrical field between two locations in space having different potentials $+V, -V$. The dashed curves correspond to the field lines. The field strength and direction is shown for two representative sets of points (near left location and in the center).

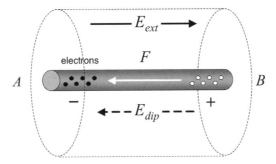

FIGURE 4.2 Displacement of electrons inside a metallic rod under an applied external electrical field E_{ext} and induced Coulomb force F. The charge separation creates a dipole with its own static field of opposite direction, E_{dip}, which cancels the applied field, leading to equilibrium.

rod extremity. After some time, the dipole field is equal in magnitude to the external field ($E_{dip} = -E_{ext}$), resulting in mutual cancellation ($E = E_{dip} - E_{ext} = 0$). Since the net field E is zero, the Coulomb force vanishes, the charge motion is stopped and the system has reached equilibrium. The same exact scenario would occur if one suddenly applied an external field (or potential difference) to the metallic rod. The result of either of these two experiments is that the electric charges in the rod are set in opposite motion to eventually form an electric dipole. If we repeat the experiment alternatively reversing the potential difference (or the electric field), the charges will oscillate back and forth, resulting in an *alternating current*.

The preceeding description prepared us to consider the principle of *antennas*, which is illustrated in Figure 4.3. Assume that the potential difference between points A and B oscillates at a rate of f times per second, with maximum values of $+V$ and $-V$, according to the sinusoidal law $V(t) = V \sin(2\pi ft)$. In this expression, f is the *frequency*, expressed in units of *cycles per second*, or *Hertz* (see Chapter 1). Because of this oscillating potential, the electric field inside the rod is also oscillatory, i.e., according to $E(t) = E \sin(2\pi ft)$, where the field strength E is called the *amplitude*. In turn, the field creates an oscillating Coulomb force, $F(t) = F \sin(2\pi ft)$, which drives the electrons back and forth along the rod. With this movement of electrons is associated an *oscillating dipole field*. The oscillating dipole field modifies the static electric field which surrounds the rod. The change is not instantaneous, since it takes some finite amount of time for the perturbation to reach any outside point. As a result, the perturbation propagates outside as an electric-field *wave*. Figure 4.3 shows that the E-wave moves away from the rod in the perpendicular direction. For clarity, the E-wave is only shown moving to the right in the plane of the page. In reality, the wave moves in all directions in the plane orthogonal to the page, its peaks and troughs forming concentric rings like a ripple on a water surface, as illustrated by the picture at top right. The E-field at distance r thus corresponds to the time-retarded wave $E(t) = E \sin[2\pi f(t - r/c)]$, where c is the *speed of light* ($c \approx 3 \times 10^8$ m/s). If we conceive of the metallic rod as a vertical wire with finite length, the above effect corresponds to the basic principle of an antenna. Each point along the wire emits an

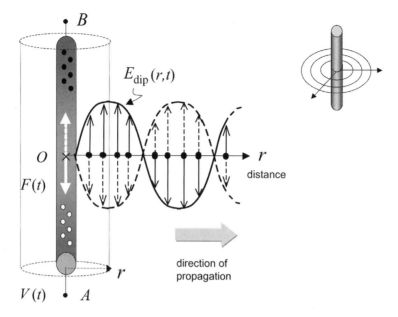

FIGURE 4.3   Generation of an E-wave, $E_{dip}(r,t)$, propagating away from a conducting rod coupled to an oscillating potential $V(t)$. The full and dashed curves correspond to the E-field amplitudes at distance r, as seen at two instants separated by a half oscillation-period. The arrow shows the wave's direction of motion. The same wave pattern is generated in all directions perpendicular to the wire's A-B axis, i.e., the electric fields $E(r,t)$ are identical in all points of the cylinder's surface at distance r and time t. The picture at top right shows that the E-wave radiates in all directions in the transverse plane (circles indicating position of peaks and troughs at a given time).

E-wave which contributes to an overall *radiation pattern*. Such a pattern depends upon the closed loops formed by the E-field lines and also on the possible antenna shapes. The complex issue of radiation pattern is addressed below in the subsection concerning antenna types.

According to Maxwell's laws of electromagnetism, a time-varying electric field is always associated with a *magnetic* field component. This magnetic field, H, is proportional to the E-field, according to $H = \sqrt{\varepsilon}E$, where ε is a constant equal (or very nearly equal) to unity in vacuum (or air). This constant can be equivalently defined through $\eta = 1/\sqrt{\varepsilon}$, where $\eta = \eta_0/n$ is called the medium *impedance* ($n = c/c'$, called the *refractive index*, is the ratio of the wave velocities in vacuum, c, to the wave velocity in the medium, c'). In vacuum, the impedance is $\eta_0 \approx 377\,\Omega$ (or $120\pi\Omega$), expressed in international units of *ohms* (Ω), see below. Because of the aforementioned proportionality, the magnetic-field wave is *in phase* with the E-wave, according to $H(t) = \sqrt{\varepsilon}E(t) = \sqrt{\varepsilon}\sin(2\pi ft)$. As shown in Figure 4.4, its direction is perpendicular to the E-field, with an orientation given by the "right-hand" rule: thumb = E, index = H and major = propagation direction. The combination of oscillating electric and magnetic fields forms the *electromagnetic*

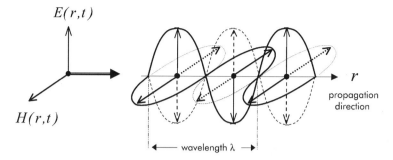

$E(r,t)$

$H(r,t)$

r

propagation direction

wavelength λ

FIGURE 4.4 Magnetic field component *H* associated with the E-wave *E*, forming together a transverse electromagnetic (TEM) wave. The full and dashed curves correspond to the electric and magnetic field amplitudes at distance *r*, as seen at two instants separated by a half oscillation-period. The wavelength λ of the EM wave, corresponding to a full TEM oscillation period, is also shown.

(EM) wave, as illustrated in Figure 4.4. Because both electric and magnetic field are in a plane orthogonal to the propagation direction, the EM-wave is referred to as *transverse electromagnetic* or TEM wave. When the E-field and the propagation direction are always in the same plane, the wave or the E-field are called *linearly polarized*. The polarization is vertical, horizontal or at 45° with respect to the ground plane. Dipole antennas always emit linearly polarized E-fields whose polarization orientation is the same as that of the antenna.

The *power* carried by an EM wave can be defined with respect to an element of surface it traverses with orthogonal incidence. The ratio of power to surface, called *intensity*, is expressed in watt/m^2 units. It can be show that the EM intensity is equal to

$$I = \frac{1}{2}EH = \frac{E^2}{2\eta} \tag{4.1}$$

where $\eta = \eta_0/n$ is the medium impedance previously defined (see Exercises at the end of the chapter).

The EM wave is characterized by a *wavelength* λ, which is defined as the distance separating two successive peaks, or

$$\lambda = \frac{c}{f} \tag{4.2}$$

The wavelength, expressed in *meters*, and the frequency, expressed in hertz (or seconds^{-1}) are two equivalent parameters defining different EM-wave domains. As described in the next subsection, the range of EM wavelengths or frequencies, which is called the *EM spectrum* spreads over orders of magnitudes. What is experienced as *light* represents a most familiar and vital form of EM waves. The sensitivity of the human eye to light corresponds in fact to a very limited portion of the optical or *lightwave spectrum*, as defined by the familiar rainbow colors. Figure 4.5 shows that the lightwave spectrum, as conventionally defined, spreads from nanometer waves

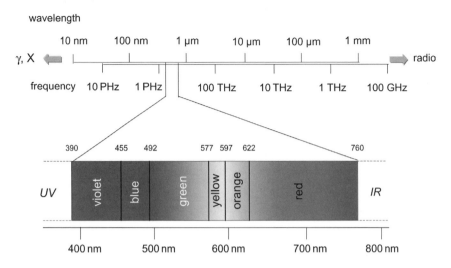

FIGURE 4.5 Electromagnetic spectrum corresponding to *optical* frequencies, and details of visible-light spectrum.

(down to $\lambda \approx 10$ nm, or $f \approx 3 \times 10^{16}$ Hz $= 30$ PHz) corresponding to the *ultraviolet* (UV) domain, to millimeter waves (up to $\lambda \approx 1$ mm, or $f \approx 3 \times 10^{11}$ Hz $= 300$ GHz) corresponding to the *infrared* (IR) domain. The frequency range for the UV domain spreads a bit over two orders of magnitude, while for the IR the range is three orders of magnitude. Frequencies higher than in the UV correspond to waves called *X-rays* (1 nm–1 pm wavelength, 1 picometer $= 10^{-12}$ m) and *gamma rays* (10 pm–1 fm wavelength, 1 femtometer $= 10^{-15}$ m). Frequencies lower than in the IR correspond to the domains of *microwaves* and *radio waves*. Between the UV and IR domains lies the short domain of visible light, i.e., from 390 nm (UV frontier) to 760 nm (IF frontier) and associated color hierarchy. Note that the eye response is not uniform according to color frequency, providing the "wrong" impression that yellow is "brighter" than blue. Interestingly, the EM spectrum of sunlight is also non-uniform. Its intensity sharply rises from the UV side, passes through a peak at 500-nm wavelength (blue) and then slowly decays towards the IR. Such a non-uniformity is explained by the physics of EM radiation from "thermal" sources, such as the sun, or incandescent lamps or candles. Telecommunications by use of lightwaves, or *optical communications*, are described in Desurvire, 2004.

4.1.2 Radio Wavebands

Although EM-waves are not so intuitive (even as described in Section 4.1.1), everyone is very familiar with *radio waves*, which belong to the low-frequency end of the EM spectrum. Indeed, most cars are equipped with radio receivers (called for short "radios") which tune to AM and FM stations. As described in a later subsection (see also Chapter 1), these two acronyms refer to *amplitude-modulated* and *frequency-modulated* radio waves, which transport signals (such as voice or music) through air. AM and FM carriers have wavelengths near 300 m and 300 cm, respectively, or frequencies near 1 and 100 MHz, as displayed on radio-station tuners. Radio lovers also enjoy *long-wave* (LW) channels which have wavelengths in the 1–10-km range, corresponding to 20–300-kHz frequencies. Apart from radio, one is also familiar with EM-wave transmission applications such as *television, mobile telephones, radars, satellites,* and a variety of other "wireless" devices.

The complete radio spectrum, which covers nine orders of magnitude, from millimeter-waves to megameter-waves (1 Mm = 10^3 km), is shown in Figure 4.6. The different radio wavebands (as conventionally defined from their wavelength domains), and their typical system applications can be listed as follows:

Extra high frequency (EHF), 1 mm–1 cm or 30–300 GHz: for radar, secure-military and satellite communications;

Super high frequency (SHF), 1 cm–10 cm, or 3–30 GHz: for satellite, fixed wireless (microwave links) communications and radar;

Ultra high frequency (UHF), 10 cm–1 m, or 300 MHz–3 GHz: for cellular mobile phones, UHF-television and radar;

Very high frequency (VHF): 1 m–10 m, or 30 MHz–300 MHz: for VHF-television, FM radio broadcasting, and air traffic control;

High frequency (HF): 10 m–100 m, or 3–30 MHz (also called short-wave or SW): for AM radio broadcasting, amateur radio and long-range air communications;

Medium frequency (MF), 100 m–1 km, or 300 kHz–3 MHz (also called medium-wave or MW): for AM radio broadcasting, maritime communications and navigation;

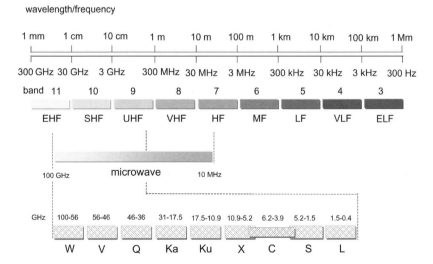

FIGURE 4.6 Electromagnetic spectrum corresponding to *radio* frequencies, designation of the main radio wavebands and frequency allocation of *microwave* subbands.

Low frequency (LF), 1 km–10 km, or 30–300 kHz (also called long-wave or LW): for AM radio broadcasting and air navigation beacons;

Very low frequency (VLF), 10 km–100 km, or 3–30 kHz: for maritime communications;

Extremely low frequency (ELF), above 100 km, or below 3 kHz: for military navigation and submarine, sonar and long-range communications.

The bands EHF to ELF are also designated by a number N, as indicated in the figure. This number corresponds to the frequency range $f = 0.3 - 3 \times 10^N$. Thus, EHF is also known as band 11, and so on to ELF for band 3.

The *microwave* domain (see Figure 4.6) concerns radio waves having frequencies from 10 MHz to 100 GHz (30-m–3-mm wavelengths). The upper fraction of the microwave domain, which overlaps with the EHF-SHF-UHF bands, is called the *microwaves*. It is itself divided into a set of sub-bands, as designated by L, S, C, X, Ku, Ka, Q, V and W, as shown in the figure. Such names were chosen at random by the military to confuse enemies. The names Ku, Ka correspond to an initial K-band, later divided into "above" frequencies (Ka, pronounce "Kay-ay") and "under" frequencies (Ku, pronounce "Kay-you"). The wavelength/frequency allocations of these microwave subbands, as well as their typical radio-system applications, are shown in Table 4.1. Note that these subband appellations exclusively concern microwaves. The lightwave domain also uses its own band definitions such as L, S, C and X. Also note that the satellite and radar domains use similar band appellations, but with relatively different frequency ranges and boundaries. For instance, C-band concerns the range 3.4–4.5 GHz for satellite and 4–8 GHz for radar, in contrast to the 3.9–6.2 GHz range shown in Table 4.1. To further complicate the picture, radars use for the K-band their own Ku-K-Ka domain split, while for satellites the split is defined as Ku1-Ku2-Ku3-Ka, with no one-to-one correspondence between the two domains.

TABLE 4.1 Conventional radio wavebands in the microwave domain and their applications

Name	Wavelength (mm)	Frequency (GHz)	Application
L	193–769	0.4–1.5	Broadcasting, 1G mobile, GPS
S	57.7–193	1.5–5.2	2G-3G mobile, LAN, WI-FI, GPS
C	48.4–76.9	3.9–6.2	Point-to-point, LAN, WI-FI, satellite
X	27.5–57.7	5.2–10.9	LH point-to-point, fixed wireless, satellite
Ku	17.1–27.5	10.9–17.5	MH/SH point-to-point, fixed wireless, older satellite
Ka	8.34–17.1	17.5–31	SH point-to-point, fixed wireless, newer satellite
Q	6.52–8.34	36–46	SH point-to-point, fixed wireless
V	5.36–6.52	46–56	Fixed wireless, future satellite
W	3.0–5.36	56–100	Fixed wireless, LAN, 4G mobile

(LH, MH and SH = long-haul, medium-haul and short-haul)

As will be shown in Subsection 4.1.4 the different radio bands differ in their ability to propagate through the atmosphere over distance, to pass through obstacles such as walls, buildings, or rain clouds, and overcome natural obstacles such as mountains and the Earth's curvature. The higher frequencies (or shorter wavelengths) experience greater attenuation. As is well-known, the HF (or SW) band is reflected by the Earth's upper atmosphere (see Section 4.1.3), making it possible to hear stations transmitting from all over the world and to communicate with each other as amateur radio operators. At higher frequencies, the UHF-VHF bands used for TV broadcasting are of much more limited range and are blocked by mountains, hence the need to have a network of repeating land stations (see a later subsection). The 0.5–2-GHz frequencies in UHF are used for *mobile cellular phones*. They have limited range, which requires a denser cellular network (see a later subsection). But the corresponding spectra are substantially wider, making it possible to share the bandwidth with a multiplicity of users, by the use of FDM, TDM, or spread-spectrum codings (see Subsections 4.2.1 and 4.2.4). The domain of *satellites* corresponds to higher frequencies, but their antennas (dish/parabolic, phased-array) are highly *directional*, making it possible to focus the EM-wave power towards specific continental areas and even dedicated land-based antennas (see Section 4.3).

4.1.3 Types of Antenna

In an earlier subsection, we described the principle of EM wave generation from an oscillating dipole. Here, we shall discuss further the issue of *radiation pattern* (how the EM wave and power are distributed into space), and some of the main types of *radio antennas*. The study of antennas and their patterns is in fact highly diversified and complex, owing to the incredible variety of antenna configurations, shapes, arrangements and designs. We shall only describe the basic concepts to provide the reader with a first and very simple introduction to this otherwise knowledge-intensive field. But after reading this subsection, the many antennas one sees while driving through the country should never look the same.

In Figure 4.1, we saw how a static E-field is generated by a dipole, with associated E-field lines. In Figure 4.3, we have shown how an E-field wave is produced by the movement of charges, or alternative current, inside a conducting wire which acts as a dipole antenna. To understand the antenna's radiation pattern requires one to look closely at the time-varying E-field lines and how their closed loops evolve over one oscillation period. Figure 4.7 shows the E-field line at five different times (1)–(5). Initially, (1) the charges are at rest with maximum separation and zero electric current, forming a static dipole; (2) driven by the antenna's potential, the charges are set in motion, generating a current, conventionally oriented from + to −, the electrons moving in the opposite direction; (3) the charges are in the same vicinity, resulting in neutral physical medium and vanished dipole; (4) the field line's extremities cross each other, as the charges separate again; (5) the initial field line now forms a closed loop with transverse diameter equal to a half wavelength $\lambda/2$. Note that the changes in field-line shape correspond to E-field radiation; hence, the E-field loop moves away from the antenna at the speed of light, c. Figure 4.8 shows the same E-field pattern formation over a full oscillation period λ. The

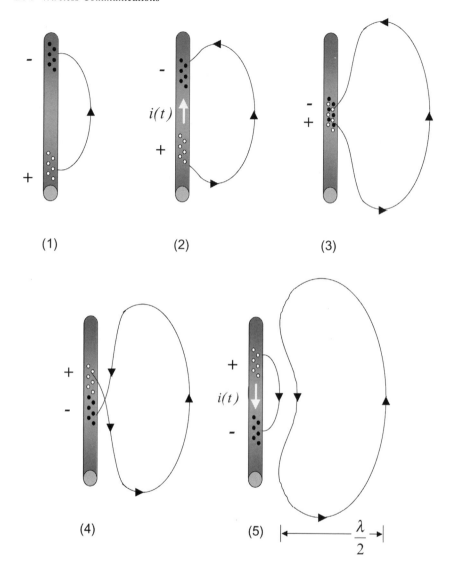

FIGURE 4.7 Evolution of a single E-field line, as electric charges move up and down a dipole antenna over a half oscillation period (see text for description).

concentric field lines correspond to different E-field amplitudes. Consistent with the previous description, the E-field vanishes at the loop borders and has maximum amplitude in the loop center, with alternating sign. As the figure shows, the radiated E-field pattern is symmetrical with respect to the antenna's vertical axis.

This represents an intuitive description of the E-field pattern generated by an oscillating dipole antenna. We might conceive a realistic dipole antenna made of an infinity of elementary radiating dipoles, with the current fed from the antenna's midpoint, to

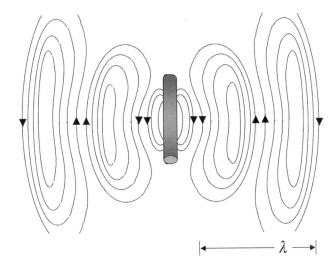

FIGURE 4.8   Electric-field lines associated with a dipole antenna, showing successive loop-formation pattern. The E-field is zero at loop borders and maximum at loop centers with alternating signs.

ensure perfect radiating symmetry. The resulting E-field is given by the superposition of all waves generated from any point along the antenna's axis. Since the corresponding waves are retarded from each other, their superposition in space is constructive or destructive, according to their mutual phase difference. The exact calculation of the resulting E-field interference pattern requires a bit of math formalism, which we shall overlook here. Suffice it to provide the result, which expresses the E-field amplitude $E(t)$ as a function of the radial distance r and the angle of observation, θ:

$$E(t) = \eta_0 \frac{i_0}{2\pi r} \sin\left[2\pi f\left(t - \frac{r}{c}\right)\right] E(\theta) \tag{4.3}$$

with

$$E(\theta) = \frac{\cos\left(\pi L/\lambda \cos\theta\right) - \cos\left(\pi L/\lambda\right)}{\sin\theta} \tag{4.4}$$

where i_0 is the alternating current amplitude, f the oscillation frequency and L the antenna's length. In this result, we recognize the time-varying component of the E-field, $\sin[2\pi f(t - r/c)]$, with a $1/r$ decay factor with distance (see following text). The second term $E(\theta)$ is referred to as the antenna's *far-field radiation pattern* which assumes distances $r \gg \lambda$. This radiation pattern provides the E-field amplitude change with observation angle θ, which is independent of the distance r from the antenna, but a function of the geometric ratio L/λ. Radiation patterns are shown in Figure 4.9, as calculated from equation (4.3) for different antenna lengths. In the first two cases, $L = \lambda/2$ and $L = \lambda$ (half-wave and full-wave dipole antennas), it is seen that the patterns are similar, exhibiting a single-lobe pattern (in direction r) having maximum E-field amplitude in the direction orthogonal to the antenna, and

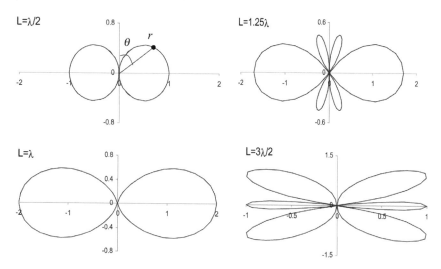

FIGURE 4.9　Far-field radiation patterns of dipole antennas corresponding to various antenna lengths: $L = \lambda/2$, $L = \lambda$, $L = 1.25\lambda$ and $L = 3\lambda/2$. In the last two cases, the field lobes are out of phase with respect to their neighbors. Note that the patterns have a revolutional symmetry with respect to the vertical axis.

zero E-field amplitude on the antenna'a axis. In the second of those cases ($L = \lambda$), however, the E-field amplitude is greater in the horizontal direction, corresponding to a more directional radiation pattern. It can be shown that this twin-lobe pattern family is common to all dipole antennas with $L \leq \lambda$. In the two other cases, ($L = 1.25\lambda$ and $L = 3\lambda/2$), the patterns are seen to exhibit additional lobes with mutually-opposite phases. These are referred to as *secondary lobes,* or *side-lobes.* Clearly, the dipole-antenna length must be chosen so that the EM energy is maximum in the horizontal transmission plane and is confined to a unique twin-lobe pattern, a condition which is satisfied only when $L \leq \lambda$.

The total EM power received at the (r,θ) point is defined according to equation (4.1), i.e., $I = E^2/(2\eta_0)$, which is proportional to i_0^2/r^2. Consistently, the radiated EM power is thus proportional to the antenna's *electric power* (i_0^2). It also decays rapidly with distance as $1/r^2$. This decay is consistent with the fact that the radiation pattern defines a unique *propagation cone* through space, within which the EM power is confined. It is easily checked that the cone's cross-section through which the EM power radiates increases as r^2. Thus the amount of EM power available at any point inside the cone must decay according to $1/r^2$. We note however that the radiated E-field only decays as $1/r$. This feature makes it possible for the EM wave to propagate over considerable distances, and hence to be captured by another receiving antenna, as discussed in the next subsection.

A useful reference for characterizing radiation patterns is the *half-power beam width* (HPBW). The HPBW is defined as the full angle within which the E-field power is between maximum and half-maximum. Since the power is given by the square of the amplitude, the angle is found at the points where the field amplitude is $1/\sqrt{2} = 0.707$ of the maximum. It is easily measured from Figure 4.9 that the HPBW is $78°$ and $47°$ for the half-wave antenna and the full-wave antenna, respectively. Note that the corresponding radiation patterns have a revolutional symmetry about the antenna's axis, meaning that the EM power is not confined in any specific *direction* of the horizontal space. Rather the pattern forms a horizontal torus with a null point on the antenna's axis (a torus looks like a New York bagel).

The preceeding description concerned the basic *dipole antenna*, which is made of a vertical metallic wire with full length $\lambda/2$ or λ, placed inside a mast. This is one of the most common types of antenna. As we have seen, the EM emission pattern of dipole antennas has a revolutional symmetry, corresponding to a twin-lobed, toroidal preferential radiation in the horizontal plane. Such antennas are well adapted to radio-signal *broadcasting*, since the EM wave is emitted with equal power in all directions of the plane. There are many radio-system applications, however, where it is a desirable feature that the EM power be emitted in a single privileged *direction*, corresponding to a *unidirectional* radiation with thinner lobe patterns. For instance, a TV satellite should broadcast the signals through a radiation lobe corresponding to the receiving region or continent. Likewise, Earth–Earth or Earth–satellite point-to-point radio links should use unidirectional antenna, but with even sharper radiation patterns or directivities. This is why a great variety of other antenna types have been progressively developed. Here, we shall restrict our description to *parabolic-dish* and *phased-array* antennas, which are found in most radio communication systems.

The *parabolic* or *dish antenna* is based upon the principle of reflecting the EM waves generated by a source antenna (the *feed*) onto a metallic surface. In general, such a reflecting surface can be anything from a plane, a corner made of two planes crossing at some angle, a section of a sphere or of an ellipsoid, or a paraboloid, as illustrated in Figure 4.10. The effect of the reflection is that the EM wave emitted in

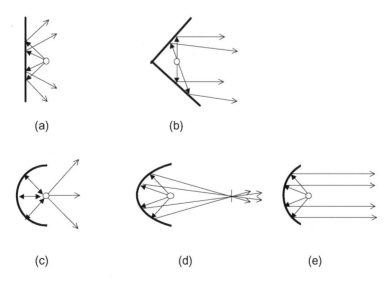

FIGURE 4.10 Different types of antenna reflectors: (a) plane, (b) corner, (c) spherical, (d) ellipsoid, and (e) paraboloid. In cases (d) and (e), the emitting dipole is placed in the surface's focal point. In all cases, the dipole axis is perpendicular to the figure's plane.

one half-space is folded back onto the wave emitted in the other half-space. As the figure shows, all reflectors improve the antenna directivity, generating a set of EM waves in a privileged angle (or cone) of space. It is seen that in all cases but the last one, the reflected rays are nonparallel and form a diverging pattern. Elliptic curves are characterized by two *focal points* or *foci*. The associated property is that all rays passing through one focus pass through the other one. It is seen that the result is a diverging pattern. Parabolic curves are characterized by a unique focus. The associated property is that all rays passing through this focus and incident to the parabolic curve are reflected in a single parallel direction, regardless of the incident angle, as shown in the figure. Thus, the parabolic reflector is the only one to steer the EM-waves originated from the focus into a single direction, forming a parallel-ray pattern.

To realize a parabolic-reflector antenna (commonly referred to as a "dish") with unidirectional wave pattern, one must therefore place the EM-dipole feed at the exact focal point. However, the waves emitted along the dish's symmetry axis reflect onto each other and can produce interference patterns. The interference is constructive if the round-trip distance $2d$ between the focal point and the parabolic surface corresponds to a wavelength, or an even number of half-wavelengths. For this reason, the round-trip distance should be an odd multiple of $\lambda/2$, i.e., $d = m\lambda/4$ ($m = 1, 3, 5\ldots$). Figure 4.11 illustrates this first design concept. A second issue is the directivity of the dipole feed itself. The directivity can be enhanced by used of a small plane or corner reflector, as previously discussed. One refers to the resulting E-field pattern as the *primary pattern*. This primary pattern should

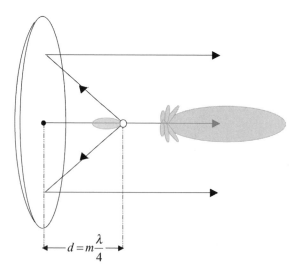

FIGURE 4.11 Principle of parabolic-reflector (dish) antenna, showing position of dipole source at focal point (open circle), relation of focal-point distance to wavelength ($m =$ odd integer) and associated radiation pattern (shaded ellipsoid pointing to reflector, left). The reflected EM-beams with resulting radiation pattern (shaded ellipsoid pointing away from reflector, right) are also shown, see also Figure 4.12.

be made sufficiently narrow as to exclusively and uniformly "illuminate" the dish surface. This can be helped by placing circular or square apertures near the feed antenna. Alternatively, the feed antenna can made from other types of devices, such as *horn antennas*. These horns consist of trumpet-like tubular openings placed at the open end of EM waveguides, such as a coaxial cable, for instance. The horn-antenna pattern is also unidirectional, and optimally single-lobed. Regardless of the feed-antenna type, however, the resulting apparatus occupies some finite cross-section in the reflected E-field beam. This blocks a fraction of the reflected waves and can therefore produce pattern distortion or unwanted secondary lobes. To compensate for this parasitic effect, the feed can be slightly displaced from the dish axis, which is referred to an *offset feed*. The feed can also be placed behind the dish, with the primary pattern passing through a center hole, then reflected by a mirror back to the dish. This is referred to as a *Cassegrain* antenna, after the same telescope principle.

Figure 4.11 also shows the *secondary pattern* generated by the dish antenna, with a typical single-lobed shape of high directivity. The calculation of this secondary pattern can be complex, unless some simplifying assumptions are used. For the purpose of this introduction, suffice it to say that an ideal parabolic-dish antenna is equivalent to a system where a beam of unidirectional EM waves is made to pass through an orthogonal circular opening or *aperture*. If D is the aperture (or dish diameter), it can be shown that the E-field pattern is given by the expression

$$E(\phi) = \frac{2}{\pi D/\lambda} \frac{J_1[\pi(D/\lambda)\sin\phi]}{\sin\phi}$$

where $J_1(x)$ is known as the *first-order Bessel function*. Such a function is very similar to a sinusoid with exponentially decaying amplitude. Note that unlike in the dipole-antenna case, the reference angle (ϕ) is taken from the dish revolution axis, which is now the preferential direction where the EM wave is radiated. Figure 4.12 shows the E-field pattern calculated from the above formula, while taking the absolute E-field value, and choosing for instance an aperture $D = 10\lambda$.

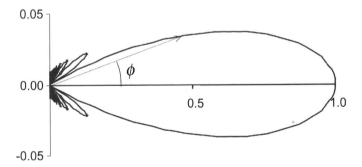

FIGURE 4.12 Radiation pattern of dish antenna, assuming an aperture-to-wavelength ratio of $D/\lambda = 10$, showing high directivity in the dish revolution axis (vertical axis expanded twenty-fold).

It is seen that the pattern is made of a long primary lobe and a multiplicity of short secondary lobes. The E-field is thus highly directive, meaning that its maximum to $1/\sqrt{2}$ amplitudes are reached within a relatively small cone of space. In this example, the cone angle (HPBW, as previously defined) is 5.8°, compared with 47–78° for the dipole antennas.

Consider a fictitious *isotropic* antenna, for which the E-field would be uniformly radiated in all directions of space. The corresponding power density at point (r,ϕ) is $P_i(r) = P/(4\pi r^2)$, P is the total E-field power. For a dish antenna with the same total power, the power density at point (r,ϕ) is $P_d(r,\phi) = PE^2(\phi)/r^2$. By definition, the *directive gain*, G, also called *directivity*, is the power ratio $G = P_d(r,\phi)/P_i(r)$, or $G = 4\pi E^2(\phi)$. The *absolute gain*, G_A, is obtained by integrating the gain over the full angle range $\phi = [-90°, +90°]$, which simply gives

$$G_A = \left(\frac{\pi D}{\lambda}\right)^2 = \frac{4\pi A}{\lambda^2} \qquad (4.6)$$

In this expression, we have introduced the dish's area, $A = \pi D^2/4$, which is also indifferently called aperture, the two formulas using diameter or area being strictly equivalent.

Usually, the absolute gain is defined in decibels, according to

$$G_A(\text{dBi}) = 10\log_{10}(G_A) \qquad (4.7)$$

The scale is referred to as dBi to recall that the power reference is that of an isotropic antenna. For instance, a dish antenna with aperture $D = 10\lambda$ has an absolute gain of $G_A(\text{dish}) = 10\log_{10}(100\pi^2) \approx 29.9$ dBi, which is close to 1,000 (30 dBi). It is also possible to take for reference the *half-wave antenna*. The absolute gain must then be corrected by the directive gain of the half-wave antenna, which (after calculation) is $G_A(\lambda/2) = 1.64$ or 2.15 dBi. Thus, with the half-wave antenna reference, the absolute gain of the dish antenna is G_A (dish) $= 29.9$ dBi $- 2.15$ dBi $= 27.7$ dB, which is close to 600 (27.8 dB). Note that the antenna gain is now expressed in dB.

This description concerned antennas having a unique dipole or aperture. As we have seen, these basic layouts generate specific radiation patterns. What happens to the radiation pattern if we put side by side an array of such antennas/apertures in a line or a plane arrangement? The answer is that the pattern will then be determined by the wave superposition of all E-field contributions from elementary source antennas. Since the sources have different locations in space and their relative phase can be independently controlled, it is possible to create an unlimited variety of interference conditions, resulting in new families of radiation patterns. These arrangements are referred to as *array antennas*, of which we shall describe a few interesting cases.

The most elementary antenna array consists in placing two dipole antennas parallel to each other in the vertical plane and separated by a distance d. If one chooses $d = \lambda/2$, it can be shown that the resulting pattern in the (r,θ) observation is the same as in the single half-wave antenna case (θ being defined from either dipole axis). The result is shown in Figure 4.13, assuming that the two antennas are in phase. Compared to the single-dipole case, the key difference is that while the pattern has the same angular shape it is not toroidal, but unidimensional, as

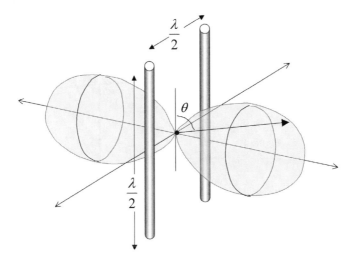

FIGURE 4.13 Radiation pattern of an array-antenna made of two in-phase, half-wave center-fed dipoles.

defined by lobes axis. Depending upon the phase relation between the dipoles, the twin-lobe axis rotates in the horizontal plane. In the case where the dipoles are out of phase, the axis is rotated by $90°$ from the direction shown in the figure. The gain calculation for this array antenna gives $G_A = 6.01$ dBi or 3.86 dB. This basic example provides a first illustration of how array antennas can be used to control and shape the radiation pattern, and therefore increase the overall antenna directivity. Larger-size antennas with improved directivity can be made by stacking arrays of dipoles in a single plane. This arrangement corresponds to the well-known "rakes" which one sees on so many roof tops. Another possibility is to introduce *parasitic elements* into the array. These are wires of length slightly longer or shorter than the source dipole. The longer element acts as a *reflector* and the shorter as a *director*. The resulting field pattern is highly asymmetric, with a privileged emission from the director's side. The *Yagi–Uda* array, named after its inventors, is a rake-antenna arrangement which includes one reflector, a half-wave (center-fed) dipole source, and a set of four directors, as shown in Figure 4.14. In an optimized configuration, the lengths of the reflector, source and directors are typically 0.475λ, 0.46λ, and 0.44λ, respectively. The reflector is placed at $\lambda/4$ from the source, while the directors are evenlyspaced with 0.31λ separations. The result is a directivity of $G_A = 12$ dBi or 14.1 dB. Since the antenna design is closely determined by the EM wavelength, it only works for a limited frequency range. The antenna's *bandwidth* can be defined as the frequency range over which such a directivity can be achieved within 3 dB. The bandwidth can be doubled if one includes a *corner reflector* (see Figure 4.10b), at the expense of directivity ($7–8$ dBi). The resulting arrangement, called *hybrid Yagi–Uda antenna*, is also illustrated in Figure 4.14. Such an antenna design is used for UHF-TV reception in hundreds of millions homes worldwide.

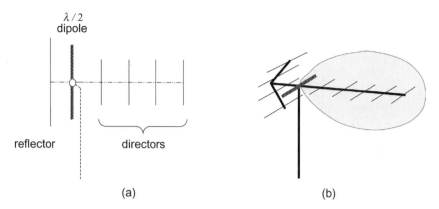

FIGURE 4.14 (a) Principle of the Yagi–Uda antenna with reflector and directors surrounding a center-fed dipole, (b) hybrid and improved design also involving back corner-reflector, with principal radiation-pattern lobe.

The principle of a *phased-array antenna* is to control both phase and amplitude of an array of dipole or apertures, in such a way as to *steer* the radiation pattern in space, like a beam. Such beam steering can be instantaneously realized by switching the phase/amplitude feeds to predetermined settings. Alternatively, the steering can be made to oscillate by varying the frequency in a saw-tooth-like pattern (frequency scanning). The EM beam is steered back and forth, corresponding to an effect of *beam sweeping*. The principle of the phased-array antenna is shown in Figure 4.15. In its unidimensional version, the antenna is made from a stack of parallel dipole elements (or any elementary EM sources) separated by a distance d.

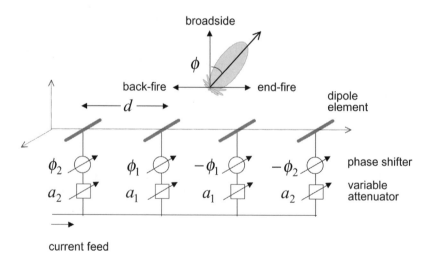

FIGURE 4.15 Principle of phased-array antenna, with preset amplitudes (a_i) and phases (ϕ_i) for each dipole source, showing beam-steering effect at angle ϕ in the resulting radiation pattern.

The current is fed from a common oscillating source. The amplitude (a_i) and phase (ϕ_i) of each dipole is mechanically or electronically controlled through variable attenuators and phase shifters, respectively. Assuming that the antenna has an even number of K sources, the calculation leads to the following radiation-pattern expression:

$$E_{array}(\phi) = 2E_{element}(\phi)A(\phi) \tag{4.8}$$

where $E_{element}(\phi)$ is the element's pattern and $A(\phi)$ is the *array factor* as defined by

$$A(\phi) = a_1 \cos\left(\frac{\pi d}{\lambda}\sin\phi - \phi_1\right) + a_2 \cos\left(\frac{3\pi d}{\lambda}\sin\phi - \phi_2\right)$$

$$+ \cdots a_K \cos\left(\frac{(2K-1)\pi d}{\lambda}\sin\phi - \phi_K\right) \tag{4.9}$$

This result shows that the array-antenna pattern is given by the elementary pattern (times two) multiplied by an array factor which reflects the effect of amplitude, phase and source separation distance. Considering that for dipole elements $E_{element}(\phi) = 1$ (no privileged direction in the antenna's transverse plane), the radiation pattern and the array factor (times two) are identical. The case $K = 2$ thus corresponds to the two-dipole array previously shown in Figure 4.13.

The effect of *beam steering* is easily explained by considering the simple case where all amplitudes are equal, but not the phases. We observe then from equation (4.9) that the array factor is maximum for the arbitrary direction $\phi = \phi_0$ when all arguments of the cosines are equal to $n\pi$, where n is an integer. For instance, with $n = 0$ we obtain

$$\begin{cases} \dfrac{\pi d}{\lambda}\sin\phi_0 - \phi_1 = 0 \longleftrightarrow \phi_1 = \dfrac{\pi d}{\lambda}\sin\phi_0 = u \\[2mm] \dfrac{3\pi d}{\lambda}\sin\phi - \phi_1 = 0 \longleftrightarrow \phi_2 = \dfrac{3\pi d}{\lambda}\sin\phi_0 = 3u \\[2mm] \text{etc.} \end{cases} \tag{4.10}$$

This result shows that when the source phases linearly increase, by an odd number of times, the previously defined quantity $u = (\pi d/\lambda)\sin\phi_0$, the E-field is maximum in the direction $\phi = \phi_0$. Thus, by the appropriate choice of phases, as defined in equation (4.10), the antenna beam can be steered to any pre-defined angle ϕ_0. Examples of array factors (or radiation patterns for pure dipole elements), as calculated from the above equations, are provided in Figures 4.16 and 4.17 for vertical ($\phi_0 = 0$) and slanted ($\phi_0 = 45°$) directions, respectively. It is seen that the beam directivity is rapidly enhanced as the number of array elements increases. One can also observe that with a number of elements $K \geq 8$, the beam can be steered to any direction between $-90°$ and $+90°$, without losing directivity. In antenna jargon, the $0°$ or vertical beam direction is referred to as *broadside*. The $+90°$ direction (to the right) and $-90°$ direction to the left are called *end-fire* and *back-fire*, respectively, see Figure 4.15.

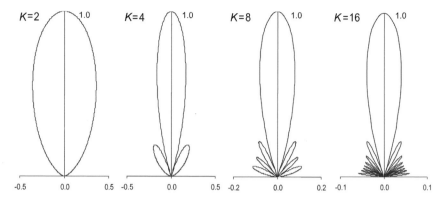

FIGURE 4.16 Normalized array factors of antenna corresponding to Figure 4.15, with $d = \lambda/2$, equal amplitudes ($a_i = 1$) and phases specifically chosen to maximize beam intensity in the $\phi_0 = 0$ or broadside direction. The different cases correspond to the number of elements $K = 2$, 4, 8 and 16.

One should note that the figures only show the array factors obtained in the upper region of the antenna. For dipole elements, the same pattern is also formed in the lower region, corresponding to a waste of EM power and directivity. To increase the directivity, the radiating elements can be chosen as unidirectional, for instance by use of reflectors or horn antennas.

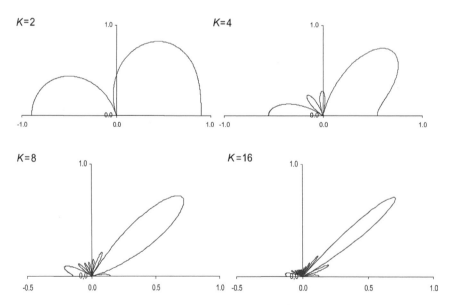

FIGURE 4.17 Normalized array factors of antenna corresponding to Figure 4.15, with $d = \lambda/2$, equal amplitudes ($a_i = 1$) and phases specifically chosen to maximize beam intensity in the $\phi_0 = 45°$ or broadside direction. The different cases correspond to the number of elements $K = 2$, 4, 8 and 16.

Another way to steer the beam is to vary the phases by *frequency scanning*. To explain how the effect works, we consider the same antenna arrangement shown in Figure 4.15, while removing all phase shifters and attenuators. We assume now that the signal current, which is fed from the left extremity, travels through the wires at some finite velocity. The dipoles located at distance 0, d, $2d$... thus receive the signal current with increasing delays. If one assumes that the current travels at the velocity of light, the phase corresponding to dipole "i" is given by $\phi_i = (i - 1)2\pi d/\lambda$. One can arbitrarily set the phase to zero in the center of the antenna, so that the configuration is equivalent to introducing phase-shifters with values set to $\pm\phi_i = \pm(2i - 1)\pi d/\lambda$ for each dipole "i" located on either side of the center. The previous result of equation (4.9) for the array factor applies. The beam has maximum amplitude in the direction ϕ_0 for which

$$(2i - 1)\frac{\pi d}{\lambda}\sin\phi_0 - \phi_i = \pm m\pi \longleftrightarrow \sin\phi_0 = 1 - \frac{m}{2i - 1}\frac{\lambda}{d} \qquad (4.11)$$

where m is an integer. If we choose for instance the configuration $m = 2i - 1$, we have $\sin\phi_0 = 1 - \lambda/d$, which defines the beam position at wavelength λ. From this configuration, a wavelength change from λ to $\lambda + \delta\lambda$ therefore results in an angle change $\delta\phi_0$ given by the relation $\sin(\phi_0 + \delta\phi_0) = 1 - \lambda/d - \delta\lambda/d = \sin\phi_0 - \delta\lambda/d$, which is the effect of beam steering. Assume for instance a vertical beam ($\phi_0 = 0$ or $\lambda/d = 1$). We have $\sin(\phi_0 + \delta\phi_0) = \sin\phi_0 \cos\delta\phi_0 - \cos\phi_0 \sin\delta\phi_0 = -\sin\delta\phi_0$ or, with the previous result:

$$\sin\delta\phi_0 = \frac{\delta\lambda}{d} = \frac{\delta\lambda}{\lambda} \qquad (4.12)$$

This result shows that the change in beam angle is independent of frequency and only dependent upon the relative frequency change $\delta f/f = -\delta\lambda/\lambda$. For instance, a relative frequency change of 10% ($\delta f/f \pm 0.1$) corresponds to a beam angle change of $\delta\phi_0 = \pm = 0.10$ rad or $\pm 5.7°$. However, a beam steering of $\delta\phi_0 = \pm 45°$ would require a frequency change of $\delta f/f = \pm 1/\sqrt{2}$, or $\delta f = \pm\lambda/\sqrt{2} = \pm 0.7f$, which represents a huge frequency deviation and is impractical. Therefore, beam-steering through frequency scanning remains limited to relatively small angles. This analysis also leads to another important conclusion: *the beam direction of phased-array antennas is relatively sensitive to frequency errors or deviations.* In the field of satellite communications, beam deviations of only a few degrees can result in large transmission loss or even outage of the communication channel, considering the very long transmission distances involved (see Section 4.3). The solution to this problem is to use frequency-dependent phase shifters.

The above description concerned antennas with single-dimension arrays. It is clear that a similar array implementation in two dimensions offers an extended range of possibilities for increasing the antenna's directivity and beam-steering ability. One of the most attractive two-dimensional implementations concerns *planar* phased-array antennas, in which the dipole sources can be integrated in large numbers and within relatively small thickness or profiles. These can be realized through a variety of integrated-circuit technologies and are commonly referred to as *patch* or *microstrip* antennas. Because of physical dimensions constraints, the range of application concerns wavelengths $\lambda < 3$ m or $f > 100$ MHz (middle of the VHF band to higher-frequency bands). The antenna is made from a rectangular metallic patch with $\lambda/2$ width, resting on an insulating dielectric layer connected to ground. An alternating current is fed on one side of the patch from the $+$ of a coaxial feed wire, generating an EM field. It can be shown that the patch radiation pattern is not very directional, corresponding to only 6 dBi. However, the patches can be integrated into large 2×2 arrays with as many as 1,000 elements, connected through microstrip transmission line feeds. The resulting phased-array antenna gains can be as high as 30 dBi at $f = 1$ GHz, for instance.

The issues related to antenna reception after atmospheric propagation of the radio waves are addressed in Section 4.1.4.

4.1.4 Radio-Wave Propagation and Reception

Being now familiar with the nature of radio waves and their emission by antennas, we shall consider here the issues of their propagation through the atmosphere and their reception by antennas.

The basic layout of a radio-communication system is shown in Figure 4.18. The system's building blocks are the *transmitter* to which a signal is fed, the *receiver* which restitutes the signal, the *transmitting* and *receiving antennas*, and the "immaterial" *radio-wave path* with its own physical characteristics. Conceptually, this layout is the same as used by Marconi in 1898 for his famous demonstration of *wireless telegraphy*, using Morse-coded messages (see Chapter 1). It also corresponds to

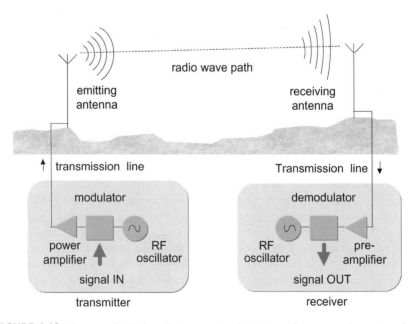

FIGURE 4.18 Layout of a basic radio communication system, from transmitter to receiver.

what we familiarly know as both radio-station or TV-channel broadcasting. Such a system can be symmetrically duplicated so that each endpoint is able to emit and receive using two different radio frequencies, realizing a duplex communication channel. The word *wireless* is now widely used to designate any type of radio communication system applied to telephony and data networking.

As seen from Figure 4.18, the transmitter and the receiver have opposite functions. The transmitter's function is to *code* the signal to be transmitted (e.g., voice) into a radio-frequency current provided by a local oscillator. Coding is made by modulating the current carrier with this (so-called) baseband signal. The modulated signal is then boosted through a *power amplifier* and fed to the emitting antenna. The antenna radiates the signal in the form of EM waves which have the same amplitude/frequency waveform characteristics as the feeding current. After traversing the atmosphere though some path pattern (see below), a fraction of the EM wave eventually reaches the receiving antenna. The receiving antenna converts the EM wave into an electrical current of identical waveform characteristics. Since the EM power reaching the antenna has been attenuated by propagation and other causes of power loss (see below), the received antenna current is generally weak, and needs therefore to be boosted to an appropriate level through a *preamplifier*. The operation of demodulation makes it possible to restitute the original baseband signal. The demodulator/oscillator layout depends upon the coding scheme, which can be either *analog* or *digital*. The basic principles of analog/digital coding and modulation, especially concerning voice, have been explained in Sections 1.1 and 1.2. A more detailed description of the coding schemes which apply to radio systems (and in particular to wireless telephone networks), is made in Subsections 4.2.3 and 4.2.4.

We shall, then, focus on the EM-wave propagation between the emitting and the receiving antenna, and all the physical effects that can be involved in the wave path.

Consider first the straight-line path joining the two antennas in free space, the antennas being at distance r from each other. This is referred to as a *line-of-sight* (LOS) radio system. In the previous subsection, we saw that if the antenna is isotropic (transmitting equal power P_{trans} in any direction of space), the EM power at distance r is, per surface unit: $P(r) = P_{\text{trans}}/(4\pi r^2)$. But if the transmitting antenna is directive, as characterized by an absolute gain G_A, the power density in the direction of maximum radiation is $P(r) = G_A P_{\text{trans}}/(4\pi r^2)$. Assuming that the receiving antenna has a surface aperture A_R, we obtain the received power

$$P_{\text{rec}} = G_A \frac{A_R}{4\pi r^2} P_{\text{trans}} \qquad (4.13)$$

If we replace the definition of G_A in equation (4.6) into this result, we finally obtain

$$P_{\text{rec}} = \frac{A_T A_R}{\lambda^2 r^2} P_{\text{trans}} \qquad (4.14)$$

where A_T is the transmitting antenna's aperture. This relation links the transmit and receive power of the two antennas according to distance, wavelength and apertures. It is called the *Friis transmission formula*, after H.T. Friis who derived it in 1946. The transmit/receive powers must be interpreted indifferently as either electrical or EM. The fact that the conversion efficiency from electrical to EM or the reverse may not be 100% is hidden in the aperture characteristics A_T, A_R. The antennas imperfections result in "effective" apertures or absolute gains that are somewhat smaller than those given by the actual physical dimensions. Some textbooks use the actual physical parameters for aperture and gain and introduce an *antenna efficiency*, $\eta_T, \eta_R < 1$, which amounts to the same result. In the definition of equation (4.14), the apertures are therefore effective parameters, otherwise the right-hand side should be multiplied by $\eta_T \eta_R$.

By deriving the above relation between transmit and receive powers, we have used a *principle of reciprocity*. This principle states that the EM power can be converted back into receiving antenna current, the same way the transmitting antenna current is converted into EM power. This principle is not strictly equivalent to applying time-reversal in the analysis of the transmitting antenna. This is because EM waves disperse spherically through space, and there is no device which can reverse the process. However, the EM-wave incident upon the receiving antenna induces an alternative current exactly the same way as the alternative current in the transmitting antenna induces EMwaves, which justifies the name of "reciprocity principle."

Let us further develop the previous results, considering a transmitting antenna which would be isotropic. Such an antenna has no privileged direction, which by definition corresponds to a gain of unity. From equation (4.6), it is possible to define the equivalent isotropic-antenna "aperture" A_i such that $4\pi A_i/\lambda^2 = G_i = 1$, or $A_i = \lambda^2/4\pi$. Replacing this definition in the Friis formula gives

$$P_{\text{rec}} = \frac{A_i A_R}{\lambda^2 r^2} P_{\text{trans}} \equiv \frac{A_R}{4\pi r^2} P_{\text{trans}} \qquad (4.15)$$

This result consistently shows that the fraction of power received from the isotropic antenna at distance r from the antenna of aperture A_R is simply given by the surface ratio $A_R/S(r)$ where $S(r) = 4\pi r^2$ is the surface of the sphere at that distance. If the receiving antenna is also isotropic, then $A_R = \lambda^2/4\pi$ and the received power is $P_{\text{rec}} = (\lambda/4\pi r)^2 P_{\text{trans}}$. This basic relation provides the power fraction that can be exchanged between two isotropic antennas. When the two antennas have directive gain (G_T and G_R), this relation becomes:

$$P_{\text{rec}} = G_T G_R \left(\frac{\lambda}{4\pi r}\right)^2 P_{\text{trans}} \qquad (4.16)$$

The *transmission loss* experienced by the EM wave between two antennas can be expressed through the decibel ratio:

$$L(\text{dB}) = 10 \log_{10} \frac{P_{\text{rec}}}{P_{\text{trans}}} \qquad (4.17)$$

The dB loss is a negative quantity, since $P_{\text{rec}} < P_{\text{trans}}$. The loss between two isotropic antennas, also referred to as *free-space loss*, is thus given by $L = 10 \log_{10} [(\lambda/4\pi r)^2] = 20 \log_{10}(\lambda/4\pi r)$. This loss is very substantial, considering that the wavelength is generally much shorter than the reception distance, i.e., $\lambda \ll r$, which is also a ground assumption for this far-field analysis. To provide an example, a 1 GHz EM-wave ($\lambda = 30$ cm) experiences over a 1-km distance a loss of 92.5 dB/km, corresponding to an attenuation factor of about $T = 5 \times 10^{-10}$ or $\frac{1}{2}$ billionth. For comparison, the attenuation of optical fibers, is 0.2 dB/km or $T = 0.955$ or 95.5%. Fortunately, antennas can be made directive and some of this loss is "compensated" by the gain product $G_T \times G_R$, which can also be substantial.

The formula in equation (4.17) can also be developed according to

$$L(\text{dB}) = 10 \log_{10} P_{\text{rec}} - 10 \log_{10} P_{\text{trans}} = P_{\text{rec}}(\text{dBW}) - P_{\text{trans}}(\text{dBW}) \qquad (4.18)$$

or

$$P_{\text{rec}}(\text{dBW}) = P_{\text{trans}}(\text{dBW}) + L(\text{dB}) \qquad (4.19)$$

where the *decibel-watt* (*dBW*) is introduced to specify that powers are expressed in reference to 1W. Thus -30 dBW, 0 dBW and $+20$ dBW correspond to powers of 1 mW, 1 W and 100 W, respectively. Since this reference is purely conventional, is also possible to express the relation in *decibel-milliwatts* (dBm), or $P_{\text{rec}}(\text{dBm}) = P_{\text{trans}}(\text{dBm}) + L(\text{dB})$. Thus -30 dBm, 0 dBm and $+30$ dBm correspond to powers of 1 μW, 1 mW and 1 W, respectively. The choice of power reference depends upon the global power range involved, the dBW being more used in radio systems and dBm in optical systems.

The result in equation (4.18) makes it possible to analyze the radio-transmission system in terms of *power budget*. One can also write from equation (4.1p):

$$P_{\text{rec}}(\text{dBW}) = P_{\text{trans}}(\text{dBW}) + G_T(\text{dBi}) + G_R(\text{dBi}) - 20 \log_{10}\left(\frac{4\pi r}{\lambda}\right) \qquad (4.20)$$

which shows that, as in a financial budget, the sum of gains and losses determines the end result. The decibel scale is therefore very practical in this respect, since the power budgeting is just a matter of additions and subtractions. However, the formulas (4.16) and (4.17) can also be used, with decibel conversion only made at the end. As an example of power budget, consider a point-to-point radio system using two dish antennas having the same aperture $A_T = A_R = 1$ m^2. The system operates in the UHF band with $\lambda = 30$ cm ($f = 1$ GHz). At a distance of $r = 1$ km, the transmission loss is, according to equations (4.15) and (4.16): $L = A_T A_R/(\lambda^2 r^2) = 1$ m$^2 \times 1$ m$^2/(0.3$ m $\times 1000$ m$)^2 = 1.1 \times 10^{-5}$, or -49.5 dB. Assuming a transmit power of $P_{\text{trans}} = 2$ W or $+3$ dBW, the received power is $P_{\text{rec}}(\text{dBW}) = +3$dBW $+ L(\text{dB}) = 3 - 49.5 = -46.5$ dBW or 22.4 μW. Such a loss budget may seem relatively high for a directive-antenna system, but even much greater values are commonplace in radio systems, especially those based on satellite relays. Further examples are provided in exercises at the end of the chapter. See more on power budgets in Section 1.5.

The above description concerned the effect of free-space transmission loss, an effect which is exclusively attributable to the spherical dispersion of the EM wave. The propagation medium does not introduce loss, as in the case of interstellar space, made of almost pure vacuum. Our familiar atmosphere may look essentially transparent to the human eye, but is not loss-free for EM waves. As a matter of fact, the existence of humidity, rain, fog, clouds, snow or dust which reduce long-range sight, shows that this is also true for visible light. Fortunately, EM waves are not blocked by heavily perturbed atmospheric conditions. This makes it possible for airplanes or ships having the right radio/radar equipment to navigate relatively safely in rainy or foggy weather.

The effect of *rainfall* on radio waves is two-fold. First there is material absorption from the water droplets. Second, the droplets cause wave *scattering*. The scattering effect corresponds to random deviation of the local wave direction, due to *reflection* and *refraction* in the droplets, which is further randomized by wind and other air turbulence. The estimation of total attenuation from absorption and scattering by rainfall is quite complex. The issue of scattering loss is discussed below. Concerning absorption by rain water, the two parameters to take into account are the atmospheric humidity (or rain-water density) and the transmission distance, r. The transmission can be modeled as an exponential law $T = \exp(-\gamma r)$, corresponding to a loss $L(\mathrm{dB}) = -\gamma\,(\mathrm{dB/km}) \times r\,(\mathrm{km})$. The attenuation coefficient, γ, depends upon two factors: (a) the E-wave polarization, i.e., linear or circular Desurvire, 2004; and (b) the elevation angle θ of the EM wave with respect to the horizontal direction. Yet, the attenuation coefficient is given by a simple closed-form relation [after ITU-R, PN838, 1994]

$$\gamma\,(\mathrm{dB/km}) = \beta R^\alpha \tag{4.21}$$

where R is the *rainfall rate* (in mm/hour) and

$$\alpha = \frac{\beta_H \alpha_H + \beta_V \alpha_V \pm (\beta_H \alpha_H - \beta_V \alpha_V)\cos^2\theta}{2\beta} \tag{4.22}$$

$$\beta = \frac{\beta_H + \beta_V \pm (\beta_H - \beta_V)\cos^2\theta}{2} \tag{4.23}$$

In the last two definitions, the signs correspond to linear polarizations parallel to the ground plane $(+)$ or perpendicular to it $(-)$. The case of circular polarization is given by omitting the terms which have the $\pm$ factor. The coefficients α, β are listed in Table 4.2 for the frequency range 1–40 GHz. As Table 4.2 shows, the coefficients have a small frequency dependence. The effect of polarization, which distinguishes the coefficients with H and V subscripts (for horizontal and vertical)

TABLE 4.2 Coefficients for calculating rainfall attenuation of radio waves

Frequency (GHz)	α_H	α_V	β_H	β_V
1	0.912	0.880	3.87×10^{-5}	3.52×10^{-5}
5	1.214	1.170	0.0012	0.0010
10	1.276	1.264	0.010	0.008
20	1.099	1.065	0.075	0.069
40	0.939	0.929	0.350	0.310

[After ITU-R, PN-838, 1994]

comes only as a small correction. Consider for simplicity the case of a point-to-point radio transmission on the Earth's surface ($\theta = 0$), with the E-wave being polarized in the linear/horizontal direction. According to the above expressions, we have $\alpha = \alpha_H$ and $\beta = \beta_H$. The attenuation coefficient is therefore $\gamma(\text{dB/km}) = \beta_H R^{\alpha_H}$. Assume a heavy rainfall with rate $R = 100$ mm/hour. It is easily calculated from the values provided in Table 4.2 that the attenuation is 2.5×10^{-3} dB/km for 1-GHz signals, while it is 3.5 dB/km for 10-GHz signals. Thus, the 1-GHz waves experience very weak attenuation while 10-GHz waves are heavily absorbed, which illustrates the fact that, under rain conditions, atmospheric transmission is frequency-selective. Note from the table that the exponent α_X ($X = $ H,V) only weakly varies (with a peak near 10 GHz), while the multiplying factor β_X strongly increases with frequency. This feature can affect satellite communications in the Ka-Ku bands (see Figure 4.6), even if the rainy atmosphere has a limited thickness (referred to as "effective rain height"), corresponding to reduced absorbing distances. See further examples in the Exercises at the end of this chapter. At frequencies above 10 GHz, two absorption bands must be considered. The first is the water-vapor resonance at 21 GHz, and the second is the oxygen resonance at 60 GHz.

We have considered so far the case of *line-of-sight* (LOS) radio systems, assuming the atmosphere to have uniform properties. In reality, the Earth's atmosphere is made of concentric *layers* having different air densities and refractive-index characteristics. Because of this, the atmospheric upper layers act as a stack of spherical mirrors, into which the radio waves (especially in the HF or SW band) can reflect or bend, making them bounce back to the Earth, as we shall describe.

For radio-wave propagation, the two main atmospheric layers of interest are called the *troposphere* and the *ionosphere*. Figure 4.19 shows their arrangement and vertical scales. The troposphere is defined as our environmental air layer,

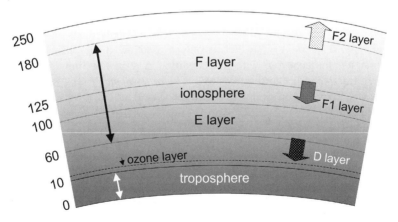

FIGURE 4.19 High-altitude structure of the Earth's atmosphere, showing troposphere, ozonosphere, and ionosphere (made of two layers E and F). The arrows show the upward and downward expansion of the ionosphere layers during 6A.M.–6P.M. daytime, forming F1–F2 and D sublayers. Altitude is shown in miles.

ranging from sea level to 10 miles (16 km, or approximately twice the Himalayas tops). The ionosphere, which ranges from 60 miles to 180 miles, is made of two distinct layers called E and F. As an interesting feature, the ionosphere splits or expands during daytime (6A.M.–6P.M.) because of the sun's radiation effect. The F layer splits into F1 (down 100 miles) and F2 (up 250 miles), while a new D layer (down 20 miles) appears under the E-layer. To be complete, the well-known ozone layer (or ozonosphere containing O_3 molecules) stands just between the D layer and the troposphere, i.e., at altitudes of 15–20 miles, as seen in the figure.

As its name indicates, the origin of the ionosphere is an effect of *ionization* of the air atoms and molecules from both cosmic and sun rays (UV to visible). In the upper atmosphere, the total solar radiation intensity corresponds to $1.37 \, kW/m^2$, which is a value referred to as the *solar constant*. The atoms and molecules found in the ionosphere are essentially H, He, NO, N_2 and O_2. The ionization results in freeing one electron from the atom or the molecule, creating a positively charged *ion*. As a result of the process, there is a *plasma* of free electrons and positive ions: H^+ and He^+ at very high altitudes (625 miles), O^+ in the F2 layer, and NO^+, N_2^+, O_2^+ in the E–D layers. This ionization results in total plasma densities in the range of $N = 10^9 - 10^{12}$ per cubic meter (m^{-3}).

At night time, the plasma density exhibits a single peak of about $N = 8 \times 10^{10} m^{-3}$ in the center of the F layer. During daytime, the F-layer splits into F1 and F2 layers, each showing equal densities generally about $2 \times 10^{11} m^{-3}$. The finite plasma densities in these layers result in a small change of their *refractive index*. For frequencies above 1 MHz, it can be shown that the index is given by the approximate formula:

$$n = \sqrt{1 - \frac{Ne^2}{\varepsilon_0 m_e (2\pi f)^2}} \equiv \sqrt{1 - \frac{80.5 \times N_{m^{-3}}}{f_{Hz}^2}} \equiv \sqrt{1 - \Delta} \qquad (4.24)$$

where $e = 1.6 \times 10^{-19}$, C is the electric charge, $m_e = 9.11 \times 10^{-31}$ kg the electron's rest mass, and $\varepsilon_0 = 8.85 \times 10^{-12}$ F/m is the vacuum permittivity. Here, we shall not explain the meaning of these parameters, and only focus on the effective index change, $\Delta = 80.5 N/f^2$, which is only a function of the plasma density N, and the signal frequency, f, with the condition $\Delta \leq 1$. The limit $\Delta = 1$, which nulls the refractive index corresponds to a *critical frequency* $f_c = \sqrt{80.5 \times N}$. Frequencies well above f_c correspond to conditions $\Delta \approx 0$ or $n \approx 1$, for which the ionosphere is essentially transparent or nonreflective. Note that the approximated model does not apply for $f < f_c$.[1]

---

[1]By definition, the wave's *phase velocity* is given by $v_p = c/n$, meaning here ($n \leq 1$) that the phase velocity is greater than the speed of light in vacuum (c). The physical wave, however, has different frequency components forming a wave packet. The packet speed is called *group velocity* (v_g). It can be shown that, in the present case, the group velocity is given by the relation

$$v_g = \frac{c}{n + \frac{4\pi^2}{n}\Delta} = \frac{nc}{1 + (4\pi^2 - 1)\Delta} \leq nc \leq c$$

showing that the wave packet travels at a velocity lower than the speed of light in vacuum.

We analyze next how the radio waves reflect onto the ionosphere, considering a single, thin-layer model. With an incidence angle ϕ with respect to the ionosphere layer ($\phi = 0$ for normal or vertical incidence), the E-field reflection coefficient is given by:

$$R_H = \left| \frac{\cos \phi - \sqrt{1 - \Delta - \sin^2 \phi}}{\cos \phi + \sqrt{1 - \Delta - \sin^2 \phi}} \right| \tag{4.25}$$

$$R_V = \left| \frac{n^2 \cos \phi - \sqrt{1 - \Delta - \sin^2 \phi}}{n^2 \cos \phi + \sqrt{1 - \Delta - \sin^2 \phi}} \right| \tag{4.26}$$

where, as before, the subscripts H and V stand for horizontal and vertical polarizations, respectively. Note that the formulas are valid under the condition $\Delta < 1 - \sin^2 \phi = \cos^2 \phi$.

It is immediately checked from the above that both reflection coefficients vanish for $\Delta \approx 0$, corresponding to layers where no ionization occurs. At vertical incidence ($\phi = 0$), the two coefficients are minimum and equal to $R = (n - 1)/(n + 1)$, or with the approximation $\sqrt{1 - \Delta} \approx 1 - \Delta/2$, $R = \Delta/4$. Let us consider two examples just above the critical frequency f_c: (a) a peak density of $N = 2 \times 10^{11} \mathrm{m}^{-3}$ (daytime peak, $f_c = 4$ MHz) and a frequency of $f = 10$ MHz (upper-edge of SW band); and (b) a peak density of $N = 8 \times 10^{10} \mathrm{m}^{-3}$ (night-time peak, $f_c = 2.5$ Mz) and a frequency of $f = 5$ MHz (lower-edge of SW band). In the first case (a), we obtain $\Delta = 0.16$, corresponding to a reflection coefficient of $R = 4\%$. In second case (b), we obtain $\Delta = 0.25$, corresponding to a reflection of $R = 6\%$, close to the previous value. As small as the reflection effect may appear, it is sufficient to make it possible for the radio signals to be detected over ultra-long distances, after bouncing upon the ionosphere. The reflection can be increased to the maximum value of 100% by using the effect of total internal reflection. This effect is illustrated in Figure 4.20 for different incidence angles ϕ. As we have seen, the reflection is maximum in the vertical

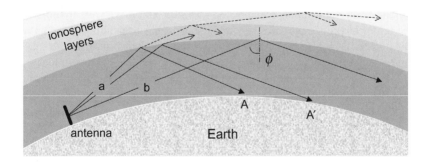

FIGURE 4.20 Effect of reflection from the ionosphere layers, causing radio waves to bounce back to the Earth. Beams directed at the largest incidences (ϕ) propagate the farthest ($A \rightarrow A'$), although with more atmospheric absorption (beams a). Beams with multiple layer reflections follow a curved path (dashed line). There exists a critical angle for total internal reflection, where 100% of the beam is reflected to the Earth (beam b).

incidence, but this configuration is useless for radio communications. Rather, the incidence should be maximized so that the wave is reflected towards the highest point possible. Because of the Earth's curvature, the highest incidences are limited to the angle range $\phi_{max} = 74-83°$, depending upon the height of the ionosphere layer involved (see Exercises at the end of the chapter). As shown by the exercise, reflection from the top ionosphere layer (approx. 250 km height) makes it possible to reach remote points on Earth as distant as 3,500 km. Even more distant points can be reached by the effect of multiple back-and-forth reflection (or "multiple-hop" propagation). In the process, the radio waves become severely attenuated, since the reflection coefficients (ionosphere and ground) are usually smaller than 100% and the effective absorption distances are significantly increased. Yet, because of the high power budgets that radio systems are capable of, SW transmission all the way to the Antipodes is possible. The principle was first demonstrated through an experiment by G Ferrié in 1902, which involved radio stations from Paris's Eiffel Tower and the island of Martinique, located 7,200 km west in the Caribbeans.

We have previously seen that the critical frequency f_c defined a limit at vertical incidence beyond which the ionosphere becomes rapidly transparent ($\Delta \propto 1/f^2$). The condition of total internal reflection for any incidence $\phi \neq 0$ is $1 \times \sin \phi = n \times \sin 90°$, or using equation (4.24), $\sin \phi = \sqrt{1 - \Delta}$, or $\Delta = \cos^2 \phi$. Such a condition determines the *maximum usable frequency* (MUF) for ionospheric transmission. The MUF is thus defined as $f_{MUF} = f_c / \cos \phi_{max} = \sqrt{80.5 \times N_{m^{-3}}} / \cos \phi_{max}$. For $\phi_{max} = 74°$, we have $f_{MUF} \approx 32.5\sqrt{N_{m^{-3}}}$. If one uses the values for night-time and day-time plasma densities, the MUF is found to be in the range $f_{MUF} = 9.2-14.5$ MHz, which falls into the HF or SW band. Note that the MUF is inherently time-varying (night and day), and is also dependent upon the intensity of solar activity (sunspot cycles) which controls the ionosphere's plasma density.

While the MUF is typically located in the SW/HF range (3–30 MHz), radio-waves at higher VHF frequencies (30–300 MHz) can also experience ionosphere reflection. The effect is then due to *meteor trails*. These meteors are small particles constantly bombarding the Earth at a rate of 10^{10} per day, typically, representing 1,000 kg daily average of pulverized mass. These particles have enough energy to ionize the atmosphere upon reaching 80–120-km altitudes. Meteor trails caused by the fall of large-size particles produce greater ionization, although on an unpredictable, sporadic basis and over relatively limited regions of space. Remarkably enough, the corresponding plasma/electron densities can reach values in the range $10^{10}-10^{14}$ m^{-3} (under-dense trail) and up to $10^{14}-10^{18}$ m^{-3} (over-dense trail). It can be shown that the path absorption varies as the inverse-cube of frequency. This feature limits applications to 30–50 MHz, representing a power budget of about 200 dB $\pm$ 30 dB for distances in the 500–1,000-km range.

Another effect causing wave reflection is known as *tropospheric scattering*. Generally, EM-wave scattering is caused by random variations of the refractive index. The irregularities of the troposphere thus make it possible to transmit radio waves over the horizon at frequencies above 30 MHz and up to the 10-GHz range. Because of the random nature of scattering, however, the transmission quality fluctuates or fades over time and signals rapidly vanish with over-horizon distances. Typical systems operate over 100–300 km at 0.8–5 GHz.

Most generally, the radio waves that are bent by the atmosphere, or are reflected onto some atmospheric layer, are referred to as *skywaves*. As we have seen, sky-waves are thus inherently able to propagate above the horizon, as opposed to LOS waves, which are blocked by the Earth's curvature. Figure 4.20 illustrates the concept of *ray bending* through a stack of atmospheric layers, which is caused by successive deflections at the layer interfaces. This effect is properly called refraction. If the refractive-index is progressively changing within a given layer (i.e., decreasing with altitude), the refraction is continuous and the path becomes a curve (see more on this topic in Chapter 2 of Desurvire, 2004). This is the case of the troposphere, which exhibits a continuous permittivity decrease with altitude at a rate of $6 \times 10^{-4}/m$. At ground level, the (relative) permittivity is $\varepsilon/\varepsilon_0 = 1.0006$, corresponding to a refractive index of $n = \sqrt{\varepsilon/\varepsilon_0} = 1.0003$. The decrease-rate value means that the permittivity is approximately unity at a 10-km height, corresponding to the index $n \approx 1$. Because of this continuous change of refractive index, EM waves launched with angles close to the horizontal plane can be sent back to the Earth, resulting in over-the-horizon propagation. This effect is referred to as a *surface wave*, in contrast with previously-described sky-waves, which are reflected or refracted from high-altitude layers. Both surface and LOS waves are called ground waves, while tropospheric-scatter and ionospheric-refraction waves are called skywaves. The different types of ground wave and skywave which have been described in this subsection are illustrated in Figure 4.21.

4.1.5 Multipath Interference

In Section 4.1.4 we saw that radio waves can propagate using different types of path. While radio beams can be made unidirectional through proper antenna design, the path is not defined by a unique propagation line. Rather, it is defined as an angular region of space within which the E-field is above certain intensity level, as described

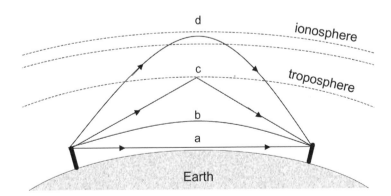

FIGURE 4.21 Different types of radio-waves, according to path: (a) line-of-sight or straight, (b) surface-wave, (c) tropospheric-scatter, and (d) ionospheric refraction. Cases (a)–(b) correspond to ground waves, and cases (c)–(d) to skywaves.

by the beam radiation pattern. Any object such as a mountain or a building found within this region reflects the wave, which leads to what is called *multipath interference*, as illustrated in Figure 4.22.

The radio waves coming from multiple paths have traveled over different distances and therefore, their phases are different. The resulting superposition can be either constructive (in-phase waves), destructive (out-of-phase waves) or anything in-between. In particular, waves can be scattered by objects such as trees and clouds, resulting in random phase changes, as opposed to deterministic. The net result is an interference pattern which depends upon the type of reflection path and scattering conditions. It is a property that the reflected EM wave has a phase opposite to the incident wave, corresponding to a π or $180°$ phase shift. Thus, direct and reflected waves that are mutually delayed by an even multiple of half-wavelengths interfere destructively, while for an odd multiple of half-wavelengths the interference is constructive. With respect to the straight path, all possible reflected paths that yield a half-wavelength delay define an elliptic region, which is referred to as a *Fresnel zone*. In three-dimensional space, the Fresnel zone is an ellipsoid. The odd-integer number n of half-wavelengths define a set of as many coaxial ellipsoids. In the plane transverse to the LOS, the Fresnel zones form a set of concentric circles.

In order to avoid destructive interference due to multipath propagation, the radio wave should propagate well within the first Fresnel zone. The radial distance between the LOS and the first Fresnel-zone boundary, can be approximated through the parabolic formula (see Exercises at the end of the chapter):

$$H \approx \sqrt{n\lambda \frac{d'(d - d')}{d}} \qquad (4.27)$$

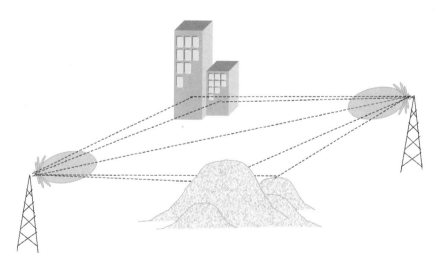

FIGURE 4.22 Effect of multipath interference of radio waves due to reflections on obstacles such as buildings or mountains.

where d is the total LOS path length, d' the distance from the origin, and $n = 1,3,5...$ is an odd integer. The first Fresnel radius ($n = 1$) at the mid-point of the link is therefore $H_{max} = H(d/2) = \sqrt{\lambda d/2}$. This radius defines a cone angle 2α within which no reflections from obstacles should occur. This angle is given by the condition $\tan(\alpha) = H_{max}/(d/2) = \sqrt{2\lambda/d}$. For the ratios $\lambda/d = 0.1$, 0.01 and 0.01, the cone angle is found to be $2\alpha = 48°$, $16°$ and $5°$, respectively. More detailed calculations show that no substantial interference occurs if the beam is clear from obstacles within 0.6 times the first Fresnel radius. This safety distance from reflecting obstacles is referred to as the beam *clearance*. Examples of clearances required for GHz communications are provided in an Exercises at the end of the chapter. Note that clearance must be absolutely respected only in the case of highly reflecting obstacles. If the reflection is weak, the resulting interference effect is also weak. However, the sum of all possible reflections (and scattering) collected within a given beam angle for all Fresnel orders can result in substantial interference, corresponding to spatial fluctuations of the signal quality and unwanted noise background. On the other hand, multipath propagation is also advantageous in areas having many obstacles within the LOS, such as towns or mountain zones. This is because reflections on buildings or mountains can create at least a possible indirect path for the radio signal all the way to the receiving antenna, while the direct LOS is blocked by these obstacles.

As we have seen, multipath interference results in spatial variations of the signal intensity level, due to the fact that waves combine constructively or destructively at different locations. The E-field resulting from this interference exhibits local holes and peaks with various depths or heights. For a fixed-antenna system, one can optimize the received signal intensity by moving the antenna around to find a spot of constructive interference (as we familiarly do with home-TV antennas). In the case of a mobile communication system, the distance between the mobile antenna and the (fixed) base-station antenna changes over time, because the person holding the handset can be inside a car, a train or an airplane. As the different Fresnel zones are traversed, the signal intensity moves up and down with a half-wavelength periodicity. This effect, which is illustrated in Figure 4.23, is referred

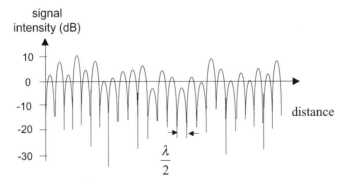

FIGURE 4.23 Effect of multipath or Rayleigh fading, showing received power fluctuations with distance traveled by mobile antenna. The vertical scale is in dB relative to the mean power.

to as *Rayleigh fading*, or *multipath fading* (Rayleigh coming from the name of the corresponding statistical distribution). The frequency of this noise depends upon both wavelength and the mobile antenna's speed (see Exercises at the end of the chapter). The fading effect is more important at ground level than in cruising airplanes, since there are more ray-path possibilities, especially in dense urban areas with tall buildings. Mobile phones are in fact designed to dynamically compensate multipath fading, bringing the signal to near-constant intensity, regardless of the motion speed.

Two other side-effects associated with multipath reflection are *Doppler shifting* and *delay spread*. The first effect concerns mobile radio systems. The Doppler shift is a change of frequency caused by the motion of the observer and his antenna. The received frequency increases when the observer moves towards from the wave source, and decreases in the opposite case. One refers to positive (upwards) shifts or negative (downwards) shifts as "blue" or "red" shifts, respectively. Here, the source is any point onto which the wave is reflected. Thus, the frequency seen by a mobile antenna moving towards a reflection point is blue-shifted, while it is red-shifted in the opposite case. The combination of multipath reflection results in random, time-varying frequency modulation. If the antenna is moving at speed V, the relative Doppler shift is defined by $\Delta f/f = \pm(V/c)$. Considering cars or trains, the ratio V/c remains extremely small (about 10^{-7} at 100 km/h speed), resulting in negligible Doppler shifts. The second effect, which concerns pulsed radio signals, is *delay spread*. Multiple reflections of a radio pulse result in random time delays (or arrival times) at the antenna point. The delays are both a function of path distance and antenna speed. For a single pulse, the result is the detection of multiple pulse echoes. For a stream of pulses, such as in digital radio, the multiple echoes randomly interfere with the incoming LOS signal, resulting in what is called *intersymbol interference* (ISI) or ISI noise.

Radio signals belonging to the microwave range (VHF-UHF-SHF) can be carried over extended distances by use of periodic *repeater stations*. One is already familiar with TV relays, these huge towers holding bundles of parabolic antennas, which usually stand on mountaintops or in the middle of the fields. Repeater stations are generally disposed every 25–50 km of LOS with the basic function to detect, reamplify and re-emit the incoming radio signals. The transmitting and receiving antennas must be highly directive for three reasons: (a) to optimize the power budget between successive stations (effect of antenna gain); (b) to avoid capturing radio signals coming from other radio transmitters (minimum radiation pattern in directions other than LOS); and (c) to avoid interference, or signal *overshoot*, between received and retransmitted signal (use of different LOS). The principle of a repeated radio link is illustrated in Figure 4.24. This figure illustrates how a slight change of LOS from one pair of stations to another makes it possible to avoid signal overshoot. It also shows that the directivity of the receiving antennas prevents capturing parasitic signals transmitted from other radio systems, as they fall into their weak pattern side-lobes. Such a zig-zag path arrangement makes it possible for several radio systems to share the ground airspace without interfering with each other, while using carrier frequencies in the same wavebands.

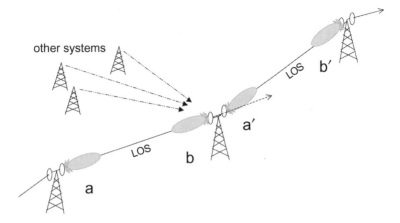

FIGURE 4.24 Principle of a repeater-based radio system. The repeater stations are disposed along a zig-zag path so that each pair (a, b) of transmitting (a) and receiving (b) antennas uses a different line of sight (LOS). The dashed arrow shows the signal overshoot from transmitter (a) into transmitter (a′). The dotted-dash arrows show interfering radio signals into receiver antenna (b), as generated from other emitting-antenna systems.

4.1.6 Effective Noise Temperature, Noise Figure and CNR

The concept of *power budget*, which defines the relation between input and output powers for transmitted radio signals, has been introduced in Section 4.1.5. We have also described different causes of signal degradation due to interference effects. Here, we shall introduce the general definition of *noise*, as associated with radio waves and microwave/electronic circuits. This makes it possible to define two essential parameters, the *signal-to-noise ratio* (SNR) and *the carrier-to-noise ratio* (CNR). Both SNR and CNR determine the overall radio-signal *quality*, a concept we are most familiar and demanding with, from our experience of tuning to AM/FM radio stations or receiving VHF-UHF TV channels. The background Section 1.4, can be useful for reading this subsection.

We have seen in Chapter 1 that "noise" can be defined as a random intensity deviation from the mean signal-intensity value $\langle I \rangle$. Most noise-generating processes are characterized by Gaussian probability distributions. One refers in this case to *Gaussian noise*. In Gaussian noise, the *standard deviation* is defined by the quantity $\sigma = \sqrt{\langle I^2 \rangle - \langle I \rangle^2}$, with σ^2 being the noise *variance*. The physicists Boltzmann and Planck showed that for hot sources (or antennas) radiating EM waves, the noise variance is related to the *absolute temperature T* (in Kelvin units), according to:

$$\sigma_{th}^2 = k_B T \frac{\dfrac{hf}{k_B T}}{\exp\left(\frac{hf}{k_B T}\right) - 1} \equiv n_{th} hf \tag{4.28}$$

which is known as *Planck's law* or *formula*. In this formula, $k_B = 1.38 \times 10^{-23}$ Joules/Kelvin, is the *Boltzmann's constant*, and $h = 6.62 \times 10^{-34}$ joules-second is the *Planck's constant*. The term n_{th} is called the "mean occupation number," as further explained below. This definition corresponds to a noise *power density*, as expressed in units of Watt/Hz or Joules (1 W = 1 J/s), which is energy. The total noise power $P_N(W)$ falling into a narrow radio band of spectral width Δf is simply given by the product $P_N = \sigma^2 \Delta f$.

Let's analyze closely the meaning and properties associated with Planck's law. From the constant units, we see that the two factors hf, and $k_B T$ both represent energies. The frequency-dependent one, hf, represents the fundamental *quantum of energy* associated with all EM waves. The word "quantum" means that EM waves can transport energy only by discrete jumps, i.e., hf, $2hf$, $3hf \ldots n \times hf$, etc... with n being an integer. These energy quanta are also called *photons*. While the term "photon" is generally associated with lightwave communications (Chapter 2 in Desurvire, 2004), giving the name "photonics," the concept actually applies to all EM waves, from radio bands to IR and visible light, all the way to UV, X-, and gamma rays. The term n_{th} thus corresponds to the average number of photons present in the EM wave radiated by a hot source at temperature T. Consistently, the EM energy is $n_{th}hf$, and the EM power is $n_{th}hf\Delta f$.

Since the photon energy increases in proportion with frequency, radio waves are the least energetic of the EM spectrum. The interesting feature of Planck's law is that the noise frequency dependence is given by the dimensionless factor $hf/k_B T$. Thus we can distinguish two categories of EM wave which correspond to the conditions of either $hf/k_B T \ll 1$ (low frequencies) or $hf/k_B T \gg 1$ (high frequencies). In the first case, using in equation (4.28) the approximation $\exp(x) \approx 1 + x$ for $x \ll 1$, we find $\sigma^2 \approx k_B T$. This result shows that at low frequencies, the EM noise is basically frequency-independent. Interestingly, the noise vanishes at absolute-zero temperature, $T = 0$. As we shall see below, because of the Big Bang origin, the universe's temperature is not zero but slightly above zero (2.7 K). This means that our universe environment is pervaded by a finite radio noise background, as opposed to being absolutely quiet or "silent." The frequency-independent noise $\sigma^2 \approx k_B T$ is commonly referred to as *thermal noise*. It corresponds to the EM noise energy radiated by stars in the low-frequency end of the EM spectrum. Figure 4.25 shows a plot of Planck's law calculated for the temperature $T = 290$ K (or +17 °C) called *room temperature* (another possible convention is 300 K or +27 °C). It is seen that at the higher-frequency end of the radio spectrum (EHF), the noise rapidly drops from the thermal limit $\sigma^2 \approx k_B T$. In the high-frequency regime ($hf/k_B T \gg 1$), corresponding to near-infrared and beyond, the noise power density exponentially decays according to $\sigma^2 \approx hf \times \exp(-hf/k_B T)$. It can be shown that the number of radiation modes at frequency f that can be excited in all directions of the universe is $M = 8\pi f^2/c^3$. At high frequencies, the total noise power is thus $M\sigma^2 = (8\pi hf^3/c^3) \times \exp(-hf/k_B T)$. Since this function exponentially vanishes, the total EM noise power, as integrated from radio-wave frequencies to infinite frequencies, is therefore finite.

The total radiated noise remains finite because as we have seen, the mean occupation number n_{th} is not uniformly distributed over the EM spectrum. The word "occupation number" refers to the extent to which an EM wave is "occupied" or "excited" as a radiation mode, which is measured in an average number of quanta

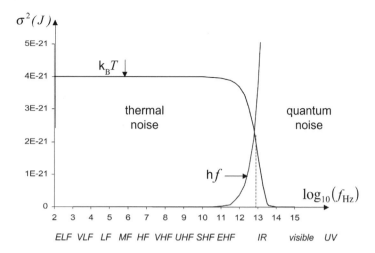

FIGURE 4.25 Noise power density σ^2 ($J = $ W/Hz) at $T = 290$ K, or room temperature, according to Planck's law, showing thermal noise limit ($k_B T$) at radio frequencies (ELF-EHF). The quantum noise $\sigma^2 = hf$, is seen to overcome thermal noise at frequencies above 6 THz.

or photons. Before the demonstration of Planck's law, it was believed that such an occupation number was uniformly distributed, meaning that the noise power in any possible radiation mode would be $\sigma^2 = k_B T$, yielding at frequency f a total noise power of $M\sigma^2 = 8\pi f^2 k_B T/c^3$. When integrating the contributions from the entire spectrum ($f = 0 \rightarrow \infty$), this hypothesis lead to the absurdity that the total noise power radiated by a hot source must be infinite! Planck's law, which is based upon the concept of energy quantization of EM waves, came up in 1900 to solve this embarrassing contradiction, initially called by physicists the "ultraviolet catastrophe." The reader is now fully aware of the importance of quantum theory to explain the physics of EM waves, which were described earlier in this section as "classical" or nonquantized EM oscillation.

Quantum theory also shows that an EM wave at frequency f is always associated with an intrinsic background noise which has the energy of one-half a photon, i.e., $\sigma_q^2 = hf/2$. This noise is referred to as *vacuum noise*, reflecting the concept that "vacuum" (as defined by absence of any matter) is not absolutely quiet. Since this vacuum noise increases linearly with frequency, its full spectral power is therefore infinite. A conceptual difficulty is that while vacuum noise generates extra uncertainty in EM power, it *cannot* be extracted by physical devices as usable energy. By means of a simple explanation, it is not possible to extract any energy from a wave occupied by one-half a photon. Taking into account the vacuum noise, we must then replace Planck's law by the new *Callen and Welton* law:

$$\sigma_{tot}^2 = \sigma_{th}^2 + \sigma_q^2 = \frac{hf}{\exp\left(\dfrac{hf}{k_B T}\right) - 1} + \frac{1}{2} = \left(n_{th} + \frac{1}{2}\right)hf \qquad (4.29)$$

Consistently, the additional quantity $hf/2$ (or more generally, hf) is called "quantum noise," as opposed to the previously described "thermal noise," $k_B T$. Figure 4.25 shows the plot of quantum noise. It is seen that quantum noise is equal to thermal noise ($hf = k_B T$) at a frequency near 6 THz, corresponding to the boundary limit between far-IR and radio, or a wavelength of $\lambda = c/f = 50$ μm. The EM spectrum is thus divided into two regions, one dominated by thermal noise (radio waves) and the other by quantum noise (IR, optical and higher-frequency waves).

The previous description leads one to the concept of *effective temperature*, T_{eff} (also called *noise temperature*) which applies to antennas or any microwave/radio devices. Such devices are thus characterized by a noise power density $\sigma^2 = k_B T_{eff}$, where

$$T_{eff}(K) = \frac{hf}{k_B} \left(\frac{1}{\exp\left(\dfrac{hf}{k_B T}\right) - 1} + \frac{1}{2} \right) \tag{4.30}$$

Actually, the above definition of effective/noise temperature, based upon the Callen and Welton law, must be completed by three other terms. These terms account for the EM-noise contributions of our "Milky Way" galaxy ($T_1 = 7 \times 10^{26}/f^3$), of the cosmic background radiation produced by the Big Bang ($T_2 = 2.726$ K), and of atmospheric effects due to absorption from water vapor and oxygen ($T_3 = 3.33 \times 10^{-27} f^3$). Consistently, the total effective/noise temperature is given by

$$T_{eff}(K) = \frac{hf}{k_B} \left(\frac{1}{\exp\left(\dfrac{hf}{k_B T}\right) - 1} + \frac{1}{2} \right) + \frac{7 \times 10^{26}}{f^3} + 2.726 + 3.33 \times 10^{-27} f^3 \tag{4.31}$$

Plots of the effective temperature vs. frequency are shown in Figure 4.26 for $T = 2.72$ K (space temperature) and $T = 290$ K (room temperature). The effective temperature given by the Callen–Welton definition plus 2.72 K (cosmic background radiation), also shown for $T = 2.72$ K, defines an absolute noise-temperature limit. As previously discussed, this limit is only weakly dependent upon frequency, at least in the "thermal regime," up to $f = 100$ GHz. It is observed that the other contributions (galaxy and atmospheric noise) significantly increase the noise temperature at both high and low frequencies, with a minimum near $f = 1$ GHz. At room temperature, the minimum is given by the thermal limit $T_{eff} \approx (k_B T)/k_B = T = 290$ K. The noise temperature is seen to be nearly constant over the frequency range $f = 250$ MHz–2.5 GHz.

Consider next a two-port microwave device or transmission line which is characterized by a power transmittance G. It can be a lossy or amplifying device, so there is no condition on the value of the transmittance G. We define the *signal-to-noise ratio* (SNR) as the ratio of signal power (P) to the noise power ($\sigma^2 \Delta f$) in bandwidth Δf. At the input of the transmission line, the SNR is thus $SNR_{in} = P_{in}/(\sigma_{in}^2 \Delta f)$, while at the output it is $SNR_{out} = P_{out}/\sigma_{out}^2 \Delta f$. By definition,

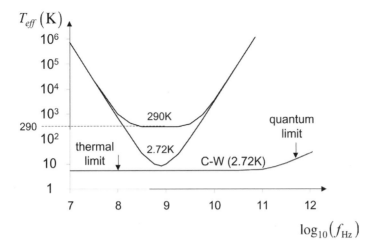

FIGURE 4.26 Effective noise temperature as a function of frequency, according to the Callen and Welton (C–W) law for physical temperature of 2.72 K (bottom), and to the full definition at 2.72 K (middle) and 290 K (top), respectively.

the transmission line's *noise figure* (NF) is the ratio:

$$F = \frac{\text{SNR}_{\text{in}}}{\text{SNR}_{\text{out}}} \tag{4.32}$$

If we substitute the expressions for SNR_{in} and SNR_{out} into the above, use the previous definition based upon noise temperature $\sigma_{in}^2 \approx k_B T$, and $G = P_{\text{out}}/P_{\text{in}}$, we get explicitly:

$$F = \frac{P_{\text{in}}}{k_B T \Delta f} \frac{\sigma_{\text{out}}^2 \Delta f}{P_{\text{out}}} = \frac{\sigma_{\text{out}}^2}{G k_B T} \tag{4.33}$$

We can further simplify this NF expression by expressing the output noise as $\sigma_{\text{out}}^2 = G(\sigma_{\text{in}}^2 + \sigma_{\text{eff}}^2)$, where $G\sigma_{\text{eff}}^2$ is the intrinsic or effective noise introduced by the transmission line. We can also associate an effective temperature to this noise, according to $T_{\text{eff}} = \sigma_{\text{eff}}^2/k_B$, which gives $\sigma_{\text{out}}^2 = G k_B(T + T_{\text{eff}})$. Substituting this new definition into the previous NF expression we finally get:

$$F = 1 + \frac{T_{\text{eff}}}{T} \tag{4.34}$$

It is seen that in all cases we have $F > 1$, meaning that the SNR is always degraded to some extent, except at the limit of a purely transparent or passive system where $T_{\text{eff}} \approx 0$. For microwave devices whose effective temperature, T_{eff}, is equivalent to the physical temperature, T, the noise figure is $F = 2$ or 3 dB. For microwave amplifiers with high gains, the NF is typically between 3 dB and 10 dB. For lossy transmission lines having a power loss $\varepsilon = P_{\text{out}}/P_{\text{in}} < 1$, it can be shown that the NF is simply given by $F = 1/\varepsilon$. The effective temperature of a lossy transmission line is therefore $T_{eff} = T(F - 1) = T(1/\varepsilon - 1)$, which is a result we shall use below. Note that $T_{eff} \approx 0$ for an ideal, loss-free transmission line ($\varepsilon \approx 1$).

For further description of the NF properties, including that of *lightwave/optical* amplifiers, and the *cascading NF formula* valid for both radio and lightwaves, the reader might refer to Section 1.5.

To complete this introductory subsection concerning noise in radio systems, it is useful to mention the definition of *carrier-to-noise ratio*, or CNR, as an alternative to the SNR. Basically, the CNR is the same concept as applied to *analog* signals (i.e., with AM, FM, or PM codings). By convention, a the difference with SNR is that in CNR the noise is expressed in terms of power density, i.e., $k_B T_{eff}$ instead of $k_B T_{eff} \Delta f$. Since the NF is a SNR-to-SNR ratio, the concepts of NF and effective temperature also apply to definitions involving CNRs. For radio-communication systems, we can therefore use either SNR or CNR, as described next, while being careful with the fact that SNR is dimensionless but CNR has the dimension of bandwidth (Hz).

For clarity, we shall consider first the (dimensionless) SNR. Assume that the transmitting and receiving antennas have apertures A_T, A_R, are separated with a distance r and use the carrier wavelength λ. According to the Friis formula (equation (4.14)), the received (analog/digital) signal power is $P_{rec} = (A_T A_R / \lambda^2 r^2) P_{trans}$, where P_{trans} is the transmitted power. If the transmitter bandwidth (Δf_T) and the receiver bandwidth (Δf_R) are mis-matched, the result should be multiplied by the ratio $\Delta f_R / \Delta f_T$. For simplicity, we consider the matched-bandwidth case, i.e., $\Delta f_T = \Delta f_R \equiv \Delta f$. The received noise power is then $\sigma^2 = r k_B T_S \Delta f$, where T_S is the effective *receiving-system* temperature, which we shall specify below. We can thus express the SNR in the form:

$$\text{SNR} = \frac{P_{trans}}{k_B T_S \Delta f} \frac{A_T A_R}{\lambda^2 r^2} \tag{4.35}$$

or equivalently

$$\text{SNR} = \frac{P_{trans}}{k_B T_S \Delta f} G_T G_R \left(\frac{\lambda}{4\pi r}\right)^2 \tag{4.36}$$

where G_T and G_R are the transmitting and receiving antenna gains (see earlier subsection).

We must now define the receiving-system temperature. As illustrated in Figure 4.18, the receiving system is made of a receiving antenna, a transmission line and the receiver itself, which includes a preamplification stage. To each of these block elements, we can associate an effective temperature, namely: T_{ra} (receiving antenna), T_{line} (lossy transmission line) and T_{rec} (receiver with preamplifier). Concerning the antenna temperature, the actual calculation is somewhat complex. It must take into account the angular distribution of the source beams (including parasitic ones such as from the Milky Way) and the radiation pattern of the antenna itself. Concerning the transmission line, we have seen above that $T_{line} = T(1/\varepsilon - 1)$, where ε is the line power loss. Taking this input loss into account, the receiver–amplifier system temperature is $T_{rec} = T_{amp}/\varepsilon$, where T_{amp} is the intrinsic amplifier temperature. The total receiving-system temperature is therefore:

$$T_S = T_{ra} + T_{line} + T_{rec} = T_{ra} + T(1/\varepsilon - 1) + T_{amp}/\varepsilon \qquad (4.37)$$

For an ideal receiving system with no transmission loss ($\varepsilon \approx 1$) and operated at room temperature, ($T \approx 290$ K), the system temperature is $T_S \approx T_{ra} + T_{amp}$, corresponding to a NF of $F = 1 + (T_{ra} + T_{amp})/290$. Thus, the SNR degradation is essentially a function of the receiving-antenna and the preamplifier effective temperatures (see Exercises at the end of the chapter for examples).

Since the CNR is usually expressed in decibels, one can write, from equation (4.36)

$$
\begin{aligned}
\mathrm{CNR(dBHz)} = {}& P_{trans}(\mathrm{dBW}) + G_T(\mathrm{dB}) + G_R(\mathrm{dB}) - L(\mathrm{dB}) \\
& - T_S(\mathrm{dBK}) + 228.6
\end{aligned}
\qquad (4.38)
$$

where $L = (4\pi r/\lambda)^2$ is the isotropic loss and $228.6 = 10 \log_{10} k_B$. In the previous formula, the system temperature is expressed in *decibel-Kelvin*, or dBK. Consistently, the CNR is expressed in decibel-Hertz, or dBHz. The conversion in decibels is simply $\mathrm{SNR(dB)} = \mathrm{CNR(dBHz)} - 10 \log_{10}(\Delta f_{Hz})$. The interest of using CNR rather than SNR is that the CNR is a reference independent of the receiving antenna bandwidth.

For satellite-based systems, where the signal makes a round trip from Earth to satellite and back, it is easy to show that the resulting CNR is given by

$$\mathrm{CNR} = \left(\frac{1}{\mathrm{CNR_U}} + \frac{1}{\mathrm{CNR_D}} \right)^{-1} \qquad (4.39)$$

where $\mathrm{CNR_U}$ and $\mathrm{CNR_D}$ represent the CNRs for the uplink and downlink paths, respectively. Since the uplink and downlink paths utilize different frequencies (note loss dependence in λ^2) and the receiving systems (satellite and Earth) have different effective temperatures, there is no reason for the two CNRs to be approximately equal. This condition would correspond to an ideal system with minimal degradation. As a matter of fact, the total CNR is dominated by the link that has the poorest transmission quality or the highest intrinsic noise.

■ 4.2 MOBILE RADIO COMMUNICATIONS

The world of *mobile radio communications* makes use of radio signals to link terminals that are not physically wired to any network and can randomly change location or are themselves in motion. This is in contrast with *point-to-point radio networks*, where the radio signals propagate through a succession of fixed antenna-repeater stations, as described in Section 4.1.5. This is also different from *radio broadcast networks*, where a fixed antenna broadcasts signals to home receivers such as TV or FM stereo, or to mobile receivers such as car radios. Since communication systems are commonly operated on a two-way or duplex mode, all terminals must be equipped with both transmitters and receivers, and the radio network must be able to create a connection link between them. Therefore, radio communications can be developed according to three basic topologies, as illustrated in Figure 4.27.

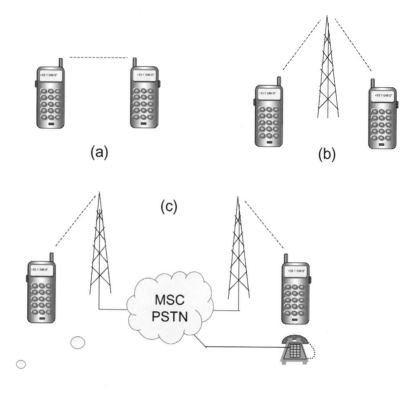

FIGURE 4.27 Three basic types of radio-communications network: (a) Private (walkie-talkie, intercom), (b) single-site station, and (c) cellular, via mobile switching center (MSC) and PSTN.

The first corresponds to the primitive form of private mobile networks, as based upon walkie-talkie/intercom equipment with multiple handsets or users. Its power and reach is usually limited, with main applications for security, police and army. Other applications concern communications in remote agricultural or geographical areas. The second type is based upon a single-site omni-directional antenna, which can also be connected to a control station for handling multiple terminals (e.g., taxi and ambulances). A typical application concerns airport traffic control, namely radio-communications between the control-tower and airplanes traversing the local airspace. The range of such networks is limited by the antenna's power, usually defining regions between 1-km and 20-km in diameter. The third type corresponds to the actual generation of *cellular* radio system. Two remote terminals are put in communication through some near-by antennas. The antennas are connected through the *public-switched network* (PSTN), as described in Chapter 2. The calls are handled from the two ends of the PSTN by *mobile switching centers* (MSC). An alternative name defining the centralized function for these centers is the *mobile telephone switching office*, or MTSO (pronounced sometimes as "mitso").

The MTSO represents the central coordinating entity, which is not only in charge of managing the calls (switching) from/to the mobile network to/from the PSTN, but also performs the operations of global powering, maintenance, supervision and billing, similarly to the *central office* (CO) of the PSTN (see Chapter 2). Since the cellular network is plugged to the PSTN, calls can be directed from fixed-telephone units to mobiles units, or the reverse, as seen in Figure 4.27.

The world "cellular" means that it is possible to connect to the network via a radio antenna only within a given circular geographical zone or *cell* (see Section 4.2.1). The words "mobile" and "cellular" are often used indifferently, although they designate two different concepts, the first for the appliance or the service, the second for the network type. The antenna premises at the center of the cell are referred to as the *base station*. A radio-network configuration having a single base station corresponds to the very origin of the *radiotelephone* service, making it possible to extend the PSTN to a local wireless area defined by the radio cell. Such a system would typically handle 400 two-way channels, meaning 400 pairs of usable frequencies (one for mobile to antenna or uplink, the other for antenna to mobile or downlink) over a region 10–15 miles wide. The principle of allocating a frequency pair for such radio links is referred to as *frequency-division-duplexing*, or FDD.

Cellular or mobile telephony only represents a first level of wireless-network service. As will be further described in this section, mobile radio networks also offer a broad range of services for data communication within *local-area networks* (LAN), *metropolitan-area networks* (MAN) and *wide-area networks* (WAN). In contrast, *fixed* radio-networks (such as point-to-point and global satellite links) are more concerned by applications of LAN-to-MAN and LAN-to-WAN data transport, in addition to classical telephone services (see Section 4.4). Since mobile telephone networks can exist both as cellular (Earth-based) and satellite, but with different constraints and technologies, we shall describe them in separate subsections. Aspects more specific to *satellite-based communications* (which also include voice- and data-transport services), will be described in Section 4.3 of this chapter. Concerning cellular telephone networks, we begin with a description of the network's architecture and constitutive elements, followed by a review of the different *generations* of mobiles, including the future *universal mobile telecommunications systems* or UMTS, as covered in the next three subsections. We then briefly describe wireless ATM (WATM) networks.

4.2.1 Cellular Telephone Networks

The making of a large-area cellular telephone network requires one to deploy an array of base stations and cells, and to connect the whole by some central management/supervision functions. For uniform coverage, the cells must be arranged according to the weakly overlapping circles pattern shown in Figure 4.28. The base stations can be managed in groups of two or more by local MSCs, making it possible to control the frequency allocation in a local subgroup of cells (see below). Cellular networks are usually represented by hexagonal tiles (Figure 4.29), although this picture masks the fact that cells are circular and actually boundless,

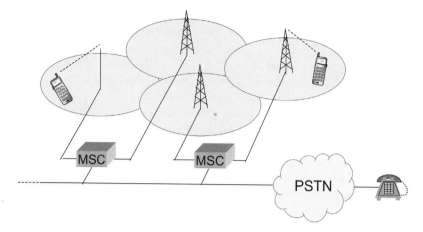

FIGURE 4.28 Layout of a mobile cellular network, showing arrangement of circular cells with their dedicated base stations, and related local mobile switching centers (MSC) connecting the network to the PSTN.

since the radio signals from the base stations are omni-directional. However, the signal power vanishes as the square of distance to the base station, which defines the unit cell as a zone of minimum transmit/receive power level. The figure illustrates the principle of *frequency-band allocation* by groups of seven contiguous

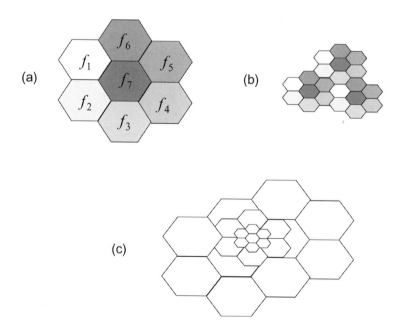

FIGURE 4.29 (a) Principle of frequency-band allocation by clusters of seven contiguous cells and (b) resulting arrangement at larger scale showing frequency reuse; (c) definition of microcell structure for denser urban area.

cells. The number of cells having different frequencies is called the *cluster size*. It is also possible to reduce the cluster size to three. Within each cell, the allocated band contains its own set of frequency pairs (or channel pairs) used for uplink and downlink communications between the mobile and the base station. When cells are arranged in a larger-scale pattern, frequencies can be *re used* since the separation distance is between two and three cell diameters (cluster size of 7) or one to one-and-a-half diameters (cluster size of 3). Typical cell diameters, which define the distance for which frequencies can be re-used without significant interference, are 20–25 km. The allocation of frequency subbands to users within a given cell is referred to as *frequency-division multiple access*, or FDMA.

Within each cell, the number of usable frequency channels is a function of the local density of mobile-phone users. Clearly, this density is greater in urban areas than in rural areas, which leads to microcell structures having smaller cell diameters, as shown in the figure. As the density of users grows over time (suburban zone expansion), the initial cells can also be split into micro-cells of smaller size. The smallest possible size of a microcell is limited to about 100 m. This is in order to avoid interference with EM noise radiated by the base station's powering equipment. However, specifically designed supplies make it possible to implement *pico-cell* networks (<10–50 m), for use in homes, offices, public areas or commercial centers. Microcell networks can also be made "portable," for use in local, one-time events or special military operations. In dense areas, the presence of tall buildings can locally create dark spots of *wave shadows*, from which no communication can be established. The effect can be overcome by implementing directional antennas which illuminate the spot from different angles. The use of directional antennas also makes it possible to define intracell areas. Indeed, three directional antennas with radiation patterns confined to $120°$ angles, for instance, can divide the cell into as many independent regions. Finally, mobile units can be spotted by *smart antennas*, whose beam angle can be directed at will (see phased-array antennas in Section 4.1.3). In cellular networks, these segmenting approaches are referred to as *space-division multiple access*, or SDMA.

The movement of mobile users through the cellular network is called *roaming*. Nowadays, roaming applies more to the concept of making mobile calls in networks owned by different operators, such as in a pan-European mobile network. As the user crosses the cell, the distance from the base station increases and the link connection can no longer be sustained. The way to solve this problem is to pass the responsibility of the connection to the closest up-coming base station, referred to as connection *handoff* or *handover*. There are two ways to achieve this, called *soft* or *hard handoff*, respectively. In the soft-handoff case, the two base stations calculate the cell-crossing time and direction, using various methods of triangulation and time-of-arrival measurements. The results are communicated to the MSC, which determines in what direction the mobile is heading and at what time the cell boundary will be reached. This gradual monitoring approach leaves the possibility of opening up a new connection with the closest "upcoming" basestation. In the hard-handoff case, the initial cell connection is interrupted at some point in time and a new connection is immediately established from the neighboring station, without interruption of the actual communication. In practice, the communication is momentarily ineffective (short silence or clipping). It can even be definitely lost because of a temporary lack of available frequency channels inside the next cell. The outcome of the handoff (soft or hard) also depends upon the *speed* of the mobile, as measured relative to the stations' antennas. Typically, systems are designed for achieving handoffs at user speeds up to 100 km/h (about 65 mph), which corresponds to highway traffic speed (note that in most countries, mobile communications are only allowed to car passengers and forbidden to drivers). Specially designed mobile networks can operate at user speeds of 300 km/h (about 200 mph) for specific use in high-speed trains, like Japan's Bullet train or European's TGV, which represents a significant customer base (business calls, suburban commuters, etc.).

4.2.2 Network Grade of Service

How many user calls can be handled within a given network cell, or more generally, by a telephone network having N channels? One must assume a channel attribution algorithm whereby a channel is attributed to incoming calls on a first-available channel basis, from channel one then channel two and all the way up to channel N. This non-trivial problem was solved by the Danish scientist A.K. Erlang. Before getting to the result, one must first define the concept of *traffic intensity*. By convention, traffic intensity is given by the product of number of pending calls during a certain monitoring period of network use. The unit of measure is the *erlang*, which corresponds to *3,600 call-seconds*, or 1 call-hour. Thus a person calling for one hour duration represents one Erlang of traffic intensity load. In North America, the unit in use is the *hundred-call-seconds*, referred to as CCS. Thus, 36 CCS = 1 erlang. The way traffic intensity is calculated is shown by the following example involving 4 channels. Assume that over a monitoring period of 10 mn, 18 calls took place with the following per-channel distribution:

3 calls occupied channel 1 for a total holding time of 8 mn;

5 calls occupied channel 2 for a total holding time of 6.5 mn;

4 calls occupied channel 3 for a total holding time of 9 mn;

6 calls occupied channel 4 for a total holding time of 6.5 mn.

The total or cumulated traffic duration is therefore $8 + 6.5 + 9 + 6.5 = 30$ mn, corresponding to an average traffic intensity of $A = 30/10 = 3$ erlangs. This result is the average total *call holding time* relative to the monitoring time. It also represents the average number of simultaneous calls handled at any time. Note that according to the definition, 3 erlangs represent 3 call-hours, or 180 call-mins. This is what we would obtain if we increased the initial 10-mn observation time six-fold (6×30 mn = 180 mn).

By definition, the observed traffic intensity is referred to as the "offered traffic," or "traffic offer." This terminology comes from the fact that from the operator viewpoint, the network use is a customer offer. The goal of the operator is to be able to handle or carry all the offered traffic, with a high success rate. One could wrongly conclude that with a 3-erlang traffic offer, for instance, only three channels are sufficient to carry it. This would be true only if there were exactly 3 simultaneous calls of

10-mn duration each, or 3 groups of 2 successive calls of 5 mn each, etc. But in our realistic example, the 18 calls come up with random times and durations, which requires 4 channels. Thus, while traffic intensity represents a useful reference measure of "number of simultaneous calls," it does not indicate how efficient is the network in carrying the offered traffic. However, the two following properties can be proven (see Exercises at the end of the chapter):

The traffic intensity, A, also corresponds to the number of calls that can be expected during the average holding time;

Given traffic intensity, A, the occupation rate of a single channel is $B = A/(1 + A)$.

The second property shows that given traffic intensity, the occupation rate of a single channel is never 100%. Consistently, it becomes close to 100% as the traffic increases ($A \to \infty$), and close to 0% as the traffic decreases ($A \to 0$). This occupation rate represents the relative proportion of time when the channel is busy or not available. Thus, it also corresponds to the channel's *blocking rate*, i.e., the relative proportion of incoming calls that are *not* accepted by the channel.

What we would need is to be able to estimate the relative proportion of calls that are being handled by the network, given traffic intensity conditions (number of incoming calls per average holding time) and number of network channels. The answer is provided by *Erlang's lost-call formula*, which defines the blocking rate according to:

$$B(N) = \frac{A^N}{N!} \left(\frac{1}{1 + A + \dfrac{A^2}{2!} + \cdots + \dfrac{A^N}{N!}} \right) \tag{4.40}$$

where A is the traffic intensity in erlangs, N is the total number of network channels and $N! = 1 \times 2 \times 3 \times \ldots \times N$ is the factorial of integer N. Figure 4.30 shows a

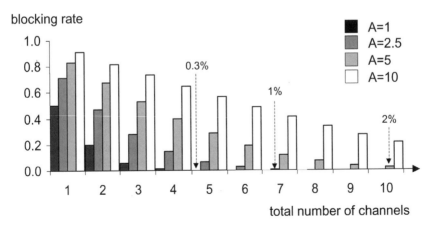

FIGURE 4.30 Blocking rate as a function of total number of network channels (Erlang's formula), assuming traffic intensities of $A = 1$, 2.5, 5 and 10 Erlangs. Some low values are given as percentages.

plot of the blocking rate versus total number of network channels, assuming various traffic intensities. At $A = 1$ erlang intensity, it is seen that the blocking rate is $\frac{1}{2}$ (i.e., 50%) if the network has one channel, 1/5 (i.e., 20%) with two channels, etc., with a value of 0.3% for five channels. This means that under these traffic conditions, five channels are sufficient to guarantee that less than 1% of the incoming calls will be lost (or >99% success rate for call connections). On the high-traffic side ($A > 1$), we observe that this low-level of blocking rate is associated with greater number of channels, for instance seven channels for $A = 2.5$ and 10–11 channels for $A = 5$, etc. We observe that for a given blocking rate, a small increase of traffic offer results in a large increase in the required number of channels. Note that the Erlang formula also works for $N = 0$ (zero channel in the network), which gives $B(0) = 1$ or a 100% blocking rate.

By definition, the upper bound of the loss rate corresponds to the network *grade of service* (GOS). Such a grade of service must be guaranteed regardless of the fluctuations in traffic intensity, from low to high usage hours. For most networks, the typical GOS is 0.01 or better (<0.01), meaning that one has a 99% chance of being successfully connected on the first try. We leave it as an exercise (see Exercises at the end of the chapter) to determine how many channels are required in a mobile-network cell in order to ensure a 0.01 GOS under certain traffic-offer conditions.

This above analysis provided the overall network blocking rate as a function of the number of network channels. We can develop the analysis further by considering a network having an *infinite number of channels*, and determine the erlang occupancy for each channel. It can be shown (see Exercises at the end of the chapter) that for a traffic offer of A, the occupancy of channel number N is given by the formula

$$O(N) = [B(N - 1) - B(N)]A \qquad (4.41)$$

Figure 4.31 shows plots of the channel occupancy for different traffic values A. It is seen that the first channel ($N = 1$) is always the most occupied, consistent with the aforementioned channel-filling algorithm. Its asymptotic occupation limit is 1 erlang for very high traffic, corresponding to the value $O(1) = [1 - B(1)]A = A/(1 + A)$. It is observed that the channels rapidly fill in with increasing traffic, consistent with the effect of occupancy saturation from lower-order channels. It is also part of the exercise to show that the full area under the diagram envelopes is equal to the traffic offer, A.

4.2.3 Early 1G Mobile Systems and Frequency Allocations

The first mobile telephone systems appeared in the late 1970s and continued to be used through the 1980s. They are referred to as the "first generation" of mobile, or "1G" for short. These 1G mobiles were all analog devices, based upon the principle of *frequency-division multiplexing* or FDM, which is described in detail in Chapter 1 (Section 1.1) and Chapter 2 (Section 2.2). Before the PSTN became

Erlangs per channel

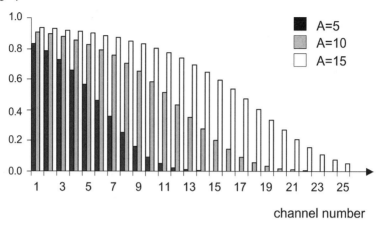

channel number

FIGURE 4.31 Individual channel occupancy, as expressed in Erlangs, for a network with an infinite number of channels. The plots correspond to three values of offered traffic, A = 5, A = 10, and A = 15. The full areas under the envelopes are equal to the offered traffic.

nearly all digital, FDM was the sole technique used to multiplex telephone calls or voice circuits, into single transmission lines. Each one-way voice circuit was physically attributed a unique carrier frequency, or channel, occupying a 4-kHz analog-modulation band (Chapter 2). The frequency attribution for handling the communication in each direction was made by the CO on a first-available channel basis. For transport between main COs, the channels were multiplexed together according to a standard FDM hierarchy, starting from small "groups" of 12 voice channels to large "hypergroups" of 2,700 channels, as described in Chapter 2.

This FDM principle is basically the same as used in early or 1G mobile networks. A first difference is that the physical connection between the network and the end-user is not a twisted-wire pair (from the CO to the fixed-telephone unit), but a wireless link (from the base-station antenna to the mobile-telephone unit). As we have seen earlier in this chapter, a wireless end-connection requires the MSC to allocate a pair of uplink and downlink frequencies. The two-way communication (wireless-to-fixed or wireless-to-wireless) is then carried over the PSTN, which initially used FDM to transport large groups of voice channels. A second difference with fixed-network telephony is that the number of end-users is not pre-determined by any wiring. Thus users "access" the radio network on a random demand (and the other way around), and channel frequencies are attributed on a cell-location basis, with *reuse* in other cell locations. Hence the name *frequency-division multiple access*, or FDMA, which represents the technique of handling analog cellular-network communications by frequency allocation and simultaneous reuse, rather than the mere frequency-multiplexing of channel capacity.

The deployment of 1G mobile was made complex because of the rich diversity of standards which different countries or blocks of countries chose to adopt. The history and geography has retained five main standards (or norms), called: *AMPS,*

Network C, NMT, TACS and *J-TACS*. The corresponding channel/frequency characteristics and main user countries are listed in Table 4.3, and can be summarized as follows:

Advanced mobile telephone system (AMPS), 824–894 MHz: Introduced in 1977 by AT&T (after Bell Labs invention of cellular communications systems) it has remained the North American standard until the turn to digital systems in the early 1990s. It was also implemented in the rest of the Americas, in the Middle-East and Pacific regions. The allowed number of mobile channels is 833.

Network C (also called C-Netz), 451–466 MHz: Introduced in 1986 by the Deutsche Bundespost (German PTT), following previous Network B generation (1971). It was later upgraded to the 900-MHz band, hence the system names C-450 and C-900), allowing for 1,000 channels instead of 222.

Nordic mobile telephony (NMT), 453–467 HZ and 890–960 MHz: introduced in 1981 by four Scandinavian operators, it took into account international aspects to allow for rapid acceptance by most European countries and others in the world. The network capacity for the two systems, called NMT-450 and MT-900 (for 450 MHz and 900 MHz) are 180 and 1,000 channels, respectively. The NMT norm was adapted by the French PTT, leading to the NMT-F system, also called *Radiocom 2000* or *R2000*.

Total-access communications systems (TACS) 890–950 MHz or 872–933 MHz: introduced in 1985 by British PTT for use in UK and Ireland, as a modification of AMPS and technically close to NMT. The norm was extended as *ETACS* to ease frequency-allocation in congested metropolitan areas such as London. The network capacity is 1,000 channels.

Japanese total-access communications systems (J-TACS) 870–940 MHz: a Japan-only version of TACS, using a slightly different frequency band. Japan also introduced the norm MCS (mobile communication systems) developed after AMPS, which appeared in two generations of MCS-L1 and MCS-L2.

Several countries developed their own national-exclusive standards after the previous ones, such as *R2000* for France, but also *RMTS* in Italy (for radio mobile telephone system, popularly called "telefonino"), *NAMTS* in Japan (for Nippon automatic mobile telephone system), Comvik ("Communication-Viking") in Sweden, and UNITAX in China and Hong Kong, for instance.

We note from Table 4.3 that by convention, the downlink frequencies (antenna to mobile) are always in the higher band. A common feature of these systems is the use of UHF frequencies in either one of two groups located near 450 MHz and 800–900 MHz, with a full width typically of 5, 10 or 25 MHz each. Historically, the trend has been to migrate to a single 900-MHz band (e.g., towards NMT or TACS in Europe, see following text). Finally, we observed that 1G systems were based upon either PSK modulation (AMPS, TACS) or FSK modulation (Network C, NMT), the principles of which are described in Section 1.1. Note that PSK and

TABLE 4.3   Characteristics of first-generation (1G) analog-based mobile systems

System	Frequency Range (MHz) Uplink	Frequency Range (MHz) Downlink	Channel Spacing (kHz)	Modulation Type	Equivalent Capacity (kbit/s)	Number of Radio Channels	Main Countries of Implementation
AMPS	824–849	869–894	30	PSK	9.6	833	**Americas**, Australia, New Zealand, Zaire, Israel, Oman, Taiwan, Singapore, Thailand, South Korea
Network C (C-Netz)	451–456	461–466	20	FSK	5.3	222	**Germany**, Portugal, South Africa
NMT	453–457 890–915	463–467 935–960	25	FSK	1.2	180 1,000	**Scandinavia**, Western Europe (except UK and Germany), Eastern Europe, Switzerland, Turkey, Morocco, Saudi Arabia, South Africa, Thailand, Indonesia
TACS (ETACS)	890–905 872–888	935–950 917–933	25	PSK	8	1,000	**UK**, Ireland, Italy, Spain, Austria, Kuwait, United Arab Emirates, Sri Lanka, Malaysia, Hong-Kong, China, New Zealand
J-TACS MCS (L1, L2)	860–870 870–885	915–925 925–940	25	PSK	0.3	1,000	**Japan** (only)

FSK are acronyms used in digital coding, so the correct terminology should be PM and FM, for analog phase and frequency modulations, respectively.

The TACS system defined four categories of mobiles, as grouped by their emitting antenna power: Class 1 for 10 W power (reserved for relatively heavy equipment, e.g., vehicle applications), and Classes 2, 3, and 4 for powers of 4 W, 1.6 W and 0.6 W, respectively (accessible to portable units of decreasing weight and size). These classes were followed by the other existing standards.

4.2.4 Global System for Mobile Communications (GSM)

The previous description of the 1G mobile standards fully illustrates the different country situations: while one could use a mobile network across entire subcontinents such as North America with AMPS, Europe was fragmented into many zones using different and mutually incompatible standards (essentially TACS or NMT). In this second case, there was no real possibility for these countries to implement a full cellular network for national coverage if the revenues were limited to national use. Roaming, the action of accessing networks owned by different operators with the same telephone unit, was restricted to specific country groups sharing the same standard, or not possible at all. By 1978, a consensus was reached to use 900 MHz with 2×25 GHz bands (uplink/downlink) as a common European reference. By 1982, the European PTTs moved further with the decision to develop a common standard, referred to as the *global system for mobile telecommunications*, or GSM, an acronym initially coming from the French as "groupe spécial mobiles." By 1992 the new-born GSM system allowed a complete, pan-European coverage, at least in 26 participating countries. Such a coverage means that any user can call, or be called, on his/her mobile from/to all these countries, regardless of national residence or local PTT ownership considerations. The approach thus introduced in Europe the concept of telecom *deregulation*, making it possible for new entities (as well as "incumbent" national PTTs) to operate the networks, at a national or broader scale. However, fixed-telephone networks would remain the exclusive privilege of the incumbent PTTs for an additional decade, until global deregulation came (1996 in the United States, 1998 in Europe).

The above provides the background context which led to the development of GSM. However, GSM also came up with a key technology improvement (as well as AMPS and NMT during the same period), which is the use of *digital coding*. The principles of digital coding (or modulation), as opposed to analog modulation, are described in detail in chapter 1 (Section 1.2). The various possible techniques for transforming analog voice signals (speech) into digital samples are also explained there (Section 1.3). Once speech channels are encoded into a digital signal, they can be multiplexed together along with other digital data. The digital multiplexing formats used in cellular systems, namely FDMA, TDMA and CDMA, are described in this subsection. The introduction in the USA, by the early 1990s, of digital technologies in wireless communications clearly defines the emergence of a *second generation* of cellular systems, as referred to as "2G". The new digital standard, called D-AMPS, used the same 824–894-MHz band. The parallel development of GSM in Europe, but with *enhanced service capabilities* (missing in previous 1G and 2G) led to the new concept of "2.5G". Such an uncertain definition represents an attempt to conceptually define some incremental rather than radical technology evolution from the original 2G. As a matter of fact, because of the introduction of enhanced services, GSM is considered by most as having marked the beginning of a "3G" phase, being radically different from 2G. Most people define 3G not only by the introduction of voice *and* data capabilities, which was already present in 2G via ISDN networks and services (see Chapter 2, Section 2.1), but also by broadband high bit-rate use. To summarize, 3G can be seen as a new era of "personal" voice-and-data cellular systems (see next subsection) with 2.5G representing the conceptual service background and 3G its broadband implementation through new standards. This is the level at which *technology* and *services* merge into single concepts, generating a flurry of novel acronyms, some of which are relatively abstract or fuzzy. To see more clearly into this 2.5–3G world, it is therefore important not to skip any step but first grasp the specific aspects and terminology of GSM systems, which we shall cover in this subsection. The evolution of GSM features and their comparison with equivalent systems now in use in the USA or Japan is also covered.

The GSM network must be viewed not as a mere wireless point-of-entry into the PSTN via a base station and a MSC, as in 1G–2G, but as a complete network-service offer that is entirely managed from the mobile unit all the way up to a central operation and maintenance office. The generic system layout is illustrated in Figure 4.32. It comprises three main sections called:

Radio section;

Network and switching subsystem (NSS);

Operations and support subsystem (OSS).

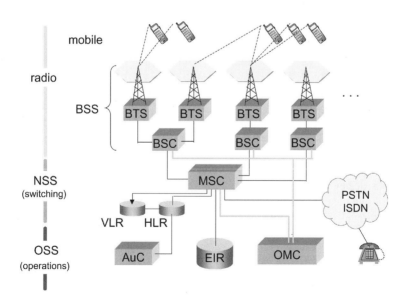

FIGURE 4.32 Layout of GSM network (see text for description).

The radio section includes the *mobile units* connecting to the network, and the *base-station subsystem* (BSS). The BSS is the subnetwork made of all the base-station infrastructure, now called *base transmitter station* (BTS) and local switching centers, called *base-station switching centers* (BSC). The BTS consists of the antenna and an 8–10-km cell. They are not necessarily close to the BSC. The BSC can also handle more than one BTS, which, in addition to facilitating handover operations between cells, also reduces the infrastructure complexity and cost.

The NSS section is the core of the GSM network. It includes the now-familiar *mobile switching center* (MSC). Unlike in previous systems, however, the MSC is able to handle a large number of BSCs. Only a few tens of MSC are required to make a GSM network of national coverage. The MSC is connected to the PSTN (and ISDN) in order to be able to handle calls between (a) mobile units and fixed units, and (b) two mobiles units with one being located outside the BSS geographical coverage or jurisdiction. As the figure shows, the MSC can interrogate two different databases. The first one is the *home location register* (HLR). The HLR contains information on mobile subscribers (network address, account, authorization number), which makes it possible to reach them on their mobiles, wherever they are in the local network. If the mobile unit traverses a different MSC jurisdiction, it is also registered in the HLR. This registration is made possible by the following. When a mobile handset is turned on, it emits a periodic signal-control tone. This tone is received by the closest BTS belonging to the local MSC jurisdiction. This makes the local MSC aware of the presence, cell location and identity of the mobile unit. This new registration is temporarily stored in a *visitor location register* (VLR) database. Once the visitor mobile is registered in the VLR, it can make outgoing calls. For security purposes, GSM handsets include a microchip card, called SIM *system identification module*. To activate the mobile unit, the user must enter his *personal identification number* (PIN), also called "PIN code," just as with an automatic cash machine. Three successive failures to enter the correct PIN result in the immediate deactivation of the SIM. This feature protects the user in the event that the handset is lost or stolen. Loss can also be reported to the operator for similar SIM deactivation, as for credit card numbers.

The third network section, the OSS, is concerned with operating and managing the network. The above authorization procedure (validation or invalidation) is handled in the OSS though an *authorization center* (AuC). The AuC validates the HLR database, as shown in the figure. Another OSS database, the *equipment identity register* (EIR) stores information consisting in a *white list* of registered units (SIM accounts) and their registered service categories, a *gray list* of units presenting dysfunctions or being un-registered, and a *black list* for denied-service units. Finally, the OSS includes an operation and maintenance center (OMC), which, apart from these top-level functions monitors the traffic load and working conditions of the MSC and the BSCs.

We describe next the GSM signal specifics. The end stage concerns the *analog-to-digital* (A/D) conversion of voice (outgoing call) or the D/A reverse (incoming call). The possible techniques for such a conversion, essentially consisting in *pulse-code modulation* (PCM) algorithms, are described in Section 1.3. As previously stated, digital mobile radio signals can be multiplexed and accessed to according to three approaches, namely: *frequency-division multiplexing* access (FDMA), *time-division multiplexing* access (TDMA) and *code-division multiplexing* access (CDMA). Let's have a closer look at each one:

FDMA: although described earlier as an analog technique (frequency division multiplexing or FDM), FDMA can also apply to digital signals. The resulting (uplink/downlink) spectrum is a comb of channels having regularly spaced carrier frequencies, but whose contents are digital signals. Such signals can be coded by ASK, FSK or PSK modulation (see Chapter 1, Section 1.2).

TDMA consists in sharing between multiple users a given frequency channel, through the technique of *time-division multiplexing* (TDM). In TDM, the digital data from N different users are time-slot interleaved, usually by groups of eight bits (1 byte with 1 bit from each channel) in order to form a multiplexed sequence of N bytes. Alternatively, the data can be time-multiplexed by blocks (m bits each), usually by groups of eight blocks to form a frame of $8m$bits, which is the GSM scheme (see below). The resulting TDM signal can be transmitted within any available frequency channel. Thus, a base station can handle several frequency channels in its uplink and downlink bands (FDMA), each of these channels being time-shared by multiple users (TDMA). This is the principle used by GSM.

CDMA is generally referred to as a *spread spectrum* transmission technique. As the name indicates, the transmitted spectrum is made broader than in conventional signals, which in particular randomizes and alleviates the deleterious effect of *multipath interference* or *fading*. Spread-spectrum transmission can be achieved by switching the carrier frequency according to a pre-defined pattern or cyclic code, which is referred to as *frequency hopping*. The user-channel hops to different carrier frequencies during transmission time, hence the name "spread spectrum." Different cyclic codes can be used simultaneously, as long as each one uses different frequency patterns at any given time in the cycle. In mobile telephony, however, time delays due to multiple reflections may cause such channels to overlap and interfere. An alternative approach consists in splitting TDM data into blocks of equal and smaller size, then reducing the blocks bit rate, and simultaneously transmitting them through all available frequency channels. This is called *orthogonal FDM* or OFDM. The CDMA is another spread-spectrum alternative where the carrier frequency is unique. Instead, the digital data to be transmitted, initially at carrier bit rate B, are *multiplied* by a binary number (code) at higher frequency B', called *chip rate*. The resulting data are thus transmitted at the chip-rate frequency, and the corresponding spectrum is broadened by the factor B'/B, called the *spreading factor*. Since many codes are possible, several CDMA signals can be transmitted through the same frequency channel. At the receiving end, the original user data is retrieved by digitally *dividing* the received signal by the corresponding code.

The digital modulation scheme that was adopted for GSM is *phase-shift keying* (PSK), as described in Chapter 1. More accurately, the scheme is a sophisticated variant of PSK which is known as GMSK, for *Gaussian minimum shift keying*. In conventional PSK, the EM phase is changed from one bit to another by a discrete step. There can be two steps of $\pm\pi/2$, as in binary PSK, or multiple steps of $\pm\pi/(2M)$, as in M-ary PSK (e.g., QPSK for quaternary PSK). In GPSK, the phase steps are continuous and calculated according to the value of all preceding bits. The calculation of the phase $\phi(t)$ over time is made according to the formula:

$$\phi(t) = \phi_0 \pm \frac{\pi}{2}\sum_k h(t - kT) \tag{4.42}$$

where "+" applies if the bit at time t has the same value as the one immediately preceding in the sequence ("−" applies otherwise), $T = 1/B$ is the bit period, and the k-summation is made over the infinity of preceding bits. The function $h(t)$ is a smooth step function which raises from zero for $t \le 1.5T$ to unity for $t \ge 1.5T$, with a value at the origin of $h(0) = 1/2$. Because of this fact, only the bits preceding in the $3T$ time interval have an influence on the phase value. This GMSK modulation scheme has interesting and useful properties. Indeed, it can be shown that if all bits are either equal to "0" or "1", the resulting phase is linearly increasing in time, corresponding to a single-frequency signal with a frequency up-shifted by $1/4T$ with respect to the carrier frequency. If bits are alternating "0" and "1," one obtains the same result but with a down-shift of $1/4T$. Thus, GMSK is a FM-like, spread-spectrum modulation scheme with a full spectral width within $1/2T$, corresponding to twice the data bit rate. The GSM system uses frequency channels spaced by 200 kHz. With GMSK, the transmitted bit rate is 270.79 kbit/s or 270.8 kbit/s. With TDMA, where the data are time-multiplexed by groups of eight user-channels, the available bitrate per user is therefore $270.8/8 = 33.9$ kbit/s. As we shall see below, however, the payload bitrate is only 24.7 bit/s, due to a relatively high amount of overhead. This 24.7-kbit/s payload also contains overhead due to signaling and error correction.

As stated earlier, GSM uses two bands of 25 MHz for uplink and downlink communications, with carrier frequencies near 900 MHz (this will be later referred to as the "GSM 900" system). The frequency ranges are 890–915 MHz for uplink and 935–960 MHz for downlink. By convention each uplink/downlink channel pair must have 45-MHZ separation. Within each band, the channels are separated by 200 kHz, which provides $25,000/200 = 125$ channels per band.

We analyze now the detailed frame structure of GSM signals before and after TDM. We recall that the GSM bit rate is 270.8 kbit/s. The basic unit for a single-user circuit is a TDMA time-slot, referred to as a *GSM slot*, and is shown in Figure 4.33. The contents are described as follows:

One header (3 bits) and one footer (11.25 bits), corresponding to a total delay between two adjacent slots of 14.25 bits/(270.8 kbit/s) = 52.6 µs. This delay is used to prevent slots emitted at different points inside the base-station cell from overlapping in time (52.6 µs $\equiv$ 16 km or two cell radii);

Two traffic fields for payload, representing $2 \times 57 = 114$ bits; the payload contains *error correction* and *encryption* (see the following);

One training field sequence (26 bits), which is a fixed pattern used for locking purposes at receiver level;

Two "stealing" flag bits surrounding the training field, which indicate whether the slot is used for carrying payload or control information.

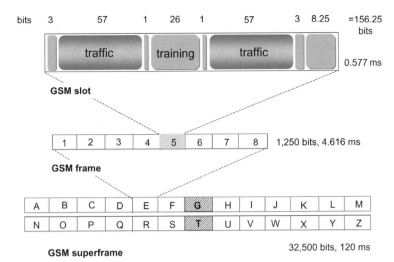

FIGURE 4.33 Top: basic structure of GSM slot, showing two traffic (payload) fields and one training field, with header and footer guard bits. Middle: Time-multiplexed frame structure, made of eight GSM slots. Bottom: Time-multiplexed GSM superframe structure, made of 24 GSM frames (A–F, H–S, and U–Z) and two stolen frames (G, T) spaced by 13 slots.

The total slot thus represents 156.25 bits and has a length of 156.25 bits/ (270.79 kbit/s) = 0.577 ms.

Eight GSM slots (or user circuits) are then time-multiplexed to form a *GSM frame*, as shown in Figure 4.33. The GSM frame size is therefore 8×156.25 bits = 1,250 bits and the duration is 4.615 ms (the exact value is 4.61538 ≡ 120/26, which has been chosen so for multiplexing hierarchy purposes, as shown later). This corresponds to a gross user-circuit bit rate of 156.25 bits/4.61 ms = 33.9 kbit/s. Since each user-circuit has only 114 bits of payload, the actual payload bit rate per user is 114 bits/4.61 ms = 24.7 kbit/s.

The GSM frames are arranged in sequences of 26 to form a "multiframe" a called a *GSM superframe*. The superframe duration is thus 26×4.615 ms ≈ 120 ms. For signaling purposes, two frames spaced by 13 frame intervals are inserted anywhere in the sequence, thus "stealing" two usable frames. The first is used for base-station signaling and the second is used to send service-text messages (e.g., caller identification). Because of this extra overhead, the actual user payload bit rate is reduced to $(24/26) \times 24.7$ kbit/s = 22.8 kbit/s. Another superframe specification consists in a 51 GSM frame sequence (51×4.61538 ms ≈ 235 ms). Greater multiframes (also called superframes by lack of exact terminology) can be formed by sequences of 26×51 GSM frames (6.120000 s), and an arrangement of 2,048 of these (12,533760 s or 3 h 28 mn 53 s and 760 ms) is referred to as a *hyperframe*.

To prevent a third party from "listening" to the user-channel (overhearing) the payload is *encrypted*. The key for decryption is held by the unit's SIM card (see more on cryptography in Chapter 3 in Desurvire, 2004). The payload is also subject to *forward error correction* or FEC (see Chapter 1, Section 1.6). The error-correction code is 456 bits long, which represents 8×57 bits, or four GSM slots of 2×57 payload bits (which explains the number 57 for the payload fields). Encryption is achieved by performing a logical function called *exclusive-OR* (XOR) on the 114 payload bits. The payload word a, referred to in crypto jargon as "plaintext," is "exclusive-ored" to a pseudo-random number b. Such a number is determined from a combination of the GSM frame number and a "session key," called K_c. The session key K_c, which is agreed between the network and the mobile unit (SIM memory), must conventionally be less than 64 bits (see Chapter 3 in Desurvire, 2004). The result of the XOR operation ($c = a.XOR.b$) constitutes the encrypted message, or ciphertext, to be radio-transmitted. If the XOR operation is repeated twice, the plaintext is recovered, i.e., $d = c.XOR.b = a$. For uplink calls, encryption is performed in the mobile unit and decryption in the BTS, and the reverse applies to downlink calls.

With both encryption and error correction, the actual voice or data payload corresponds to a bit rate in the vicinity of 10 kbit/s. Voice and data must be processed differently, based upon the fact that, unlike voice, data cannot be subject to further compression and partial loss. Using adaptative-differential PCM techniques (ADPCM) for speech coding with 10–13-kHz sampling rate (Chapter 1, Section 1.3), reasonable sound-quality levels can be achieved. The software used for the A/D or D/A conversion is called a *codec* (for coder/decoder). The codec sampling rate (i.e., the output bit rate) is a function of the required quality. For voice, the offered quality levels correspond to either 13 kbit/s (full rate, or enhanced quality) or 6.5 kbit/s (half rate, low quality). For data, the offered rates are either 21.4 kbit/s (raw), or 14.4 kbit/s (enhanced) or 9.6 kbit/s (regular). Encryption and FEC correspond to various amounts of overhead, making the resulting bit rate up to the maximum value of 22.8 kbit/s, as previously determined. For all options, encryption introduces an overhead of 1.4 kbit/s (or 0.7 kbit/s for half-rate voice). This leads to the following possible configurations:

Raw data without FEC: we have 21.4 kbit/s (data) + 1.4 kbit/s (encryption) = 22.8 kbit/s, which is the configuration of minimal overhead (6.1%);

Enhanced data with FEC: the FEC requirement is 6.6 kbit/s, totaling 14.4 kbit/ s (data) + 6.6 kbit/s (FEC) + 1.4 kbit/s (encryption) = 22.4 kbit/s;

Regular data with FEC: the FEC requirement is 11.1 kbit/s, totaling 9.6 kbit/s (data) + 11.1 kbit/s (FEC) + 1.4 kbit/s (encryption) = 22.1 kbit/s;

Full-rate voice with FEC: the FEC requirement is 8 kbit/s, totaling 13 kbit/s (codec) + 8 kbit/s (FEC) + 1.4 kbit/s (encryption) = 22.4 kbit/s.

The first option without FEC, giving a capacity of 22.8 kbit/s is not practical, since data cannot be safely transmitted with the ever-present risk of error corruption. Concerning data, the available GSM bandwidth comes therefore as two main options: 14.4 kbit/s and 9.6 kbit/s. In GSM-network terminology these two options are referred to as TCH/F14.4 and TCH/F9.6, respectively (TCH = traffic channel, F = Full rate).

From the second and fourth above examples it is seen that the typical GSM over-head (including FEC) is 42% for both voice and data. Note that people usually refer to the overhead as defined by the traffic size (Figure 4.33), which for a GSM slot represents 117 bits out of 156 bits or $117/156 = 73\%$ efficiency. Although such traffic is heavily loaded with FEC and encryption (in addition to some amount of frame signaling), it represents a valuable data reference since FEC guarantees high SNRs and low data error-rates, and encryption provides a high degree of privacy. For GSM, we retain the standard data capability (corrected and encrypted) of $B_{\mathrm{GSM}} = 14.4$ kbit/s. This is a useful reference for comparison with the alternative American and Japanese systems, as described in the following text.

In America, the digital version of AMPS, called D-AMPS, is also indifferently called ADC or NADC (for *American or North-American digital cellular*), or USDC (for *U.S. digital cellular*). It is also called TDMA for short, as opposed to CDMA, but this denomination can be misleading since GSM is also TDMA with a different framing system. There are four main differences between USDC (as we shall call it) and GSM:

The frequency channels are separated by 30 kHz, as opposed to 200 kHz (GSM)

It uses DQPSK (differential QPSK), which is a bit more complex than GMSK of GSM, since

It is a 4-level modulation (see Chapter 1);

The slots are not the same for uplink and downlink;

The slot bit rate is 48.6 kbit/s, as opposed to 270.8 kbit/s (GSM).

The slot structures and the multiframe USDC hierarchy are both illustrated in Figure 4.34. The main difference of USDC with GSM (apart from those listed above) are the following:

The slot length is 320 bit, corresponding to 6.67 ms duration;

Uplink and downlink traffic is made of 260 bits, out of 320 bits, representing 81% efficiency (compared to 73% for GSM);

The slots are arranged by groups of three (USDC frame), which corresponds to an effective user rate of $48.6/3 = 16.2$ kbit/s; however the actual traffic is given by 16.2 kbit/s $\times 81\% = 13.16$ kbit/s (or 13 kbit/s for short);

The multiframe is made of 16 frames including one for signalling.

After inclusion of FEC, the 13-kbit/s data rate is reduced to the effective user rate $B_{\mathrm{USDC}} = 9.6$ kbit/s, compared with the GSM capability of $B_{\mathrm{GSM}} = 14.4$ kbit/s.

The Japanese standard for digital mobile was developed after D-AMPS/USDC, but with reverse compatibility with its own analog J-TACS system. The result is the *personal digital cellular*, or PDC system, also called *Japanese digital cellular*, or JDC system. In this system, the modulation is DQPSK, like USDC. The channel separation is 25 kHz, compared with 30 kHz (USDC) and 200 kHz (GSM). The raw bit rate is 14 kbit/s, compared with 16.2 kbit/s (USDC) and 33.9 kbit/s (GSM). Finally, the effective user rate after FEC is $B_{\mathrm{PDC}} = 9.6$ kbit/s, which is the same as in USDC.

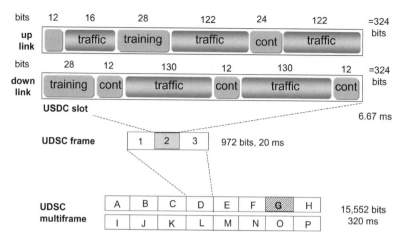

FIGURE 4.34 Top: basic structure of USDC slot (D-AMPS), corresponding to uplink and downlink channels, showing header, control and traffic (payload) fields. Middle: Time-multiplexed frame structure, made of three USDC slots. Bottom: Time-multiplexed superframe structure, made of 16 USDC frames (A–F and H–P) and one stolen frame (G).

As previously stated, GSM introduced a new set of unprecedented or "enhanced" services, which do not reduce to traditional telephony and data communications. Such services can be categorized into four categories of "speech," "data," "short messages," and "supplementary." To list some representative examples in each of these categories:

Speech services: mobile telephony (mobile to PSTN and mobile-to-mobile); "ubiquitous" roaming inside and outside one's country; call-privacy protection through encryption; voice messaging (voice mailbox); and emergency calling (e.g., 112 as a unique number for Europe);

Data services: facsimile; telex; videotext; and mobile connections to *integrated-services data network* (ISDN), to *circuit-switched public data networks* (CSPDN), to *packet-switched public data networks* (PSPDN) and to X.25/frame-relay networks (see Chapter 2);

Short-text services: caller line identification (CLI) which shows the calling number on the unit display; mobile-to-mobile paging (short alphanumerical text messages, also called *short-message service* or SMS); and broadcast messaging (e.g., weather, highway-traffic information, breaking news, location-based city maps...);

Supplementary services: call password; call waiting; call forwarding; call-forwarding on subscriber busy; call forwarding on no reply; call back; multi-party or conference calls; and, most generally, PBX services (see Chapter 2).

The evolution of GSM from the initial concept of a 900-MHz carrier with two 25-MHz bands, and the parallel developments of digital mobiles in the United

States and Japan, led to the so-called transition from 2.5 to 3G (or the full emergence of 3G), which is described in the next subsection.

4.2.5 From 2.5G Towards 3G Mobile Systems

The previous description represents a first phase of the GSM, as based upon the choice of a 900-MHz carrier with two 25-MHz bands, which is the GSM 900 product offer. For reasons of compatibility with the other bands centered near 450 MHz and 800 MHz (see Table 4.3), several variants of GSM systems were also developed, called in accordance the GSM 450 (either 450–467.5 MHz or 478–496 MHz), and the GSM 800 (824–894 MHz). By late 1990, the concept of *personal communication networks* (PCN) emerged, also called DCS-1800 for *digital communication system* at 1,800 MHz. The PCN or DCS-1800 was initially conceived for applications in the 1,800-MHz band (1710–1785 MHz), as a natural evolution phase from the pending 900-MHz generation, which was thought not sufficient to handle the future capacity demand. A similar label was that of *personal communication services*, or PCS, which evokes the service rather than the product aspect. Compared to previous GSM and other digital-cellular services, the PCS had additional specifications for *cost reduction* (access to larger numbers of users), *low-power operation* (smaller handsets) and *high-density coverage* (picocells of only a few meters) in suburban areas. Another argument for doubling the carrier frequency *f* was to reduce the antenna size (namely by half); such an improvement, however, implies the reduction of cell size, recalling that free-space loss is in $1/f^2$, which is consistent with denser mobile networks.

As it turned out, the PCS/DCS-1800 initiative was overcome by the rapid and successful deployment of the GSM 900. By early 2001, over 60% of worldwide cellular systems were based upon the GSM standard. Such progress reduced the competing DCS-1800 to a mere conceptual extension of GSM. Under the PCS conceptual approach, the GSM also developed its own new standards for the 1,800-MHz and 1,900-MHz bands, extending its family to five operating bands. In the USA the PCS market developed nevertheless, to complement the already well-settled AMPS, USDC (digital version of AMPS) and the expanding family of GSM 450–1,900. Apart from their differences in band allocations and coding/framing standards, the GSM systems of Europe, USDC of American and PDC of Japan offered similar service features. For this reason, the term "PCS" was extended to define this mobile-system generation as a whole, with the associated concept of *2.5G* to evoke an intermediary evolution step. But, short of more precise definitions, it has also been argued that PCS represents the beginnings of 3G, whether or not all expectations had actually been met in terms of broadband capability and standards implementation. Such an argument is however supported by the introduction into the PCS family of two major GSM service upgrades which are destined to give rapid Internet access. These are called:

High-speed circuit-switched data (HSCSD), and

General packet radio service (GPRS).

The rationale for developing HSCSD and GPRS is two-fold. First, the available data rates of 9.6–14.4 kbit/s (called TCH/F9.6–TCH/F14.4) were too slow for Internet applications. Second, the burst-prone nature of GSM traffic caused relatively long connection set-up times and made inefficient use of the system resource, which kept the connection prices too high. The solutions to this problem were (a) to introduce the concept of *packet* in order to tie together several frames and increase the user bit rate, and (b) to efficiently distribute the system bandwidth over the traffic by implementing optimum packet- and circuit-switched algorithms. Since GSM uses eight data slots per frame, it is possible to increase the user's bandwidth by tying together two, three, four slots or more, yielding the possible capacities:

TCH/F9.6: 19.2 kbit/s, 28.8 kbit/s, 38.4 kbit/s and up to 76.8 kbit/s;

TCH/F14.4: 28.8 kbit/s, 43.2 kbit/s, 57.6 kbit/s and up to 115.2 kbit/s.

The HSCSD standard allows for tying up to four channels, i.e. 38.4 kbit/s or 57.6 kbit/s, with the possibility of asymmetric allocation (e.g., $N = 1-2$ slots uplink and $N' = 3-4$ slots downlink). For symmetric use, the uplink and downlink time slots are also paired: each terminal transmits and receives N slot sequences alternatively. The HSCSD system architecture remains circuit-switched, as in conventional GSM, which minimizes the changes in system architecture and traffic management. However, three major drawbacks have so far delayed the implementation of HSCSD. The drawbacks are: (a) for the mobile units, the increase in power consumption and overheating; (b) for the user, the increase in connection price (four-fold and up to eight-fold in comparison to the half-codec rate; and (c) for the operator, the saturation of the cells and network capacity.

The second type of GSM upgrade, GPRS, follows an approach similar to HSCSD but based upon packet-switching. Since the Internet usually represents long connection times for HTML file transfers but containing idle periods for reading web pages, it is sensible to allocate the connection on a packet-switched rather than a circuit-switched basis. As described in Chapter 3 with ATM, packet-switching makes it possible to efficiently share the same physical link by means of *statistical multiplexing*, which fills up gaps between packets of a given channel. As result, the connection cost is no longer based on the actual duration of the call session, but on the number of packets actually transmitted. Such a tariffing approach is therefore perfectly adapted to the case of Internet connections. It is also sensible that voice be carried over packets, rather than over conventional time slots, but this requirement may heavily impact the design and complexity of mobile units. Because both voice and data should make optimal use of the time slots through circuit- and packet-switching, but not necessarily on the same level of service, GPRS distinguishes three service *grades*, defining the possibilities of:

1. simultaneous voice call and active Internet connection (surfing, browsing, e-mail);

2. simultaneous voice call and Internet connection, the connection being maintained without data transfer (voice call to be terminated otherwise);

3. either voice call or internet connection.

The uplink and downlink bandwidths can be chosen according to a wealth of options to select from the previously listed rates, i.e., $N \times$ TCH/F9.6 and $M \times$ TCH/F14.4. Instead of selecting an exact number of time slots for both uplink and downlink, it is sensible to define the bandwidth option as a maximum guaranteed number of slots, P, to share between the two links. As a result, GPRS offers no less than 29 slot classes, which are summarized in Table 4.4. The classes are either *full duplex* or not, depending upon the condition $P = N + M$ or $P < N + M$, respectively. The full-duplex operation means that all assigned slots are used, noting that it does not necessarily mean equal bit rates for the uplink and the downlink. The classes 15 to 18 can be distinguished from the others by the fact that the terminals (mobile and base-station) must be able to transmit and receive at the same time. Indeed, in these options the total number of slots to be transmitted and received, $N + M$, is greater than eight, which is the total number of available slots in a given frame. The maximum capacity in *both* directions is given by class 18, i.e., 115.2 kbit/s for uplink *and* downlink. In contrast, class 29 provides the maximum capacity for uplink *or* downlink, i.e. in a single direction during a given frame duration. It is seen from the table that the 29 options correspond to a wide choice of uplink and downlink capacity options, but the number of transmitted slots (defining the cost) is only 2, 3, 4, 5, 6, 8, 10, 12, 14, 16, defining a scale of one to ten. As previously mentioned, the 10–16 range is reserved for terminals having dual-mode transmit/receive capability. We observe that in spite of the technology upgrade of GPRS, the GSM capacity is limited to some 115 kbit/s, which remains well below the speed offered by wired technologies, whether based on microwave or lightwave transmissions. GPRS, which was introduced in 1999–2001, still appears to be one of the best technologies at this time for high-speed *mobile* Internet access. This should not be confused with *fixed wireless* Internet access, which is described in another subsection The full development of 3G mobile (UMTS) and future 4G generations are discussed in Sections 4.2.6 and 4.2.7.

4.2.6 Universal Mobile Telecommunications System (UMTS) and cdma2000

As discussed earlier, the phases 2G and 2.5G correspond to the introduction in cellular systems of *digital coding* (GSM, USDC, PDC) and *enhanced voice-and-data services* (HSCSD, GPRS), respectively. The concept of 2.5G is associated with a first deployment phase of *personal cellular services* (PCS). The concept of 3G is essentially one of broadband implementation (up to 2 Mbit/s), in addition to extending PCS with *multimedia* (voice, *interactive video* and data) capabilities. More than just value-added GSM, 3G should represent a "seamless" offer, meaning that it should work irrespective of the radio-access technology details, of the network infrastructure, of the country/geographic location, of the environment (rural to urban to home and office), and of the type of service or application. The spirit of 3G is also that of a convergence between traditional telecommunications, multimedia/entertainment and the Internet, which have so far developed independently.

Because the previous GSM-based technology is inherently limited to 100-kbit/s capacities, new digital standards for broadband cellular technologies had to be developed. Understandably, such standards would have to be backward-compatible with the

TABLE 4.4 Slot classes for GPRS, as defined by available uplink (top line) and downlink (left column) capacities in kbit/s and number of attributed slots

Uplink → Downlink ↓	14.4	28.8	43.2	57.6	72	86.4	100.8	115.2
14.4	1 1/1 (2)	2 2/1 (3)	4 3/1 (4)	8 4/1 (5)				
28.8		3 2/2 (3) 5 2/2 (4)	6 3/2 (4) 9 3/2 (5)	10 4/2 (5)		19 6/2 (8)		24 8/2 (8)
43.2			7 3/3 (4) 13 3/3 (6)	11 4/3 (5)		20 6/3 (8)		25 8/3 (8)
57.6				12 4/4 (5) 14 4/4 (8)		21 6/4 (8) 22 6/4 (8)		26 8/4 (8) 27 8/4 (8)
72					15 5/5 (10)			
86.4						16 6/6 (12) 23 6/6 (8)		28 8/6 (8)
100.8							17 7/7 (14)	
115.2								18 8/8 (16) 29 8/8 (8)

The numbers $X\ N/M\ (P)$ correspond to the class number (X), the number of slots in the uplink (N) or in the downlink (M), and the maximum allowed number of slots ($P \leq N+M$). All classes are full duplex, expect when underlined ($\underline{X N/M\ (P)}$). Classes 21–22 and 26–27 differ by slot-latency options.

existing ones, which complicates the innovation process and the definition of a unique trend. As a matter of fact, and expectedly so, the main operators from Europe, the United States and Japan are already set for different preferences and their own road-maps. To further complicate the picture, the background 3G standards were defined in the early 1990s, that is long before the Internet was so well known and popular as it is now. In 1992, the ITU formulated a project known as FPLMTS for *future public land mobile telecommunications system*, which was re-named for convenience IMT-2000, for *international mobile telecommunications* 2000. The number 2000 was both a reference to the future and to a 2,000-kbit/s – 2,000-MHz capacity-band target. The outcome of the proposal took the form of two main different strategies, which were called "UMTS" and "cdma2000," as described in the following text.

Despite strategy differences, one common technology trend was the transition from TDMA to CDMA. As we saw earlier, CDMA makes better use of the bandwidth and is more resistant to multipath interference. The second trend is the use of a higher-level coding such as *8-ary PSK*, also called *octogonal phase-shift keying*. Increasing the number of modulation levels from one (PSK) to eight or four (QPSK) to eight (see Sections 1.2 and 1.7) increases the number of bits/ symbols (or bit/s/Hz) efficiency by three-fold or two-fold, respectively. As we shall see, however, it is sensible that the new extended bit rate hierarchy be defined not only with respect to GSM slot classes but also ISDN. The trend for *wideband* CDMA has thus been named *W-CDMA*.

Let's make a brief tour of the different options leading to 3G (see also Figure 4.35 for visual reference):

EDGE (*enhanced data rates for global-GSM evolution*): As the name indicates, it is a proposed natural evolution of GSM/GPRS. It is the only approach towards 3G which does not consider CDMA, but keeps to TDMA. It is

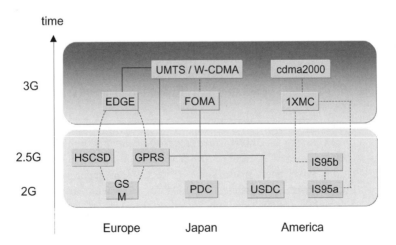

FIGURE 4.35 Road-map from 2G to 3G, with different intermediate system technologies. Dashed lines correspond to smooth upgrades, full lines correspond to full system-technology change.

divided into two implementation phases called *EDGE Phase 1* and *EDGE phase 2*.

- EDGE Phase 1 is also referred to as *enhanced GPRS* or E-GPRS, as based upon packet-switching. The objective of E-GPRS is a maximum user capacity of $8 \times 48 \, \text{kbit/s} = 384 \, \text{kbit/s}$, using 8 time slots with 8-PSK modulation. Other approaches vary the slot capacity according to FEC requirements (8-PSK being more sensitive to error), from 22.4 kbit/s to 59.2 kbit/s, leading to a maximum of 473.6 kbit/s. Note that these capacities include the codec/data payload plus FEC and encryption, which does not exactly compare with the GPRS slot capacities (TCH/F9.6 and TCH/F14.4);

- EDGE Phase 2 represents the same capacity goal with circuit-switching, as a form of *enhanced HSCSD*, or E-HSCSD. This capacity may not be available anywhere in the cell (or available only near the basestation), which makes the EDGE approach a difficult one, if not accepted as being a true 3G solution by Mbit/s-capacity standards.

UMTS (*universal mobile telecommunications system*): Initially conceived by 1992 as a *multienvironment* cellular system (rural, suburban, urban, home office) with up to 2 Mbit/s capacity-per-user, allowing for interactive multimedia and other broadband (Internet-based) services. Some key features of UMTS are the following:

- The term "W-CDMA" (*wideband CDMA*) is also indifferently used to designate UMTS although it refers to a coding technology, not a network or a service. The bandwidth is 5 MHz, which is 25 times broader than that of GSM. In conventional CDMA, the data from each baseband channel are multiplied by a code having a higher bit rate (chip rate), which expands the spectrum. The codes in use for this technology are referred to *gold codes* (as also used in GPS). In W-CDMA, the chip rate is made to change according to the available power, unlike in the conventional approach. The bandwidth spreading factor is thus varied from 4 to 128 times, the greater spreading corresponding to the stronger signals. Note that the other 3G CDMA technology, cdma2000 (see below) proceeds from a different and more conventional approach.

- Following the IMT-2000 initiative, the UMTS was taken over by the task forces known as 3GPP (*third-generation partnership project*) and the *UMTS Forum*. Different proposal versions and experimental implementations have been made since 1999 (v99 in Japan) to 2002 (v5 in Europe) with v6 expected by 2004 for worldwide implementation. In 1999, Japan developed an intermediate standard called *freedom of mobile multimedia access* (FOMA). Note that in UMTS, timing synchronization in the base stations does not rely upon the global positioning system (GPS), which is a service under the control of the U.S. Department of Defense, and on which Europe and Japan have made the choice not to depend. The UMTS network topology and specifics are described in detail further below.

cdma2000 is the other alternative standard to UMTS. The rationale is to further develop the previous (and only available) CDMA system called *cdmaOne* or *IS-95a*, then *cdmaTwo* or *IS-95b*.

- It is both useful and interesting to provide at first a technical background of the IS-95a-b systems, which are essentially different from the TDMA-based GSM. The CDMA technique is also more conventional than that used in W-CDMA/UMTS, as we shall explain. For spread-spectrum conversion, the system uses *Walsh codes*. These constitute a family of 64 numbers of 64-bit length that are mutually orthogonal ($a.AND.b = 0$). This makes it possible to retrieve any channel from the CDMA aggregate by the identical code multiplication. Thus 64 frequencies are used for each user-circuit transmission, unlike for TDMA, corresponding to a bandwidth *spreading factor* of 64. The codec or user bandwidth is 13 and 14.4 kbit/s for voice and data, respectively, corresponding after FEC to 28.8 kbit/s (uplink) and 19.2 kbit/s (downlink). The IS-95a system was developed for both 800-MHZ and 1,800–1,900-MHz bands. In a way similar to pre-empting more than one time slot in TDMA, the user-circuit capacity can be upgraded by using more than one code in CDMA. Thus in *IS-95b* (or *cdmaTwo*), up to 64 codes could be used in theory, corresponding for instance to 64×19.2 kbit/s = 1.22 Mbit/s. In practice, only 15 codes can be safely used, yielding 15×19.2 kbit/s = 288 kbit/s, which is exactly 2.5 times the theoretical value achievable by GPRS (115.2 kbit/s). A further upgrade with 32 codes yields 32×19.2 kbit/s = 614.4 kbit/s, which is also quoted as a reference target.

- The *cdma2000* project thus represents an ultimate 3G development of IS-95b. The proposed generations of cdma2000 are referenced to by the labels $1 \times$ MC and $3 \times$ MC for "first" and "third" generation multicarrier. An equivalent terminology is $1 \times$ RTT or $3 \times$ RTT (for *radio transmission technology*). The cdma2000 $1 \times$ MC (or $1 \times$ RTT) uses a single CDMA carrier with 1.25-MHz width. The cdma2000 $3 \times$ MC (or $3 \times$ RTT) uses one 3.75-MHz carrier for uplink and three 1.25-MHz carriers for downlink. The theoretically achievable capacities are 384 kbit/s ($1 \times$ MC) and 4 Mbit/s ($3 \times$ MC), respectively. Other alternatives based upon a single 1.25-MHz channel are referred to as *cdma2000 1xEV-DO* or *cdma2000 1xEV-DV* for "data-only" or "data and voice." For the downlink channel, the corresponding capacity potentials are 2.4 Mbit/s and 5.2 Mbit/s, respectively. For $1 \times$ EV-DO, the corresponding uplink capacity is 302.7 kbit/s. (No number is available at this time for $1 \times$ EV-DV.) Note that for modulation formats, the $1 \times$ EV approach is based upon either 8-PSK (average reception), or 16-QAM (clear reception). Poor reception prevents multilevel implementation, which reduces the capacity performance to that of QPSK. Finally, we note that cdma2000 relies upon GPS for timing synchronization in the base stations, unlike W-CDMA/UMTS.

Deploying the three types of 3G systems, EDGE, UMTS/W-CDMA and cdma2000 required the regulatory authorities and governments to allocate new frequency bands and sell corresponding exploitation licences to operators. Recalling that the 2G–2.5G mobile (PCS/PCN) systems already occupy the bands 450 MHz, 800–900 MHz and 1,800–1,900 MHz, there are two possible solutions: (a) reusing the 2G bands for 3G upgrade; and (b) allocating new bands in the EM spectrum.

Although looking like a rational or economical choice, the first solution is a very complex one because it would have to be progressive both in terms of geographical deployment and relative 3G penetration. The coexistence between the two system types also causes a problem of "pollution" by radio signals in the opposite-type receivers. New rules for collective and efficient *spectrum management* by the different mobile operators and local authorities must then be agreed upon, with negotiations in terms of resource sharing and commercial deployment. The second solution, i.e., using new EM bands for 3G, is apparently more straightforward. But a completely new network must be developed, with relatively limited possibility of cell reuse since the cell sizes for 3G are smaller. A second constraint is that the EM spectrum is already occupied by other radio services and agencies. Considering the UHF (300 MHz–3 GHz) band, here are some of the constraints (as also illustrated in Figure 4.36):

In most countries, the region 420–806 MHz is used for land-based TV broadcasting. Should 3G become a substitute to analog TV, or satellite-TV take

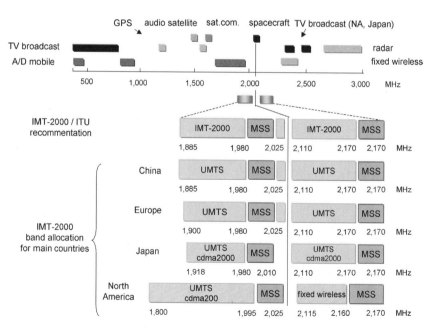

FIGURE 4.36 (top) UHF spectrum and current band allocations for 2G mobile communications, satellite communications, TV broadcasting, GPS/spacecraft navigation and radar; (bottom) IMT-2000/ITU frequency-band recommendation for 3G, including mobile satellite services/systems (MSS).

over, this band could be progressively exploited, but this is a far-sighted and hypothetical vision;

In the frequency interval above 800–900 MHz and under 1,800–1,900 MHz, the 1.2-GHz and 1.6-Hz bands are used by the *global positioning system* (GPS); the 1.5-GHz band is used by satellite digital-audio broadcasting; the 1.62-GHz band is used by mobile-satellite communications (Iridium, Globalstar for uplink);

In the rest of UHF, between 2,000 MHz and 3,000 MHz, the 2.02–2.11-GHz band is used for spacecraft navigation and communications; the 2.12–2.16-GHz band is used in North America for fixed wireless communications; the 2.29–2.40-GHz band is used for fixed wireless; the 2.3-GHz and 2.5-GHz bands are used in North America and Japan for land-based analog-TV broadcasting, respectively; the 2.49-GHz band is used by Globalstar for downlink; finally, the 2.7–3.4-GHz band is used for radar applications. The following text discusses these applications in more detail.

The remaining gaps left in this already busy EM spectrum sky were therefore selected by IMT-2000 for allocating the 3G frequencies, as illustrated in Figure 4.36. It is seen that uplink/downlink 3G frequencies should occupy two slots centered near 1.9 GHz (width 115 MHZ) and 2.1 GHz (width 60 MHz) located on both sides of the 2.0–2.1-GHz spacecraft band, respectively. In addition, two smaller bands of 30 MHz each have been reserved next to each of these slots for *mobile satellite services/systems* (MSS), as shown in the figure. North America has already fully exploited the 1,800–1,900-MHz band for PCS (2.5G), and already has attributed the 2.1-Hz band for fixed wireless applications. Therefore, the implementation of 3G in North America will be a matter of reusing the 1,800–1,900 band, according to the approach discussed earlier, in contrast with other countries.

It is to be expected that a UMTS network should have some similarities with the GSM one, but it should also be designed to handle different traffic types, i.e., circuit-switched or packet-switched, without congestion. Conveniently, the UMTS terminology to designate network elements (NE) is not too different from the GSM, apart from inevitable new acronyms. Figure 4.37 shows the NE layout of UMTS and the respective names. One can recognize globally the familiar arrangement of a GSM network (see Figure 4.32 for comparison), but with the following key differences:

The first network zone is that of the mobile units which are referred to as *user equipment* or UE; the UE is made up of the *mobile equipment* (ME) and an *UMTS subscriber identity module* (USIM); the SIM "smart card" has the same functions as in previous GSM, consisting in user identification, authentication and encryption;

The second network zone is the UTRAN, for *UMTS radio access network*. The UTRAN is made up of a set of *radio-network subsystems,* or RNS; each RNS includes two base stations and a controller; the base stations are now called *node B* (a 3GPP terminology); two nodes B are being connected to a single *radio network controller* (RNC); the RNCs are connected together for the purpose of assuming certain common roles during user

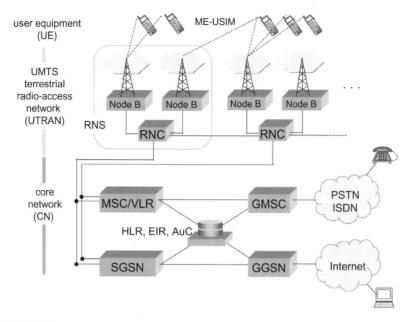

FIGURE 4.37 Layout of UMTS network (see text for description).

connections: *serving* RNC (SRNC), *drifting* RNC (DRNC) and *controlling* RNC (CNRC); these are elaborate management and logical functions within the UTRAN access points (soft and interfrequency *handover* being an example).

The third network zone is the *core network* (CN); as the figure shows, the CN is organized on two parallel structures, to which the RNCs are connected; the entry nodes are

1. a GSM-like *mobile switching center* (MSC), as directly associated with a *visitor location register* (VLR); it is concerned with circuit-switched traffic;

2. a node called *serving GPRS support node*, or SGSN; it is concerned with packet-switched traffic;

the MSC and SGSN entry nodes share a common database of *home location register* (HLR), and *equipment identity register* (EIR), which are associated with an *authorization center* (AuC); consistently, each or these node connects to either a circuit-switched network (PSTN, ISDN) or a packet-switched network (e.g., Internet); these connections require two different gateways, called GMSC for *gateway MSC* and GGSN *for gateway GPRS support node*, respectively.

To complete the picture, it is worth listing the names of the different software interfaces involved between the different elements above. They are called:

Cu interface: Internally links the hardware of the ME (handset) with the USIM (smart card);

Uu interface: Handles the W-CDMA radio connection between UEs of any manufactured brand to the nodes B;

Iub interface: Ensures connection between node B and its RNC;

Iur interface: Ensures communications between different RNCs, in particular the transition from drift-RNC to service-RNC during soft handover (see above);

Iu interface: Connects UTRAN to the core network (CN).

Conveniently, the above interfaces have common standards which allows independent manufacturers to develop their own versions of mutually compatible UMTS systems and subsystems.

4.2.7 3G Services and Beyond 3G

In a previous subsection, we listed the basic "enhanced" services offered by GSM, as a 2.5G phase. Through UMTS or cdma2000, does 3G introduce any new category of added-value service? It is a debate, since some consider that most services actually needed are already in place with 2.5G, and 3G has not yet met its high expectations. But such a perspective could be only momentary and also fails to recognize the new features introduced by wide-band operation in cellular networks. One needs then to step away from the network perspective and technological details to address what one would really like to get from an advanced mobile service. To quote some of the new features:

Interactive video: The same service as already available from fixed-units connected to the Internet; the challenge is to popularize a real portable "video-phone" and make it as easy to use as in the previous fixed and mobile "telephone" generations. For color displays, a technical issue is image resolution and quality, which are criteria somewhat opposite to the intrinsic size limitations of a handset.

Voice recognition: To achieve hands-free mobile connections (call dial-up and pick-up) with features such as authentication and simple menu commands; the possibility of dictating short alphanumeric messages, analogous to quick e-mail (see further below).

Short messaging services (SMS): Already existing in GSM with 160-character text size, but cumbersome for regular e-mail users, yet already extremely popular in the teenage population; with 3G, possibility via Internet to attach HTML files (e.g., e-card, JPEG pictures, documents, etc.); multimedia upgrades of SMS are called MMS; also, new possibilities for *instant messaging* (like *paging*) which can have many useful applications for personal and business use; SMS represents a clear potential for voice recognition and synthesis technologies, which eliminates the hassle of dialing text characters one by one on tiny keyboards and scrutinizing messages from small-size displays; this also opens perspectives to hearing- and vision-impaired users.

Mobile commerce: Not so much the idea of purchasing items from the handset as to be able to charge transactions without entering credit-card information;

the user is thus able to connect, via the Internet, to his/her personal *e-account, e-subscription* or *e-wallet* for authentication, authorization, charging/depositing and any validation purposes.

Wireless Internet access (WAP): The possibility to download and display HTML website pages, using a new compressed format; the number of format-compatible URLs (*uniform site locators* or Web site addresses) is however restricted and the display quality is inherently reduced; it is nevertheless a useful means to access flight/train schedules, stock information and bank account balances; WAP is the precursor of the fully-fledged *webphone*.

Electronic messaging: The possibility to access or consult one's electronic mail (e-mail) from a mobile unit; this feature is already within GPRS capability (MMS) and known as *i-mode*, a service offered by DoCoMo in Japan; in contrast with WAP, i-mode is always "switched on," without any dialing requirement. See Section 4.4 on *wireless LANs* (WLAN) and Subsection 4.2.7 on *wireless ATM* (WATM).

Wireless device-to-device connections: Also known as "*Bluetooth*" makes it possible for devices such as computers (fixed and laptop), printers, faxes, headsets and cellular handsets to be connected via a radio link (frequencies are 802.11 MHz and 2 GHz within a few meter range. A key application is mobile Internet access from a laptop computer, via a cellular phone (no more wires and wall plugs!). See Section 4.4 on WLAN; see also next item;

Internet-controlled appliances: Some envision future *home appliances* such as refrigerators, washers, heating, surveillance cameras, etc... being connected to the Internet, via fixed or short-range mobile (Bluetooth) radio links. The purpose is that of implementing remote control/activation and maintenance (installation, upgrade, repairs, trouble-shooting, etc.) of any of these appliances; it could also concern *vital portable appliances* such as pace-makers and other health-monitoring devices, and retinal displays, a futuristic view of *personal/portable virtual reality,* as already used by specialized army forces.

Internet-controlled cars: Another hint of the long-term future, which extends the concept of Internet-controlled appliances to individual transportation; applications range from maintenance, inspection-check and warning systems (an even more intelligent dashboard than the current generation), safety (prevention of accidents, collisions and loss of control), security (user authentication, surveillance, de-activation/tracking/location of stolen vehicles), efficiency (on-site map downloading, automatic navigation, traffic-path optimization). Another application is driving "take-over" on parallel highways, which makes sense for long cross-country trips and even more so for day-to-day commuting. During that time, one could safely watch news, movies, go through one's e-mail and make business calls.

This vision of 3G services is not so new, if one recollects the doomed videophone of the 1970s. It is also pervaded with a certain amount of hype, which fails to recognize the inherent technology limitations, or even worse, the practical view of lab implementation to field installation, with high returns for the operator and low connection cost for the customer. The only difference between the two views is that our current one is by and large closer to reality, with a new mindset from both private and business users. This is mainly attributable to the invasion of the Internet and mobile communications in both private and business environments. Such considerations then raise the issue of whether 3G technologies will ever be adequate to handle such an ambitious promise. In this respect, it is worth briefly describing the concepts associated with 3.5G, all the way up to 4G. According to any expectation, these upcoming generations are not defined with any more precision and accuracy than 3G. However, one can summarize some of their differentiating aspects and outreach according to the following:

3.5G: the next incremental move from 3G, schematically defined as the transition from initial narrowband technologies (a few 100 kbit/s) to full wideband implementation (2 Mbit/s to 20 Mbit/s); associated with the concept of *personal digital assistants* (PDA), conceptually evolved from the PCS and PCN view, and that of the *personal access network* (PAN) which concerns the home environment (Bluetooth); the PDA would offer a range of preference-focused services such as personalized Web browsing, individual address-notebook and book-keeping, speech and handwriting recognition (already introduced in Palm-pilot devices) and many other personalized-service features; some analysts prefer to define 3.5G as "evolved 3G" or "enhanced 3G," and any future developments as "mature 3G" or "beyond 3G (B3G)" systems.

4G: the radical transition of cellular systems and services to ultra-wideband operation at 100 Mbit/s (full mobility) to 1 Gbit/s rates (low mobility); converging technologies and service compatibility with wireless LANs; for mobile services, such an evolution towards higher bit rates goes together with further chip integration and processing speed, and improved power efficiency; introduction of *software-defined radio* (SDR) where SW algorithms control the generation and tuning of frequencies, as opposed to built-in HW definition; introduction of *orthogonal frequency-division multiplexing* (OFDM), a cousin and alternative to CDMA (see Section 4.2.4 on GSM); independently of technology performance, 4G could also be thought of as a phase of cost-effectiveness both in terms of terminals conception and network operation; the multiplication of ever-new service offers should drive new trends for service attractiveness and user-friendliness.

It is clear that at this current stage of 3G deployment, these previous considerations concerning 3.5G and 4G technologies and services represent only projections and speculations for the future.

4.2.8 Wireless ATM (WATM) Networks

The Layer 2 protocol called *asynchronous transfer mode* (ATM) was developed to meet the needs of multiservice delivery of voice, video and data in *fixed packet-switched networks* (see Chapter 3). This is in contrast with other Layer 2 protocols such as Ethernet or X.25, which were developed to address the needs of data transport in local-area networks (LAN). Unlike layer 3 TCP/IP, which operates in a *connectionless* mode, ATM is *connection-oriented*, which makes it more (if not best) suited to *real-time* multimedia applications. But ATM switches do not have the higher capabilities of IP routers, which is a limitation for carrying dense and randomly varying traffic. In Chapter 3, we described the principle of *mapping IP over ATM*. Such a mapping, where IP datagrams are encapsulated into ATM cells, makes it possible to benefit from the advantages of either protocol, the network including both ATM switches and IP routers, according to a predefined service zone. Since the approach proves to be beneficial for fixed wired-networks, it must also be true for *fixed wireless* (MMDS/LMDS, see Section 4.4) and *mobile/cellular* (3G) networks. Hence the concept of *wireless ATM* or WATM.

One could conceive of a WATM to be made of a standard core ATM network with peripheral radio access for fixed or mobile stations. While such a simplified a picture is accurate, there are a few hidden issues involved in radio access, and which are of important consequence for practical implementation. A first issue is the large differential in bit rates, namely from the 155-Mbit/s ATM offer down to the capacities of 30–45 Mbit/s (LMDS) or 2–20 Mbit/s (3–3.5G) available from (most advanced) fixed or mobile radio systems, respectively. Another issue, which is characteristic of radio-system environments, is the control of packet-cell delay, error-rate and loss. Because ATM was initially developed for fixed wired-networks, it is intrinsically more suitable to *wireless local loops* (WLL) where terminals (nodes) have fixed positions. The ATM protocol, which defines virtual connections and paths, is not adapted to cellular/mobile networks where the terminal nodes are moving. Thus, some network protocol enhancements had to be developed in order to address the issue of mobility, particularly considering the three functions of *soft handover*, *location management* and *call routing*. The situation becomes even more complicated if the WATM service classes and requirements must remain strictly the same as in standard ATM ones. Here, we shall focus on the issues which cover the larger picture and are specifically associated with *mobile* WATM, which is what is usually meant by WATM.

One of the key differences between ATM and cellular networks is that, in the first case, cells are routed without path hierarchy or any mediation of control centers. In contrast, cellular systems use centralized *mobile switching centers* (MSC), *network switching subsystems* (NSS) and *core network* (CN) service gateways. Among the centralized functions are the *home/visitor location registers* (HLR, VLR) for user localization and identification, and authorization/billing. Thus, a dedicated *ATM mobility server* (AMS) could be implemented to replace the MSC with a double mission. The first is to reroute calls that are addressed to the local cellular network jurisdiction into that same network and control handovers, as the MSC does; the second is to act as a gateway to the backhaul ATM network. Since ATM connections between two nodes are based upon the establishment of *virtual connections* and *virtual paths* (see Chapter 3), the paths must be modified as a mobile node moves from one base station (BS) cell to the next.

Three methods to achieve soft handover in WATM networks are possible. These are called: (a) *dynamic rerouting* (changing the path dynamically); (b) *virtual-connection-tree handover* (reconfiguring the virtual paths according to a preset connection-tree hierarchy); and (c) *cell forwarding.*

The last solution, *cell forwarding*, is the most straightforward and does not need AMS. It consists in concatenating together the initial path (which lead to the first BS) with the new path (which leads to the new BS). The cells destined to the new BS node thus follow an elongated path which includes all the previously addressed BS nodes. The principle of cell forwarding and virtual-path handover is illustrated in Figure 4.38. The disadvantage of cell-forwarding is that path elongation increases the delays and the queuing requirements at ATM switching nodes. This is in addition to the common possibility that the elongated path loops onto itself, meaning that cells can travel more than once over a given physical link, returning to the same nodes as they might have come from, which is inefficient. A solution to this problem is to have pre-established virtual connections with optimized path allocation. In the so-called *Bahama handover* scheme, the virtual paths (also called branches in this use) are periodically reconfigured to the optimum scheme. To be even more effective, the approach uses the *virtual-path identifier* (VPI) for network addresses rather than the physical BS address. The advantage of cell forwarding is that no AMS is required, the switching intelligence being distributed among the nodes, including base stations as well.

The approach of *virtual connection-tree handover*, as mediated by an AMS, makes a more efficient use of the network. The principle is illustrated in Figure 4.39. The downlink cells are thus routed to the new BS, utilizing the shortest-available new virtual path/branch. A drawback is the possibility that

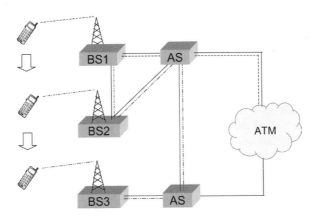

FIGURE 4.38 Principle of call-handover through *cell forwarding* in wireless ATM (WATM) networks, made of radio base stations (BSi) and ATM switch (AS) nodes. The solid lines indicate the network's physical links. The dashed line indicates the initial virtual path followed by the ATM cells as the mobile station first connects to base station BS1. The dot-dashed line and the double-dot dashed line indicate the elongated virtual path followed by the ATM cells as the mobile stations connect to the base stations BS2 and BS3, respectively.

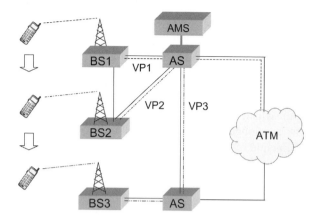

FIGURE 4.39 Principle of *virtual connection-tree handover* in WATM networks. The solid lines indicate the network's physical links. The ATM cells first follow the virtual path VP1 (dashed line), then VP2 (dot-dashed line) and VP3 (double-dot dashed line), as progressively re-configured by an ATM mobility server (AMS), while the call is handed over from base-station BS1, to BS2 then BS3 (AS = ATM switches).

downlink cells that were routed to the old BS prior to the handover but not yet transmitted have to be rerouted or be lost; another possibility is that uplink cells coming from the new BS arrive *before* the uplink cells coming from the old BS. Both cases requires cell re-ordering. A solution to alleviate these problems is to use *handover delimiters*. Such delimiters are sent by the mobile unit to signal a call start to the new BS/branch and a call stop to the old BS/branch. Upon receiving the start delimiter, the AMS buffers and queues the cells until the end-delimiter is received, then routes the sequence to the new branch.

Finally, *dynamic rerouting* can also be implemented, which requires one ATM swiching node to redefine virtual paths during the handover according to preset rules and in real time. One technique, referred to as *nearest common node* (NCN) handover, consists in rerouting the traffic through the node that is closest to the two BS involved. This requires the mediation of *zone managers* (ZM), which are nodes in charge of managing a cluster of neighboring nodes. The attribution of NCN is given by the local ZM, and the NCN role is passed on in case of multiple handover sequences. The approach assumes that the latency and link-transmission delays experienced during the handover are negligible compared to the delays associated with the radio path. If the link is direct, NCN handover is similar to cell forwarding. In the case of indirect links, the ZM initiates a new path after searching and identifying the NCN. Both (old and new) downlink paths are held active until the handover is stabilized and duly completed. Cells that could have been duplicated are discarded by the receiving station. Nonreal-time traffic (see ATM service classes in Chapter 3) can allow some buffering at both BS (downlink traffic) and mobile terminal (uplink traffic) until the process is completed. Clearly, the three approaches for WATM handover have their own advantages and drawbacks, and their effectiveness is a matter of both network topography and resource (node density, number of radio access points, network traffic load) and service-class requirement (e.g., bit rate, cell-loss rate, timing relation, payload type, network percentage usage).

The above provides only a flavor of the extra complexity involved in WATM networks, which is not found in either wired-ATM or cellular networks. A key difference between WATM and cellular networks such as GSM or UMTS, is that there is no central hub to take care of routing optimization, registering home and temporary "visitor" users, and managing operations. Since IP can be seamlessly mapped over ATM (see Chapter 3), WATM represents yet another channel for broadband wireless access (BWA, see Section 4.4), as compared to previously-described 3G/4G mobile. It is not clear yet whether the "mobile ATM" approach will some day take the upper hand over the emerging IP-based networks with wire-less-access capability (currently known as WAP and i-mode). See more on this topic in Section 4.4 on "Fixed wireless networks." The reason to develop wireless IP access in the "fixed wireless" context is dictated by the emerging concepts of *wire-less local loops* (WLL) and *personal area networks* (PAN), in which "mobility" takes new meanings, dimensions and values.

▪ 4.3 SATELLITE-BASED COMMUNICATIONS

The field of *satellite communications* is now 40 years old, dating back from the 1962 launch of the first international telephone satellite, Telstar. This satellite had the capacity of 12 telephone circuits. These circuits could be used for international phone calls over the Atlantic during half-hour slots and a few time-slots per day because of its fast revolution period relative to the Earth. The following decades of progress in electronics, computers, analog and digital radio and signal processing, directive-antenna design, solar batteries, rockets, and a wealth of space technologies have set the grounds for the vast array of modern *satellite services*, which are not limited to telecommunications, as first reviewed in the first subsection. We shall consider next some basic engineering of *satellite orbits*, which helps in understanding the difference between orbit types, their shapes and their respective distances to the Earth. Then we will focus on the main subject, which is *satellite telecommunications*.

4.3.1 Types of Satellite-Based Network Services

Satellite-based network services can be categorized into five general types (Figure 4.40):

Fixed satellite services (FSS): Establish a communication link between two fixed "gateway" Earth stations; one of the earliest applications is *international telecommunications*, which has been progressively taken over by *submarine optical-cable networks* (see following text and also see Chapter 2 in Desurvire, 2004).

Mobile satellite services (MSS): Divided into *aeronautical* (AMSS), *maritime* (MMSS) and *land-based* (LMSS) services, as designated by the type of

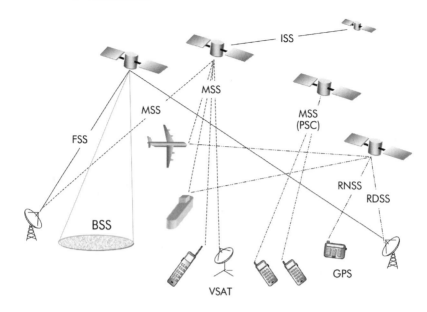

FIGURE 4.40 Different types of satellite-based services: fixed-satellite service (FSS), broadcast satellite service (BSS), mobile-satellite service (MSS) with aeronautical, maritime or land-based mobile units, MSS for personal satellite communications (PSC), radio-navigation satellite service (RNSS), radio-determination satellite service (RDSS) and intersatellite service (ISS). GPS = global positioning system, VSAT = very-small aperture antenna terminal.

mobile antenna; establish a communication link between mobile units and fixed Earth-stations, or directly between mobiles. In AMSS and MMSS, the receiving units are mobile, which requires tracking the satellite position (phased-array antenna); in the case of land-based units (LMSS), the units can be used for broadcasting (e.g., TV/radio live reporting for sports events, from remote countries or war zones), which usually requires a portable dish antenna; its can also concern *personal satellite communications*, where constellations of low-orbit satellites make possible direct communications between portable handsets (*enhanced*-MSS, or EMSS); as discussed below, this field has been overshadowed by the sweeping development of *land-based cellular systems, which are described in the previous section*;

Broadcasting satellite services (BSS): The domain of *world radio and TV broadcasting*; the radio/TV signals are emitted by Earth stations directed toward the satellites, which in turn broadcast them to predefined country/continental zones; the receiving units can be fixed (with well-known dish antennas) or mobile (in the case of mobile units, the antennas can be a portable dish.

Radio-navigation satellite services (RNSS): A unidirectional communication system which provides the geographical position of a land-based mobile unit for navigation purpose; examples are the *global positioning system*

(GPS) and its Russian or European versions, GLONASS or Galileo (see more on GPS and other global positioning systems in Subsection 4.3.2); RNSS also exist for aeronautical and maritime applications;

Radio-determination satellite services (RDSS): A two-directional version of RNSS, where satellites also acquire information on the position of mobile units, and provide it to both units and monitoring Earth stations; an example of RDSS is the system *Iridium* (see below);

Intersatellite services (ISS): Used for establishing communications links between different satellites in GEO-GEO, GEO-LEO and LEO-LEO orbiting configurations (see below for definitions), using both radio and lightwave signals; an example of radio-ISS is provided by the Iridium system;

Earth-exploration satellite services (EESS) and *meteorological satellite services* (METSS): Used for general scientific, public and military purposes.

Figure 4.40 illustrates the broad variety of satellite-based services. In this section, we shall focus on the two first service categories, i.e., FSS and MSS, which are more directly relevant to the telecommunications field. The next subsection concerns the engineering basics of satellite systems and Earth-to-satellite links. We further discuss global positioning systems, which are associated with certain types of MSS systems and terminal units. The aforementioned services and in particular the *constellation*-based systems are also described.

4.3.2 Engineering Basics of Satellite Orbits

The fact that objects such as satellites can be launched into space and set to rotate indefinitely around the Earth no longer captures the imagination. Yet, the principle of satellites is the same as the one making it possible for our solar system to exist and sustain itself instead of collapsing. It is also the familiar principle of the Earth rotating around the Sun or the Moon around the Earth, with revolution periods of about one year (365.25 days) or one month (27.32/29.53 days, depending upon definitions), respectively. Thus, satellites can be viewed as manmade "moons," which can be made to occupy fixed or variable positions in the Earth's sky. In this subsection, we shall make an introductory tour of the engineering basics of satellite orbits, where they come from and how they can be designed for workable systems. Although we need to introduce lots of formulae, definitions and astronomic jargon, there are no complex notions involved. It only takes some effort to analyze step by step the mechanics of the satellite trajectory, and its three-dimensional characterization. In the following, we shall consider one after another all the aspects involved in the basic knowledge of satellite orbits.

Gravitation

The trajectories followed by satellites around the Earth (like planets around the Sun) are referred to as *orbits*. An orbit is the object's closed path dictated by mechanics (the science of motion), according to its initial conditions of location and speed, when submitted to the Earth's *gravitational field*. It is worth recalling Newton's law of gravitation, as expressed for the Earth's case. This law defines the force F which attracts, towards the gravity center of the Earth, any object located at distance r and having a mass m. It simply writes

$$F = -\mu \frac{m}{r^2} \tag{4.43}$$

where $\mu = 3.986 \times 10^{14} \ \mathrm{m^3/s^2}$ is the Earth gravitational constant and the sign "$-$" recalls that the force is attractive. The units for F,m,r are *Newton* ($1 \ \mathrm{N} = 1 \ \mathrm{kgm/s^2}$), kilogram (kg) and meter (m), respectively. On the Earth's surface, the mean distance to the Earth's center at the Equator and sea level is $R_E = 6{,}378$ km, which gives, for a mass of $m = 1$ kg (2.2 *pounds*), a force of $F = 9.798$ N or 9.8 N. This attractive force is felt by human beings as their *weight*; the weight is different from the mass, even if popular use confuses the two. Indeed, while the mass is a universe invariant, the weight depends upon the distance r to the Earth's center and to the mass of the Earth, which is contained in the parameter μ. We observe that the gravitational force rapidly decreases with distance. For instance, at a distance corresponding to 3.33 Earth radii ($3.33 \ R_E = 21{,}238$ km), this force is one-tenth as strong.

Circular Orbits

Since objects naturally fall to Earth, setting a satellite into the right orbit must require specific initial conditions. In the general case, launching an object at a certain distance of a gravitational field with initial direction and speed can lead to different possible trajectories, as shown in Figure 4.41. This is a simple problem solved by mechanics. The result of the analysis is that the object trajectory, as defined by the radial distance r versus observation angle θ (see Figure 4.41) is given by

$$r(\theta) = \frac{p^2/\mu}{1 + (p^2/\mu)q \cos(\theta + \phi)} \tag{4.44}$$

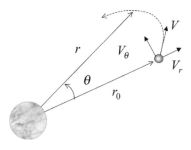

FIGURE 4.41 Launching an object in space at distance r from the center of a gravitational field, with initial speed/direction condition (V) which projects into radial (V_r) and transverse (V_θ) components. The resulting trajectory, which depends upon these initial parameters, is shown as a dashed line.

where p, q, ϕ are constants given by the initial launching conditions. As Figure 4.41 shows, the initial speed, V, breaks into a radial and a transverse component, V_r and V_θ. The transverse speed is related to the *angular speed*, ω, according to $V_\theta \equiv r\omega$. Note that the unit for ω is *radians/s*. The astronomer Jean Képler, who first did this analysis, discovered that the constant p is related to the angular speed and radial distance according to

$$p = \omega r^2 = \omega_0 r_0^2 \qquad (4.45)$$

where the "0" subscripts indicate values at the initial launching point. Assume next that we can achieve the condition $q = 0$. From equation (4.44), the radius of the trajectory is constant (independent of the angle θ) and equal to: $r \equiv r_0 = p_2/\mu$. This condition corresponds to a perfect *circular* trajectory or orbit. Substituting the value of p from this definition, we get the required angular speed, which we now call Ω_0:

$$\Omega_0 = \sqrt{\frac{\mu}{r_0^3}} \qquad (4.46)$$

The time it takes for the object to return back to the initial point or to complete a full *orbit revolution* is therefore given by

$$T_0 = \frac{2\pi}{\Omega_0} = 2\pi \sqrt{\frac{r_0^3}{\mu}} \qquad (4.47)$$

which is referred to as the *revolution period*. We note that both the angular speed yielding a circular orbit and the corresponding revolution period are independent of the satellite's mass. This is at least true for manmade satellites, whose relative masses do not disturb the gravitational field, unlike the Moon with respect to the Earth.

Since we know how to achieve circular orbits, we can manage to make the revolution period equal to the Earth's rotation rate. If the satellite is launched in the Equatorial plane in the same rotation direction as the Earth, it should then occupy the same point in the sky (or equivalently, move synchronously with the Earth, as if it were attached to it by an invisible cable). The Earth rotates about itself in one 24-h day, more accurately in $T_S = 23$ h 56 mn 4.1 s. This period is called *sidereal day*, the adjective meaning an absolute reference with respect to the stars. Note that the sidereal day is shorter than the conventional 24-h day by about 3.6 mn. It is a simple exercise (see the end of the chapter) to show that the satellite's distance giving $T_S = T_0$ is $r_0 = 42{,}163$ km, corresponding to an altitude of 35,785 km (circular-orbit case). Such satellites are called *geosynchronous*, and their orbit is referred to as a *geosynchronous Earth orbit* or GEO. If the orbit plane is the Equator, the satellite is apparently immobile in the sky, which is referred to as *geostationary* satellite orbit (GSO). If the orbit plane is different from the Equator, the satellite is still geosynchronous (GEO) but not stationary. The same is true for noncircular orbits having a revolution period equal to the sidereal day. In either or both cases, the apparent motion in the sky takes a shape similar to the number "8" with the upper loop being located the Northern Hemisphere and the lower one in the Southern Hemisphere. The loop's amplitude depends upon the orbit-plane "inclination," while its oscillation period depends upon the orbit's "eccentricity" (see the following definitions). Interestingly, the GEO (synchronous and stationary) concept and orbit design were first proposed in 1945 by the science-fiction writer Arthur C. Clarke, author of the famous *2001, a Space Odyssey*.

Since the revolution period varies according to $T \approx r_0^{3/2}$, satellites launched at distances closer to the earth than $r_0 \approx 42,000$ km have shorter periods or revolve faster, hence making more than one revolution per day (see Exercises at the end of the chapter).

Elliptical Orbits

This analysis concerned *circular orbits*. As discovered by Képler, the general case corresponds to *elliptical* orbits. Ellipses are like flattened circles having two centers each called a *focus* (or foci for plural). The center of gravity occupies one of the two foci, and the revolving object or planet thus revolves according to an elliptical trajectory, like the Earth around the Sun.

The fact that orbits are most generally elliptical is showed by rewriting equation (4.44) in the form

$$r(\theta) = \frac{a(1 - e^2)}{1 + e \cos (\theta + \phi)} \tag{4.48}$$

where a is the ellipse's *major-axis diameter* and e its *eccentricity*. The meaning of these two parameters, and their relation to other geometric parameters of the ellipse, are shown in Figure 4.42 and the caption. In astronomy terminology, the closest and farthest points to one of the foci are called *perigee* (or *perihelion*) and *apogee*, respectively. The angle θ, measured between the perigee and the satellite's position, is referred to as *true anomaly* (see the following discussion on "anomaly" angles).

We must now relate the physical parameters p, q, μ to the ellipse parameters a, e, by identifying equation (4.48) and equation (4.44), i.e.:

$$\begin{cases} a(1 - e^2) \equiv p^2/\mu \\ e \equiv qp^2/\mu \end{cases} \tag{4.49}$$

We shall specify now the relation between the constants q and ϕ and the initial conditions r_0, V_r (see Figure 4.41):

$$\begin{cases} \sin \phi = \dfrac{V_r}{pq} \\[2mm] \cos \phi = \dfrac{1}{q}\left(\dfrac{1}{r_0} - \dfrac{\mu}{p^2}\right) \end{cases} \tag{4.50}$$

To simplify the analysis, we can assume that the satellite is launched from the point at distance r_0 with zero radial speed, i.e., $V_r = 0$. This condition gives $\sin \phi = 0$, or $\phi = k\pi$ ($k = 0, 1, 2 \ldots$) or $\cos \phi = \pm 1$. In turn, we get from

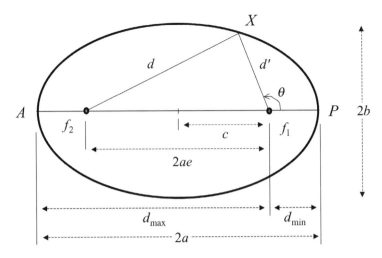

FIGURE 4.42 Characteristics and properties of ellipses with eccentricity $e = c/a$ and major-axis diameter $2a$: the distance between the focal points f_1, f_2 is $2ae$, the minor-axis diameter is $2b = 2a\sqrt{1 - e^2}$, the sum of distances from any point X of the ellipse to the foci is $d + d' = 2a$, the minimum and maximum distances to a given foci are $d_{\min} = a(1 - e)$ and $d_{\max} = a(1 + e)$, respectively. The two points closest to $f_1(A)$ and farthest from $f_1(P)$ are called apogee and the perigee, respectively. The angle θ is called true anomaly.

previous definition:

$$q = \pm\left(\frac{1}{r_0} - \frac{\mu}{p^2}\right) \tag{4.51}$$

Recall next that the constant p is proportional to the initial angular speed ω_0, according to $p = r_0^2 \omega_0$. One can then express this angular speed with respect to that giving a circular orbit (Ω_0) according to $\omega_0 \equiv \eta \Omega_0$ where η is a real positive number. The case $\eta \neq 1$ therefore corresponds to conditions departing from the circular orbit. Replacing the definition in equation (4.46) for Ω_0 into equation (4.52) yields

$$q = \pm\frac{1}{r_0}\left(1 - \frac{1}{\eta^2}\right) \tag{4.52}$$

and

$$\begin{cases} a(1 - e^2) = \dfrac{p^2}{\mu} = r_0\eta^2 \\[2mm] e = q\dfrac{p^2}{\mu} = \pm\left(\eta^2 - 1\right) \end{cases} \tag{4.53}$$

From this last result, we can rewrite the orbit definition in equation (4.48) in the form

$$r(\theta) \equiv r_0 \frac{\eta^2}{1 + (\eta^2 - 1)\cos\theta} \tag{4.54}$$

with $\theta' = \theta$ or $\theta + \pi$ (depending from which foci the angle is measured). When $\theta' = 0$, we have $r = r_0$. Thus r_0 represents the distance between the focus and the perigee, which is also the minimum possible distance from the satellite and the focus point, as called d_{min} in Figure 4.42. At the point of apogee ($\theta' = \pi$), we have $r = r_0\eta^2(2 - \eta^2) = d_{max}$, which corresponds to the maximum possible distance between the satellite and the focus point.

In order for the previously defined trajectory to be an ellipse, however, the quantity $\eta^2 - 1$ must represent an *eccentricity* and therefore be comprised between 0 and unity ($e = 1$ representing the limiting case of an ellipse with zero minor-axis diameter). This condition yields $0 \le \eta < \sqrt{2}$, corresponding to angular speeds $0 \le \omega_0 < \Omega_0\sqrt{2}$. Figure 4.43 shows plots of the elliptical orbits obtained for different values of the parameter η, assuming $r_0 = 1$ in all cases. It is seen that, according to expectation, initial conditions with $0 \le \eta < \sqrt{2}$ yield either elliptical ($0 < \eta < 1$ or $1 < \eta < \sqrt{2}$) or circular ($\eta = 1$) orbits. What happens for higher angular speeds corresponding to $\eta > \sqrt{2}$? The above trajectory formula is still valid but only for angles such that $1 + (\eta^2 - 1)\cos\theta' > 0$, or equivalently $\cos\theta' > -1/(\eta^2 - 1)$, using the fact that now $\eta^2 - 1 > 0$). For $\eta = \sqrt{2}$, this condition is $\cos\theta' > -1$, which is always verified except for the angle $\theta' = \pi$. As the angle approaches this limit, the distance becomes infinite or $r \to \infty$. This means that the trajectory is no

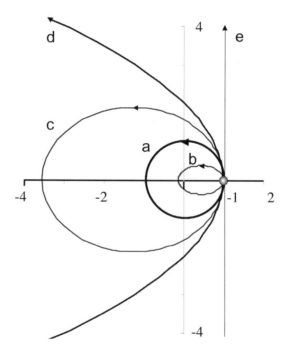

FIGURE 4.43 Different types of orbit corresponding to launching angular-speed conditions $\omega_0 = \eta\Omega_0$: (a) circular, or $\eta = 1$; (b) and (c): elliptical with $\eta = 0.5$ or $\eta = 1.25$, respectively. The case (d) correspond to $\eta = \sqrt{2}$, which is the limiting case of open or a deorbiting trajectory. The case (d) corresponds to an infinite initial angular speed.

longer closed, but open. The "satellite" moves indefinitely away from the launching point $(2r_0, \theta = 0)$ towards the asymptotic direction defined by $\theta' = \pi$. This case is illustrated in Figure 4.43, which only shows the trajectory in the vicinity of the gravitation center. Just under this limit ($\eta = \sqrt{2} - \varepsilon$), the trajectory is a closed elliptical orbit with major-axis diameter $a \approx r_0/(2\varepsilon\sqrt{2})$, as can be checked by using equation (4.53) in this approximation, which increases to infinity as $\varepsilon \to 0$. We can therefore call the $\eta = \sqrt{2}$ limit a "deorbiting" condition, since the trajectory becomes open and the satellite escapes the gravitational capture. Above the $\eta = \sqrt{2}$ limit, the asymptotic direction angle is given by $\cos \theta' = -1/(\eta^2 - 1)$. For infinite initial angular speeds ($\eta \to \infty$, or $\omega_0 \gg \Omega_0$), we have $\cos \theta' \to 0$, or $\theta' \to \pi/2$, which shows that the object escapes the gravitational field in the perpendicular direction (see Figure 4.43), as if the field has no effect on the trajectory.

Note that in satellite systems, the launch into final orbit is made from firing a booster at the point of apogee, and not at the perigee. This is because the required orbital speed is minimal at the apogee, contrary to the perigee, as will be shown below. Thus, the acceleration force, fuel consumption and fuel weight are also minimized and cost effective.

Three-Dimensional Orbit Definition

Once we have determined the orbit's shape and parameters (and the location of the satellite within this orbit), the next step is to accurately define the orbit's orientation with respect to the Earth's Equatorial plane. This is a tricky three-dimensional definition problem, but it becomes simple if one introduces the three angle parameters shown in Figure 4.44. The figure shows that the orbital and Equatorial planes have a relative angle defined by the *inclination i* and intersect at the Earth's center (since the Earth's center occupies a focus of the ellipse). The intersection points between the orbit and the Equator are called *ascending* and *descending nodes*, respectively. The angle ψ between the line of nodes and the perigee direction is called *argument of perigee* (or perihelion). The longitude angle φ of the line of nodes in the Equatorial plane is called *right ascension* of the ascending node. Thus the set of angles i, ψ, φ completely define the three-dimensional orientation of the orbit with respect to the Equator. Such a definition must be completed by specifying the direction of the satellite's rotation, i.e., the same as the Earth (called *prograde*) or opposite to the Earth (called *retrograde*).

Unlike the true anomaly, θ, which defines the satellite's position in the orbit path, the angles i, ψ, φ are theoretically fixed parameters. This is true in the ideal view of an Earth-satellite system with no external or internal perturbations to modify its stability. Such ideal conditions correspond to *unperturbed* or *Képler orbits*. External perturbations can be listed as follows:

> The small combined gravity pull from the Sun and the Moon, called *lunisolar attraction*. Such a perturbation represents about 10^{-7} times the Earth gravity at orbit altitudes $<10,000$ km. It results in an oscillating elevation angle (latitude) of $\pm 0.02°$/week;

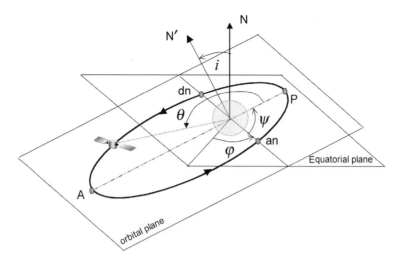

FIGURE 4.44 Parameters used to completely define orbits and satellite location. The orbit plane is at inclination i from the Equator plane. On the orbit plane: P = perigee, A = apogee, an = ascending node, dn = descending node, ψ = argument of perigee, and θ = true anomaly (satellite position in orbit). On the Equator plane: ascending (an) and descending (dn) nodes, and φ = right ascension of ascending node.

The radiation pressure exerted by the solar wind; it can be shown that the force due to solar pressure is $F = 9.2 \times 10^{-6} S$ (N/m^2) where S is the satellite's surface oriented towards the Sun; such a force corresponds to an acceleration of $\gamma = 9.2 \times 10^{-6}$ m/s^2 per unit mass (kg) and unit surface (m^2); the transverse component of this acceleration perturbs the longitude corresponding for instance to a maximum of $\pm 0.002°$/week ($m = 100$ kg, $s = 1$ m^2);

The effect of nonuniformity between land mass and ocean basins; in average, this default can be viewed as the Equator being slightly elliptical, with a 150-m difference between the major and minor axis lengths; this ellipticity results in weekly speed variations up to ± 0.1 m/s, corresponding to $\pm 0.08°$/week.

Internal perturbations of the orbit concern the drag from the Earth's atmosphere, which causes the satellite speed to slowly decrease over its operational lifetime, leading to eventual "end of life" de-orbiting in the absence of corrective speed adjustments. But the main internal perturbation is caused by the fact that the Earth is not perfectly spherical. Rather, it is "oblate," meaning that it is flattened about the poles' axis. The reason of such an oblateness is the centrifugal force exerted by the rotation of the Earth on its geological constituents, which behave like high-viscosity fluids. As a result of this centrifugation effect, the Equatorial radius is 21.4 km greater than the polar radius, representing a $1/288.257 = 0.3\%$ relative increase. This asymmetry creates a perturbing *torque* effect, which causes the line of nodes to rotate in the direction opposite to the satellite's rotation direction. For this reason, the effect is called *nodal regression*. It can be shown that at each revolution the angle defining the longitude of the line of nodes (right ascension φ) has moved by the negative quantity:

$$\delta\varphi = -3\pi J_2 \left(\frac{R_E}{r}\right)^2 \cos(i) \tag{4.55}$$

where $J_2 = 1.0823 \times 10^{-3}$ is a constant reflecting the flattening of the gravity field and i is the orbit inclination. It is seen from the expression that the regression is maximal for polar orbits ($i = 0$) and for Equatorial orbits ($i = 90°$). Converted in terms of daily rate, the nodal regression is:

$$\delta\varphi/day = -3\pi J_2 \left(\frac{R_E}{r}\right)^2 \frac{\cos(i)}{T} \times 86{,}400 \tag{4.56}$$

where T is the orbit period in seconds (see the following definition) and $86{,}400 = 24 \times 3{,}600$ is the number of seconds per day. An application of nodal regression is the *Sun-synchronous orbit*. In this case, the line of nodes (and orbital plane) rotate at the same rate as the Earth around the Sun, presenting the same angle throughout the year cycle (see Exercises at the end of the chapter). The purpose of Sun-synchronous orbits is the avoidance of *Earth eclipses* which disrupt the satellite's power activation from the Sun and cause detrimental thermal shocks (see below). Another purpose is the possibility of monitoring Earth locations through the year with the same satellite–Earth–Sun angle configuration, which is useful for weather and military surveillance applications.

The flattening of the Earth also perturbs the argument of perigee, ψ. Its change over time is called *precession of perigee* (also "rotation of the apsides"). It can be shown that the precession (daily) rate, also called *apsidal rotation rate*, is

$$\delta\psi/day = \frac{3\pi J_2}{2} \left(\frac{R_E}{r}\right)^2 \frac{5\cos^2(i) - 1}{T(s)} \times 86{,}400 \tag{4.57}$$

This precession is seen to vanish for critical inclinations such that $\cos(i) = \pm 1/\sqrt{5}$, corresponding to $i = 63.435°$ or $i = 116.565°$, respectively. These critical angles are used for applications of highly-elliptical orbits where the perigee's argument must be kept constant, as in HEO (Molnya) orbit types (see below).

Two exercises at the end of the chapter provide numeric examples of nodal regression and precession of perigee. As we shall see next, with the knowledge of these orbit characteristics, called *orbital elements*, it is possible to precisely predict a satellite's location at any time of history, starting from its day of launch. However, the related predictions must be corrected over time in order to take into account the previously described perturbations, even if they average out.

Key Orbit Parameters

The preceding analysis shows that it is possible to completely "design" a satellite orbit according to specific and well-known launching conditions from a point in space. As we have seen, the launching conditions (speed, magnitude and absolute direction) determine an ellipse with major axis $2a$, eccentricity e and orientation angles i, ψ, φ with respect to the Earth's Equatorial plane. From a communications-system perspective, however, the main design parameters of interest reduce to the following three:

1. The *minimal* and *maximal distance to the Earth*, which define a connection round-trip time, called *latency*;

2. The *revolution period*, which globally determine satellite-access times or *slots*;

3. The *inclination* of the orbit relative to the Equator, which defines the pattern of regions traversed each day by the satellite during one or several revolutions.

The minimal and maximal distances are given by the relations $d_{min} = a(1 - e)$ (perigee) and $d_{min} = a(1 + e)$. Since the Earth occupies one of the foci, the minimal and maximal distances between the satellite and the Earth's surface are obtained by subtracting the Earth's radius r_E. The minimal and maximal latencies are thus

$$\Delta T_{max, min} = 2 \frac{a(1 \pm e) - R_E}{c} \tag{4.58}$$

where $c = 3 \times 10^8$ m/s is the speed of light. In the case of GEO satellites, we have $e \approx 0$, $a \approx 42,000$ km and $R_E \approx 6,400$ km, which gives $\Delta T = 0.237$ s or approximately $\frac{1}{4}$ second. Such a small latency is noticeable as a source of annoyance when international phone calls transit via GEO satellites, at least in normal conversations where two parties try to exchange a few words at a time (see below on the issue of GEO latency).

The next design parameter of interest for satellite systems is the revolution period. In the case of elliptical orbits, it can be shown that the period is simply given by the relation

$$T = 2\pi \sqrt{\frac{a^3}{\mu}} \tag{4.59}$$

which with our notations becomes

$$T = 2\pi \sqrt{\frac{r_0^3}{\mu}} \left(\frac{1}{2 - \eta^2} \right)^{\frac{3}{2}} = T_0 \left(\frac{1}{2 - \eta^2} \right)^{\frac{3}{2}} \tag{4.60}$$

where (to recall) we have $0 \leq \eta < \sqrt{2}$ and T_0 is the period corresponding to a circular orbit. When the orbit period is an exact fraction of a day, i.e., $T = 24\,\text{h}/n$ ($n =$ integer), the orbit is referred to as an nth *resonant* orbit.

We saw earlier that the angle θ between the satellite's position and the perigee is called the *true anomaly*. In the general case of elliptical orbits, the satellite's speed varies with position (see next), therefore the true anomaly does not increase linearly with time, unlike in a circular-orbit case. As a matter of fact, the function $\theta(t)$ is defined by a transcendental equation, meaning that it takes an iterative numerical resolution process to determine $\theta(t)$ for a given time t. In order to simplify the description, a parameter called *mean anomaly* (or "phase") was introduced. The mean anomaly is simply defined as:

$$M(t) = 2\pi\frac{t}{T} = 2\pi\frac{t}{T_0}(2 - \eta^2)^{\frac{3}{2}} \tag{4.61}$$

where t is the time elapsed from the crossing of the perigee. Hence, we have $M(0) = 0$ at perigee, $M(T/2) = \pi$ at the apogee, and since M is an angle, $M(T) = 2\pi \equiv 0$. The mean anomaly thus corresponds to the true anomaly of a fictitious satellite with circular orbit having the same period T. The mean anomaly increases linearly with time, unlike the true anomaly. Note that in this field's jargon, the time t corresponding to a satellite location in the orbit is called "epoch" or "epoch time." For the reader eager to know the relation between mean and true anomaly: $\cos\theta = (\cos E - e)/(1 - e\cos E)$, where E is called *"eccentric"* anomaly. The eccentric anomaly is the solution of the transcendental equation:$M = E - e\sin E$, given the mean anomaly M at epoch t.

We now have at hand all the elements to fully characterize an orbit and the position of the satellite within this orbit at a given epoch. These elements make it possible for official agencies to edit *prediction reports* for the day-by-day position of any registered satellite with respect to the ground. The reference records used by space agencies (such as NASA) follow what is referred to as the *two-line element* (TLE) format. The TLE is in fact made of three coded lines, made of one title line followed by two 69-character lines, looking like:

```
XXXXXXXXXXX
1_NNNNNU_NNNNNAAA_NNNNN.NNNNNNNN_+.NNNNNNNN_+NNNNN-N_+NNNNN-N_N_NNNNN
2_NNNNN_NNN.NNNN_NNN.NNNN_NNNNNNN_NNN.NNNN_NNN.NNNN_N.NNNNNNNNNNNNNNN
```

In this code, line 0 is an 11-character title identifying the satellite by name. The code definitions corresponding to the two 69-character lines 1 and 2 are provided in Table 4.5. As seen from the table, this report contains complete information regarding the satellite and its orbit elements. As an illustration, consider the following example of TLE:

```
ISS (Zarya)
1_25544U_98067A_ _ 03012.50098380_(other parameters)
2_25544_ 51.6357_82.3009_0005171_26.7855_163.1305_15.58940125236716
```

Following Table 4.5, this record concern the *International Space Station* (ISS), named after its first element, Zarya. The data indicate that the ISS was launched in 1998 as the 67th launch event in that year. The record's epoch is year 2003, day 12 (January 12th) at universal time 0.500998380 (12 h 01 mn 25 s). Its orbital elements are $i = 51.6357°$, $\varphi = 82.3009°$, $e = 0.0005171$, $\psi = 26.7855°$. The mean anomaly at this epoch is $M = 163.1305°$. The number of revolutions per day is 15.58940125, corresponding to one revolution per 0.064146145 days, or one revolution in period $T = 1.53950748$ h $\equiv 1$ h 32 mn 22.226 s (about 1 h 30). The rest of the data (except the last checksum digit), $n = 23,671$, is the

TABLE 4.5   Code definition corresponding to two-line element (TLE) record of orbits

Line 1		Line 2	
1	Line number (=1)	1	Line number (=2)
3–7	Satellite registration number	3–7	Satellite registration number
10–11	2-digit launch year (e.g., "98" for 1998)	9–16	Orbit inclination i (degrees)
12–14	Launch number in year (international)	18–25	Right ascension of ascending node φ (degrees)
15–17	Designation of piece (e.g., piece "A")	27–33	Orbit eccentricity e (e.g. ".01" for $e = 0.01$)
19–20	2-digit epoch year (e.g., "03" for 2003)	35–42	Argument of perigee ψ (degrees)
21–32	Epoch day (see caption*)	44–51	Mean anomaly M at epoch (degrees)
34–43	Parameter related to speed	53–63	Number of revolutions per day
45–52	Parameter related to acceleration	64–68	Revolution number at epoch since launch
54–61	Other parameter	69	Checksum (**)
63	Ephemeris type		
65–68	Element number		
69	Checksum (**)		

The left column indicates the character rank in the string. (*) Epoch day is defined as Julian calendar day followed by hour-fraction of day at universal standard time (GMT), e.g., 158.60763888 for June 7th (day 158) at 14:35 GMT. (**) Modulo 10 sum of all characters including blanks, periods, plus (value = 0) and minus (value = 1) signs.

revolution number as counted from launch date. The total ISS time in orbit is thus $23{,}671 \times T_{\mathrm{h}}/24\,\mathrm{h} = 1{,}518.4$ days or 4.157 years, or 4 years and 57 days. This brings the theoretical ISS launch time to November 16th 1998, which is close to the real date of November 20th, 1998. The small error (0.2%) can be explained by slight changes in the orbit parameters and other small perturbation effects during this long 4-year course. Note that the orbit period of $T = 1.53950748\,\mathrm{h} \approx 1/15.6$ days is close to the case of a 16th-resonant orbit. This is a convenient rotation rate for synchronizing work and rest activity schedules between different crews.

Satellite's Relative Speed

While the revolution period represents a reference design parameter, it does not indicate at what speed the satellite is moving at a given point of its orbit, unless the orbit is a circular one. With elliptic orbits indeed, the satellite's speed (V) is not uniform, unlike in the circular- case. Referring to Figure 4.41, the total speed is given by $V^2 = V_{\mathrm{r}}^2 + V_{\theta}^2$, where V_{r} and V_{θ} are the radial and angular/tangential speeds, respectively. We already know the value of V_{θ} since it is given by the relation

$V_\theta = r\omega \equiv r(p/r^2) = p/r$. It can be shown from detailed analysis that the total satellite speed reduces to the simple equivalent formulae:

$$V^2 = V_r^2 + V_\theta^2$$

$$= p^2 q^2 \sin^2 \theta + \left(\frac{p}{r}\right)^2$$

$$= \frac{2\mu}{r} + p^2 q^2 - \frac{\mu}{p^2}$$

$$= \frac{2\mu}{r} - \frac{\mu}{a} \tag{4.62}$$

where (to recall) $q = 1/d_{min} - \mu/p^2$ and $d_{min} = a(1 - e)$. In this result, the term $pq \sin \theta$ corresponds to the radial speed (V_r), which vanishes at both perigee and apogee ($\theta = 0$ and $\theta = \pi$), and is maximum for $\theta = \pi/2$ and $\theta = 3\pi/2$. The angular/tangential speed ($V_\theta = p/r$) is maximum or minimum at the perigee ($r = d_{min} = a(1 - e)$) or apogee ($r = d_{max} = a(1 + e)$), respectively. Thus, one can conceive of an elliptical orbit as a trajectory starting from the perigee with maximum speed ($V^2 = 2\mu/d_{min} - \mu/a$), decelerating all the way to the apogee to reach minimum speed $V^2 = 2\mu/d_{max} - \mu/a$, and accelerating back to the perigee. In the case of circular orbits, we have $e = 0$ and $d_{max} = d_{min} = a$, therefore the speed is uniform with magnitude $V = \sqrt{\mu/a}$, where a is the orbit's radius. Note that in this equation, V corresponds to the absolute speed of the satellite relative to the Earth's center. Considering next the satellite speed *relative to the Earth's surface*, which is of interest for satellite-service applications, we must take into account two effects (a) the Earth's own rotation about the pole axis, and (b) the inclination of the orbit with respect to the Equator. This effect of earth rotation depends upon the *latitude* λ of the observation point, as defined by the elevation angle above the Equator (e.g., for New York, $\lambda \approx 42°$ and for the polar circles $\lambda \approx \pm 66.5°$). It is simply established that the tangential speed at latitude λ is $V_E(\lambda) = 2\pi r_E \cos \lambda/T_S$. On the Equator ($\lambda = 0$), the speed is $V_E(0) = 2\pi r_E/T_S = 2\pi \times 6{,}378$ km/ $(23$ h 56 mn 4.1 s$) = 0.465$ km/s or $1{,}674$ km/h. As surprising as it may sound, the Earth is indeed turning this fast about itself: people located on the Equator cover an absolute distance of nearly 40,000 km per day! At the latitudes of New York or of the polar circles, this tangential speed is reduced by the factors $\cos 42° = 0.74$ or $\cos 42° = 0.40$, corresponding to $V_E = 1{,}240$ km/h or $V_E = 669$ km/h, respectively. We are now able to evaluate the satellite's speed with respect to a point on Earth located just under its position (the satellite is said to be at this point's *zenith*). We shall assume that the satellite rotates in the same direction as the Earth (i.e., counterclockwise when looking from the North Pole), which leads to three basic cases:

1. *Equatorial orbit*: the relative satellite speed is the difference $V_R = V - V_E = \sqrt{2\mu/r - \mu/a} - 2\pi r_E/T_s$;

2. *Orbit with inclination of angle 'i' with respect to Equator*: the relative satellite speed is $V_R = V - V_E \cos(\lambda)\cos(i) = \sqrt{2\mu/r - \mu/a} - 2\pi r_E \cos(\lambda) \times \cos(i)/T_s$;

3. *Polar orbit* $(i = 90°)$: no effect of Earth rotation, i.e., the relative satellite speed is $V_R = V = \sqrt{2\mu/r - \mu/a}$.

An illustration of basic satellite period and speed calculations, which concerns the interesting example of the ISS, is provided in an exercise at the end of this chapter.

Types of Orbit

The previous analysis has shown that designing a satellite system is first a matter of choosing an orbit type (circular, elliptic, geostationary), then defining its parameters such as eccentricity, minimum and maximum distance to the Earth (maximum and minimum satellite altitude), and the revolution period, as linked with the previous ones. Finally, the orbit must be defined within orientation plane, which we have defined as the inclination (plane angle with the Equatorial plane). The orbit's inclination can be anything between Equatorial $(i = 0)$ and polar $(i = 90°)$. Figure 4.45 summarizes different orbit-choice possibilities. One can distinguish four types of orbits after their altitude ranges, and with respect to two intermediate layers called *Van Allen belts* (see the following explanation):

1. *Low-Earth orbits* (LEO): Altitude range 100–1,500 km, any inclination from Equatorial to polar, typically circular (or elliptic with very low eccentricity);

2. *Middle-Earth orbits* (MEO): Altitude range 5,000–15,000 km; any inclination from Equatorial to polar;

3. *Highly elliptical Earth orbit or high-Earth orbit* (HEO): Altitude range 500–40,000 km;

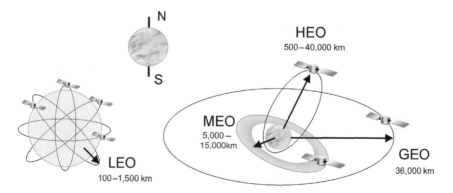

FIGURE 4.45 Different Earth-orbit types: low (LEO), middle (MEO) and geostationary (GEO) in equatorial orbit planes, and highly elliptical (HEO). The corresponding altitude ranges are indicated for each type.

4. *Geosynchronous Earth orbit* (GEO): Elliptic or circular orbit which has a mean revolution period of $T_0 = 23$ h 56 mn 4.1 s, corresponding to the exact Earth's rotation period (sidereal day); the particular case of *geostationary* satellites (GSO) corresponds to circular orbits located in the Earth's equatorial plane; their orbit radius and altitude are 42,163 km and 35,785 km, respectively;

The 1,500–5,000 km altitude gap between LEO and MEO satellite types is justified by the need to avoid the so-called *inner or lower Van Allen belt*. The second 15,000 km upper limit for MEO altitudes is also defined by the *higher or outer Van Allen belt*. The two Van Allen belts, so-named after their discovery by James Van Allen in 1958, are regions of intense radiation due to the collision of high-energy charged particles with the Earth's atmosphere. These particles, of cosmic-ray and solar origin, are trapped by the Earth's magnetic field and forced to follow the field lines according to helical trajectories converging to the Earth's magnetic poles. The resulting plasma forms two concentric, doughnut-shaped clouds in the plane normal to the Earth's magnetic axis, which is slightly offset from the poles ($11°$). The extent of the two belts depends upon the particles' energies. These range from 100 keV to 10 MeV (the *electron-volt* is an energy unit with value 1 eV $= 1.6 \times 10^{-19}$ joules). The inner Van Allen belt is essentially made of protons and spreads over 650–15,000 km altitudes. Its peak intensity is reached near 5,000 km. The inner belt also contains heavy ionized nuclei such as O, He, N and C with energies up to 50 MeV. The outer Van Allen belt is mainly formed of electrons and spreads over 13,000–50,000 km altitudes, with peak intensity at 19,000 km. To be complete, there exists a third Van Allen belt in between the inner and outer ones, which is closely linked to solar activity (11-year cycles). Given the altitude ranges for LEO, MEO and HEO satellites, we see that satellites are exposed to some extent to the Van Allen belt radiation. This radiation has most damaging effects on both semiconductor electronics and optical devices. It also causes electronic background noise, digital detector errors, and deleterious electrostatic discharges. Finally, it is harmful to astronauts in extravehicular or unshielded activity. This is part of the reason why manned space-craft (e.g., Space Shuttle, ISS) are made to orbit at relatively low altitudes such as 250–400 km, and also why there are no satellite orbits within the two Van Allen belts.

The four orbit types, LEO, MEO, GEO and HEO have different characteristics and service applications.

Consider first *GEO systems*. As previously described, GEO satellites occupy fixed or nearly fixed points in the Earth's sky with a synchronous Equatorial orbit. The fixed case corresponds to circular orbits located in the equatorial plane, called geostationary (GSO). The other case corresponds to GEO satellites with any elliptical and inclination parameters, which makes them describe "figure 8" trajectories in the sky. This property of Earth synchronicity offers the key advantage of a fixed or nearly fixed radio coverage over huge continental scales. As a matter of fact, the maximum theoretical intercity distance is about 18,000 km, as defined by the Earth's horizon, representing a bit less than one-half of the Earth's circumference. Since radio rays that are nearly tangential to the Earth have a thicker atmospheric layer to traverse, they are more absorbed by, or susceptible to rain, ice or dust than the Equatorial rays (see Section 4.1). The near-tangential rays must also stay clear of obstacles such as mountains or buildings. Over-looking such differences in signal quality away from the optimal Equator region, only three GEO systems suffice in theory to cover all points on Earth, excluding the polar regions (see Exercises at the end of the chapter). A second advantage of GEO systems is that ground-based antennas do not have to track the satellites, i.e., they have fixed or permanent orientations. These antennas should be highly directional, with a radiation pattern or aperture sufficient to capture the GEO satellites having "figure 8" motion. These two key advantages are in contrast with LEO, MEO and HEO systems, which have more limited geographical coverage and need antenna track-ing. Note that portable transceiver units for LEO systems only use basic dipole anten-nas. These antennas are oriented at a pre-defined angle (or can be tilted to the right position), which alleviates this tracking requirement, at the expense of a poor directive gain (see Section 4.1). Given all these limitations and antenna-link prerequisites, the area effectively covered by individual GEO satellites is called the "*footprint.*"

Having considered the advantages of GEO systems, we can list some of their main drawbacks:

The need for powerful launchers, e.g., rockets such as ESA's Ariane or *space transportation systems* (STS) like NASA's Space Shuttle, to put them into the high altitude of 36,000 km; this requires a complex series of space man-euvers, such as:

(a) Positioning into intermediate transfer orbits,

(b) Need of two apogee motor firings (AMF),

(c) Motor ejection at perigee,

(d) Orbit-plane reorientations, and

(e) Corrections and final satellite-axis reorientation.

In contrast, LEO satellites can be placed into their nominal orbits through a single launching operation (followed by the necessary reorientation); another disadvantage is that the highly remote location of GEO satellites prevents the possibility of maintenance (or repair) though STS missions, unlike with non-GEO systems.

The need to correct small deviations from the fixed point; indeed, small pertur-bations (see earlier description) cause the satellite to drift away from its nominal or stationary position; the required correction of 0.1° or 0.05° in lati-tude or longitude, called *station keeping*, can be made by on-board firing thrus-ters (or boosters), which have inherently limited fuel capacity; a way to alleviate this station keeping requirement is to chose GEO orbits with small eccentricities and inclinations; as previously described, the satellite then follows "figure 8" loops; with a minimal number of station-keeping corrections (i.e., on a weekly or bimonthly basis) the loop excursions about the nominal point can then be kept within the capture angle of the Earth-station antenna;

The periodic advent of *Earth eclipses*; indeed, the Earth periodically crosses the satellite–Sun line of sight, which momentarily interrupts the satellite's exposure to the Sun's heat and puts it into the shade; the satellite must be designed to avoid deleterious thermal shocks due to nonuniform heat distribution between the illuminated and shaded zones; it is simple to show that the maximum eclipse time per day is given by $\delta t = 2 \tan^{-1}(R_E/h) \times 24\,\text{h} \times 60\,\text{mn}/360° = 2 \tan^{-1}(6{,}378\,\text{km}/42{,}164\,\text{km}) \times 24 \times 60/360 = 68.8\,\text{mn}$ or 70 mn. A more detailed calculation shows that such eclipses occur daily during a 42-day period centered on the two equinox dates, with a 30 mn duration at ± 20 days; note that for GEO satellites, the northern and southern sides are alternatively in the shade for six months and illuminated for another six months at a relatively low $23° \, 27'$ angle;

The signal *latency*, corresponding to the round-trip time between two Earth stations, which is about 0.24 s, as shown earlier in this subsection. As some readers may have experienced during international calls going though GEO systems, such a latency is not compatible with the rate of normal conversations; it also causes synchronicity problems with packet-communications protocols such as the Internet (Chapter 3);

The uneven signal coverage, which gives an advantage to near-Equatorial latitudes; even at intermediate latitudes (such as Tokyo, New York, Paris or London with $\lambda = 35-50°$), the line of sight (LOS) must stay clear of buildings such as skyscrapers. At higher latitudes, mountains and even hills can also block the LOS; as previously mentioned, the atmospheric absorption also increases rapidly with latitude, due to longer atmospheric paths, which necessitates dish antennas of larger size or aperture;

At higher latitudes, the event of *sun outages*, which correspond to conjunction events where the Sun passes close to or right behind the GEO satellite during the two equinox periods; these conjunctions cause the Sun radio noise to overwhelm the signal, provoking a temporary interruption of the communication link; such outages can last between 1 and 10 minutes, depending upon the dish size and latitude, and occur up to several times during each equinox; the Sun angle with respect to the Equator varies according to $u = 23° \, 27' \sin[2\pi(t - \tau)/365]$ where t is the day of the year and $\tau = 0$ or 182 for the northern and southern hemispheres, respectively; from this formula, simple trigonometry and calendar analysis make it possible to predict the conjunction dates and their duration as a function of the receiving antenna latitude (as pointing to the GEO satellite);

The exhaustion of available GEO spots: in an equatorial orbit, there should be no more than one GEO satellite per $2°$ angle in order to avoid interference between different systems; this gives $360°/2° = 180$ spots; to date, already 150 are occupied!;

Orbit at the end of their operational lifetime, GEO satellites permanently remain in orbit since they are not subject to atmospheric drag, unlike with other satellite types; this creates a zone of "space junkyard" humorously referred to as *Clarke's belt*; such junk satellites present collision hazards vis-à-vis GEO satellites being in their orbit placement and correction phase, or vis-à-vis any spacecraft traversing the region, however small the collision probability; their existence still requires tedious bookkeeping.

We consider next the case of *HEO systems*. The rationale for highly elliptical orbits is to provide a broad Earth coverage, as in a GEO system, but with orbit planes other than that of the Equator. One of the rationales is the coverage of polar or high-latitude regions which are not accessible to GEO systems. While the perigee is an altitude of a few hundred kilometers, the apogee is at 25–40,000 km altitudes. As seen earlier, the satellite's speed goes through a maximum at the perigee and a minimum at the apogee. The satellite thus spends more time travelling to the apogee and moving back from this point than circling the Earth to pass through the perigee. As a result, the satellite's coverage is similar to a GEO, especially when close to the apogee, except for the interruption caused by the perigee passing. However, the position of the perigee and hence, of the orbit-axis orientation, are not stable in time. Indeed, we have seen that the perigee undergoes a precession which is a function of the orbit's inclination i, equation (4.56). Interestingly, the functional dependence is $\delta\psi(i) \propto 5\cos^2(i) - 1$, which means that there are two critical inclinations $(\cos(i) = \pm 1/\sqrt{5})$ for which there is no precession effect. As previously mentioned, these critical inclinations are $i = 63.435°$ or $i = 116.565°$, respectively. The original concept of precession-free HEO was implemented by Russia in 1965 with the Molnya satellite system. The development of this system was motivated by the need to have a nearly fixed national coverage of the USSR territories, which spread across 11 time zones in the Northern Hemisphere. In the Molnya system, the inclination angle was $i = 63.435°$; the apogee and perigee altitudes were $h_A = 39,957$ km and $h_P = 548$ km, respectively. This gives a major-axis length of $2a = 2R_E + h_A + h_P = 53,261$ km and an eccentricity of $e = 1 - (R_E + h_P)/a = 0.74$. The revolution period is $T = 2\pi\sqrt{a^3/\mu} = 4.32 \times 10^4 = s = 12.0$ hours, which shows that the Molnya satellite completes two revolutions per day. The perigee being situated in the Southern Hemisphere, the satellite thus spends the majority of its time in the Northern Hemisphere. Because of the high eccentricity of the orbit, the satellite actually spends 11 hours per revolution in the northern part, which shows that the corresponding satellite coverage is interrupted only 2 hours per day. The name of this system has been kept since to designate this HEO class, as referred to "molnya orbits." Typically, HEO systems have altitude ranges between 500 km (perigee) and 25,000 km (apogee), with high eccentricities of $e = 0.74$. As in the Molnya system, the key advantage of HEO systems is to cover vast continental areas with orbit inclinations different from the Equatorial ones (as we have seen with GEO systems, the number of available spots is now quite reduced).The coverage time per revolution is close to that provided by a GEO, hence the designation of "*super-synchronous*" for HEO satellites. Two satellites can be placed in a single HEO orbit at opposite points $(\theta, \theta + \pi)$ in order to ensure full-time coverage by 12 h turns, providing an uninterrupted "virtual GEO" service. The latency is however dependent upon the satellites positions, and is maximum (about $\frac{1}{4}$ s) at the apogee point. The minimum latency is at the perigee, but this point is normally outside the operation window. The effective minimum latency corresponds to the satellite being located on the orbit's minor axis. Since the ellipse-axis ratio is $b/a = \sqrt{1 - e^2}(e = 0.74)$ this latency is 0.67 times $\frac{1}{4}$ s, or about 1/6 s. While there is no perigee regression for $i = 63.435$, the effect of nodal regression remains. As one exercise at the end of the chapter shows, the corresponding regression rate is -0.442 deg/day, or $180°$ in 1.1 year. It is therefore not possible to obtain a permanently fixed LEO orbit, even if the nodal regression is relatively small. While LEO systems are in principle easier to launch compared to GEO ones, the issues of orbit design, tracking and operation/maintenance are obviously far more complex.

We consider next *MEO systems*. As previously discussed these satellites stand clear from the inner and outer Van Allen belts. These belts define a slot of altitudes between 5,000 km and 15,000 km. The advantage is to have a compromise between GEO and LEO systems, i.e., short latencies associated with relatively broad coverage, or "footprint," of the continental zones traversed. For circular orbits, the latency is constant and given by $\Delta T = 0.237 \times a_{MEO}/a_{GEO} = 0.06 - 0.12$ s (or 0.1s for 11,000-km orbit altitudes). Note that this latency should be multiplied by the number of satellite hops required to communicate between two Earth stations (see more on latency below). The MEO revolution period is $T = 2\pi\sqrt{a^3/\mu} = 1.2 - 3.1 \times 10^4$ s $= 3.35 - 8.6$ hours. It is easily checked that the altitude that provides $T = 6.00$ hours (4 revolutions/day) is 10,355 km. Such satellites are said to be *semisynchronous*. In an exercise provided at the end of the chapter, it is shown that it requires a minimum of 9 to 12 MEO satellites to cover a full Earth circumference, the number depending upon the altitude range (the lowest number being for 15,000-km altitudes). To reach any point located in the antipodes, the number of required satellite hops can be up to three, which increases the latency by $0.2-0.3$ s. Consistently with the earlier-described Molynia orbit concept, the use of elliptical orbits with high eccentricity makes it possible to increase the altitude (footprint) and to slow down the satellite speed by the apogee (longer coverage time before handover); however, these advantages are introduced at the expense of increased latencies and system-management complexity.

In addition to mobile telecommunications applications, MEO systems are mostly used for meteorology, surveillance, military, mapping and other scientific applications. Two illustrative applications of MEO telecommunications systems are *ICO* (for *intermediate circular orbit* telephone communication system), and *Odyssey* (see Tables 4.6 and 4.7 in Section 4.3.3 for characteristics). Both were designed for orbital altitudes of 10,355 km (semisynchronous coverage) with a *constellation* of 10 satellites (the word "constellation" meaning an organized set of satellites ensuring complete Earth coverage). A main nontelecom application of MEO satellites is the well-known *global positioning system* (GPS), which belongs to the RNSS service class. Most generally, the purpose of a positioning system is to accurately determine a terminal's location on Earth by means of satellite signals, which also provides a unique/universal time reference. The two main principles used are:

1. Measuring the Doppler frequency shifts in two frequency bands (150-MHz/ 400-MHz) from a passing satellite; which gives the change of terminal/satellite distance over time; the position is determined from the orbit parameters, with only a 2-dimensional resolution; this is the principle of the first positioning system called NNSS (*Navy navigation satellite system*) or *TRANSIT*;

2. Measuring time delays between the terminal and four synchronized satellites; the relative delay measurements ($|t_2 - t_1|$, $|t_3 - t_2|$, $|t_4 - t_3|$) are converted into radial distances, which determine the intersection point of four spheres, resulting into an exact, 3-dimensional position.

The last principle is the one used in the U.S.-operated NAVSTAR/GPS (for *navigation system with time and ranging*/GPS), popularly known as "GPS," since its launch in the early 1970s. A similar system was developed between 1982 and 1996 by the Russian Federation, called *GLONASS* (*global navigation Sputnik/satellite system*). Recently, Europe decided to implement its own global-positioning system, *Galileo*, to be fully operational in 2008 with exclusively civilian control and advanced service features (see more details below). Note that only GPS and Galileo offer (or will offer in the final case) free-of-charge services for civilian applications.

It is interesting at this point to further describe these global-positioning systems, since they are becoming an integral part of many aspects of modern wireless telecommunications. In these systems, satellites are called *space vehicles*, or SV. One can briefly summarize their technical features and service classes as follows:

The U.S.-owned (Department of Defense) system, *"GPS,"* is based upon a constellation of semi-synchronous SV (11 h 56 mn-period) distributed into six orbit paths with inclinations of $35°$ ($55°$ latitude), relative node longitudes of $60°$, and altitude of 20,350 km This constellation arrangement makes it possible for up to $5-8$ different SV to be visible from any point on Earth. The GPS offers two services classes, called *standard positioning service* (SPS) and the *precise positioning service* (PPS). The SPS is available and free to any public user and has a resolution accuracy of 100 m (horizontally) and 156 m (vertically), with 340-ns time accuracy. The PPS, which is only available to US military and other authorized users, has a resolution of 22 m (horizontally) and 27.7 m (vertically), with 200-ns time accuracy. As previously mentioned, GPS uses two carriers, one called L1 (1,575.42 MHz), the other L2 (1,227.60 MHz), which are exclusively for "downlink" purposes. The carriers L1 and L2 are used for SPS and PPS, respectively. The L1 carrier phase is modulated by a 1-MHz/1,023-bit *pseudorandom noise* (PRN) for spectral broadening and resistance to interference. Each SV has its own PRN as an individual identification number (there are up to 32 such numbers, which are memorized in the GPS terminals). The L1 and L2 carriers are then modulated with a precise 10-MHz code sequence. Finally, system parameters (such as orbit characteristics, clock corrections and almanac of neighboring VS) are encoded into a "navigation message" on top of L1 at 50-Hz clock rate. The data are sent by the SV every 30 s in the form of 1,500-bit data frames containing five 300-bit sub frames. The data contain information on the frame's emission time (SV clock), clock correction and parity/CRC overhead. The times of arrival (TOA) and rangefindings are determined in the terminal by automatic search and alignment with the PRN of one of the hovering SV. Several range measurements are carried out with different neighboring SV (up to five, for redundancy and further corrections). The identification of these SV is made straightforward from the almanac data obtained in the first acquisition. The resulting information is not only the terminal position but also its speed and the corresponding *universal coordinated time* (UCT) of the measurement. Improved positioning accuracy can be provided by the *differential GPS*, or DGPS. The DGPS uses a land-based station, usually coastal, which should be located not more than 30 km from the terminal. From the knowledge of this station's absolute position, the GPS data can be further corrected, leading to SPS accuracies of 1–10 m. Highly sophisticated differential-GPS should theoretically provide resolutions in the mm–cm range.

The Russian-owned (Russian Space Agency) system, *GLONASS*, consists in 24 satellites distributed into three orbits at 25.2° inclination (64.8° latitude) and 120° apart in node longitude, with altitude of 19,100 km and 11 h 15 mn revolution period. Such a configuration gives at least 5 satellites visible from any point on Earth. The system uses FDMA with 48 different "downlink" frequencies spaced by 0.5625 MHz and distributed into two L bands: L1 (1,602.5625–1,615.5000 MHz) and L2 (1,240–1,260 MHz). The two service classes correspond to *standard-precision* (SP) and *high-precision* (HP) navigation signals. The SP service corresponds to accuracies 57–70 m (horizontal) and 70 m (vertical), in addition to terminal-velocity resolution of 15 cm/s. The HP service increases the positioning accuracy to 10–20 m, which, like GPS, can be further enhanced by differential techniques. Improvements of the GLONASS system are under way towards a 2004 generation. Note: The European Geostationary Navigation Overlay Service (EGNOS) is currently operational and combines the resources of the U.S.-owned GPS and Russian-owned GLONASS positioning. The EGNOS system uses GEO satellites and an array of land-based stations which improves accuracy to within 5 m, compared to about 20 m with previous PPS (source: www.esa.int).

The future European-owned (European Space Agency) system, *Galileo,* should be operational in 2008. It will consist of a 30-satellite constellation, to be launched between 2004 and 2007, in three orbits with 34° inclination (56° latitude), altitude of 23,616 km (above the upper Van Allen belt, unlike the two previous systems) and revolution period of 14 h 5 mn. The signaling uses binary-PSK (BPSK) with two E bands for downlink: E5 (1,188–1,215 MHz) and E6 (1,260–1,300 MHz). Two uplink frequency bands at 1.3 GHz and 5 GHz are also included for ensuring two-way communications. Note that the downlink bands are close to that of GPS, which makes possible *interoperability* between the two systems via *dual-mode* receivers. A novelty is that Galileo satellites will have on-board atomic clocks providing nanosecond (10^{-9} s) time accuracy. The free-of-charge service should provide positioning resolutions better than 30 m with 100-ns timing accuracy. The two higher service classes for professional or military applications specify a 20-ns clock accuracy and a 6-m or 4-m resolution, respectively. Other features of Galileo include guaranteed/uninterrupted service with improved reliability, traceability and liability, enhanced positioning accuracy (even in the free-service class), full Earth coverage including at high latitudes, and GPS-interoperability. This in addition to new monitoring capabilities, including that of urban development, environment, agriculture, geography, road/rail transport, and emerging services concerning rail/road transport (vehicles/container tracking, fleet and cargo management) and maritime/air traffic control and surveillance, to quote a few.

We consider finally the case of *LEO systems*. The 100-km–1,500-km altitude slot of LEO is the closest to the Earth that can be allowed. This is because the higher altitude is inside the inner Van-Allen belt (but sufficiently far from its peak intensity region, and the lower altitude just above the region where atmospheric drag could cause rapid slow-down and final de-orbiting. Because of their close proximity to the Earth, it is relatively easy (and cost-effective) to launch and position LEO satellites into orbit. It is also possible to communicate with LEO satellites through low-power, nondirective/dipole antennas, which can be integrated in portable terminals. The maximum latency for LEO systems is between 10 ms and 25 ms, depending on the orbit, which is at least ten times shorter than in GEO systems. In practice, LEO systems are made of satellite *constellations*, as described in details below. Because one satellite is always passing in the vicinity of the terminal, latency is reduced to a value close to the round-trip time at zenith point, e.g., $\Delta T_{min} = 2(a - r_E)/c = 2h/c$, representing minimum latencies of 6.6 ms to 10 ms for orbits of altitudes $h = 1,000$ km and 1,500 km, respectively.

Because of their relatively low orbital altitudes LEO satellites have much shorter revolution periods, which complicates handover issues. Namely, the period is given by $T = 2\pi\sqrt{a^3/\mu}$, or with, $a = 6,478$–$7,878$ km, $T = 5.2$–6.9×10^3 s $= 1.45$–1.95 hours. The LEO satellites thus progress in the sky at angular speeds of $360°/T_{mn} = 31$–$40°/10$ mn (neglecting the $\pm \cos(\lambda)\cos(i) \times 0.25°/$mn correction due to the Earth's rotation), which means that one to several handovers should be implemented during the course of a normal telephone communication. The handover problem is alleviated by steerable phase-array antennas which compensate the satellite's motion relative to the ground for a certain duration (see below).

Here, the concept of *handover* applies to define the transition of a connection between a fixed (or slowly moving) ground-based terminal and two rapidly moving satellite stations. This is the opposite situation of cellular telephony (Section 4.2), where the base station is fixed and the terminal is mobile. In constellations, the equivalent of a unit cell is the circular area covered by the satellite and is called a *spot*. As we have seen, spots are moving on the ground at a 3.5–4°/mn. A series of satellites belonging to the same orbit is called a *satellite belt*, which also corresponds to a *service belt*. This belt generates a series of successive spots which sweep the area under the orbit (Figure 4.46). As the figure illustrates, the spots must have a minimum overlap for soft-handover purposes. Between two adjacent service belts, the spots must also overlap so as to form a tile pattern (see Figure 4.46). This requires synchronicity between the two belts as well as an accurate positioning of each satellite within one belt, respective to the two neighbors. An alternative approach consists in using belts with alternative rotation directions (counter-rotating belts). In this case, the spot overlap must be somewhat increased in order to avoid creating temporary gaps in the coverage (see figure). Note that this picture is valid only above a certain circular region on Earth (e.g., the

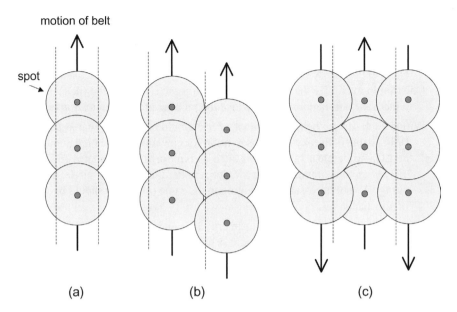

FIGURE 4.46 Arrangement of overlapping areas (spot) covered by satellite constellations: (a) within an isolated service belt; (b) in two adjacent corevolving service belts; and (c) in three adjacent, counter-revolving service belts. The darks pots show the satellite position in the spot and the dashed lines show the area effectively covered or serviced.

Equator) where the service belts are locally parallel; in the other regions, the belts form an angle and eventually cross at some point (e.g., the poles). This shows that the task of uniformizing and optimizing spot-overlap patterns, hence the full constellation coverage of the Earth's surface, is much more complex than for GEO or MEO systems. It must also be dictated by considerations of continental geography to favor the most-populated areas.

Independently of coverage optimization issues, it is a basic exercise to first determine the number of SV required to cover the Earth with a LEO constellation (see Exercises at the end of the chapter). Assuming 10% overlap between spots, the result shows that for 100 km/1,000 km/1,500 km altitudes, the number of SV required to cover one Earth circumference is between 20/7/6, each covering areas with maximum inter-city distances of 2,300 km/6,700/8,000 km, respectively. The total number is between 400/49/36. These figures just indicate a minimum SV requirement for LEO constellation design. It is clear that a constellation that would have *several hundreds* of SV orbiting at 100-km altitudes is not a practical approach. Choosing instead the highest altitudes (1,000–1,500 km) restricts the SV number to less than 50. But several factors lead one to increase this requirement by 30%–100%, corresponding to 64–98 satellites (1,000 km) or 48–72 satellites (1,500 km). One can list a few justifications for such an extra margin, all being mutually consistent:

At such a scale and manufacturing quantity, SV design must be simplified to a maximum extent, trading performance (traffic-handling capacity) with unit production cost, which is the opposite of GEO-satellite design approach;

In order to rapidly implement the constellation and to reduce initiation costs, SV payloads should be grouped into single rocket launchers (e.g., by 2, 4, 8 or even 12); since a launch success rate is not 100% (by however little), the SV design must be cost effective;

One should maximize the number of SV within each service orbit (belt) in order to facilitate soft handover and reduce the number of communication circuits;

Within each service belt, the number of SV should be set such that at least one SV is visible in the sky at an observation (or elevation) angle high enough for the LOS to be clear of obstacles such as buildings or mountains (e.g., 15–20°); when the communicating SV drops under this elevation angle, a neighboring SV (not necessarily belonging to the same orbit belt) should be visible for soft handover;

LEO satellites are subject to atmospheric drag and tend to deorbit after a few years, in spite of possible periodic orbit corrections (which require booster fuel); SV spares are thus needed for backup purposes and periodic replacements through the system's lifetime.

For these different reasons, the main LEO constellations (800–1,500 km) have been designed with between 48 and 80 SV (e.g., *Globalstar*, *Iridium*, *Skybridge*), except for *Teledesic*, which has a whopping target of 840 units (though now reduced to "only" 288). Such constellation systems are described in the next subsection. It is however interesting to make a basic system-cost evaluation considering "just" the satellite and the launching aspects. For satellites, the reference figures are 25,000\$ to 75,000\$ (25–75k\$) per kilogram of SV payload, depending upon technology, lifetime, power and circuit-handling capacity. For launchers, the cost is about 15 k\$/kg for LEO and 50k\$/kg for GEO. Based upon these approximate figures, it is straightforward to evaluate the minimum cost of a constellation, according to different system-design possibilities and SV-payload mass (see Exercises at the end of the chapter). For LEO systems, the result is in the order of 1–4B\$, which turns out to represent some 50% of the actual costs for corresponding system designs (see table in the next subsection). It is beyond the purpose of this exercise to get to real numbers, since there are many other and complex cost factors involved (technical/business analysis, R&D, launch trials, fixed and operational costs, loan interest, insurance, licenses, etc.), as is the case in all global telecom systems.

4.3.3 Satellite Telecommunications

The advent of *satellite constellations* for telecom applications is the outcome of a long history of satellite-based, global services, from fixed to mobile applications. This history consists in several generations where "*comsats*" (communication satellites) constantly grew in mass and circuit-handling capability (electronic speed) and terminal antennas decreased in size. In the first twenty-year period (1960–1980), satellite payloads increased from 40 kg to 150 kg and up to 700 kg. The Earth-based dishes were 30 m in diameter (!), which is a huge size, considering that they had to be mounted on special masts offering minimum angular freedom for tracking-optimization purposes with latitude and time. The first commercial GEO satellite, known as "the Early Bird" (professionals also call satellites "birds"), was launched in 1965 as the first of the INTELSAT (*international telecommunication satellite [organization]*) family.

The 1980 period represented a second generation of FSS systems, using 10-m dish antennas with 1,000-kg payloads. The 1990s introduced the first *mobile* satellite systems (MSS) with 1–2-m down to 0.5-m dish-antenna sizes and 1,500-kg satellite payloads. This advance opened the perspectives of wide-coverage digital networks free from the limitations (at the time) of land-based radio networks. Because of their reduced antenna size, these early MSS systems were called *VSAT* (*very small aperture terminals*).

Figure 4.47 shows the basic VSAT network topology. It is a star network characterized by a central *hub*, connecting together several *closed users groups* (CUG) of VSATs through a unique GEO satellite by means of a large, 5–11-m dish antenna. The GEO satellite acts as a passive "repeater" for all traffic between the hub and the VSAT units. The hub comprises a *network control center* or NCC (which is not necessarily located by the hub-antenna premises), and the *baseband equipment* (HBE) corresponding to the different CUGs. Each CUG's traffic is independently controlled by a *subnetwork control center* (SNCC) via its own HBE (note that a CUG is a functional cluster of VSAT units which do not necessarily belong to the same geographical area). Finally, the SNCC are connected to the local network system (PSTN, ISDN, X.25). As the figure illustrates, for any VSAT (A) to connect to another VSAT (A′ or B, inside or outside a given CUG), two round-trips to the

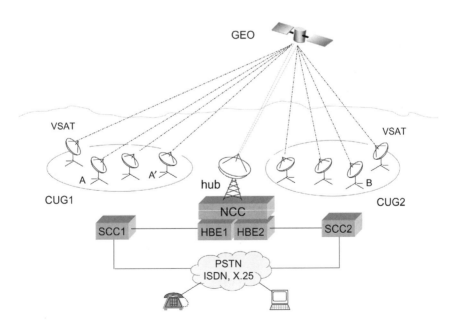

FIGURE 4.47 Basic VSAT network topology (GEO = geostationary satellite, VSAT = very small-aperture antenna terminal, CUGi = closed unit group "i," NCC = network control center, HBEi = hub baseband equipment for CUGi, SCCi = subnetwork control center for CUGi).

GEO are necessary, which amounts to a minimal delay of about 0.5 s. The two uplink/downlink frequency bands are 6/4 GHz and 12/14 GHz, and the access protocols are TDMA or FDMA (CDMA being less advantageous for power considerations). A VSAT network is able to handle as many as 16,000 terminal users (called *client installations*), with service classes up to 256/512 kbit/s capacities (voice, compressed-video and data). New VSAT-based commercial services applications in aircraft, ships and land vehicles are provided by INMARSAT (*international maritime satellite organization*) and its satellite family.

Because of the two round-trips to the GEO relay, these early VSAT networks are not so well adapted to truly interactive video and other real-time services. An improved approach consists in locating the hub functionality in the GEO satellite itself, thus reducing the delay to only one round-trip time.

The *constellation* concept, based upon LEO, MEO and hybrid LEO + MEO systems, consists in realizing a global world coverage by use of a relatively large number of SV. The SV occupy N different orbits having $180°/N$ node longitudes, as illustrated in Figure 4.48. There is a wide flexibility in the choice of orbit inclinations, which are also referred to as *orbit planes* (corresponding to the service belts). If the orbit inclination is 90°, the orbit planes are perpendicular to the Equator and all intersect at the poles, as in (a) in the figure. In practice, the orbit inclinations are chosen either slightly under 90° or within a 45°–90° range. With inclinations close to 90°, the orbits then intersect near the polar regions but the intersection points are different for each orbit pair. The closer the orbit inclination to 45°, the longer the SV remains above the most populated continental regions and away from the less-populated polar regions. Some configurations use 45–50° planes with an additional 90° or polar plane for complete world coverage, including regions well above the polar circles.

In the case of "*Walker orbits*," the orbit planes are paired with longitudes centered on $180°/2N$ angles; two Walker subconstellations can be interleaved making

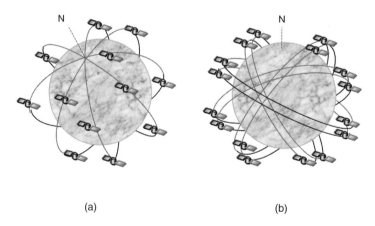

(a)   (b)

FIGURE 4.48 Principle of satellite-constellation networks, showing orbit planes or service belts (here 4, with 4 satellites each) covering the full Earth's surface: (a) using polar orbits at 90° inclinations, or (b) using Walker-type paired-orbits at 45° inclination (paired orbit planes drawn parallel for clarity).

the paired-orbit planes cross each other at different longitudes (see (b) in Figure 4.48); such a configuration makes it possible to create homogeneous and periodic patterns for satellite positions relative to ground, referred to as *orbital repetitiveness* (see following discussion on *Skybridge*).

Satellite constellations, familiarly called "*big LEO*," came in the picture in the early '90s. They represent a third generation for satcoms, or a second one for MSS, with mobile access from light-weight, hand-held terminals having 20/30cm-long dipole antennas, hence the service name of *enhanced*-MSS or EMSS. A novelty in some big-LEO systems is the use of *phased-array antennas* to continuously steer the beams onto fixed spot locations, thus compensating the SV motion relative to the ground, before handover to another passing SV is required. Recall that a key advantage of LEO systems is the short latency of 10–20 ms, to compare with 600–120 ms or 237 ms with MEO or GEO systems, respectively.

Other large satellite systems concern HEO, GEO and hybrid GEO + MEO satellite arrays. The features common to constellations and HEO/GEO arrays is their global coverage and on-board switching capabilities, and in some cases, the use of *inter-satellite service* (ISS) links for routing. GEO-based systems are located in or near the Equator plane. Recall that they cannot be spaced by less than 2° in the sky, which limits the total number of possible *satellite slots* to $360/2 = 180$. An advantage, however, is that the same frequencies can be re-used every two slots, which is the principle of *space-division multiplexing* (SDM) already discussed with GSM networks.

Lots of ambitious constellation programs have been proposed and launched by consortia from USA, Europe and several other countries. Some are now fully operational, or awaiting licensing, near-future deployment, or further financing, some are even bankrupt. The number and diversity of these systems is quite impressive, as reflected by the data shown in Tables 4.6 (LEO, MEO and LEO + MEO) and 4.7 (GEO and GEO + MEO), which only represent a selection of the main programs. The two tables provide a variety of system characteristics, ranging from the number of satellites, orbit parameters, frequency and data rate to the type of services (e.g., voice, video, data, messaging, paging, fax, e-mail, asset-tracking and broadband communications). The planned system/SV lifetime, as determined by SV deorbiting (LEO), fuel-reserves for station-keeping (GEO) and other aging considerations are indicated. Note that in LEO systems, SVs can be periodically replaced, unlike in GEO systems. Unfortunately, the complete set of data for each system is not always available, and some might be subject to change according to reconfigurations, program mergers, new frequency/band licenses or reuse, bandwidth upgrades and service upgrades. Therefore, the data shown are only indicative, and it might be useful to refer to the corresponding program Web sites to know their actual evolution. Here, we shall briefly describe some of the most representative LEO, MEO and GEO systems, with a selection of their interesting facts and features (see also tables and Web references for other details):

Globalstar (LEO, alt. 1,414 km): Launched in October 1999 at a cost of $2 billion by a consortium led by Loral, and Qualcomm; includes 48 satellites (plus 8 spares) in 8 planes of 6 (plus1 spare) satellites each; satellites were set into orbit using the United States, Russian Federation and Chinese rockets,

TABLE 4.6 A selection of main LEO/MEO constellation systems and their characteristics

Constellation or System	Orbit Type, Altitude (km)	Inclination (°)	NP	NS	W (kg)	Frequency (GHz)		Data Rate (kbit/s)			Satellite Lifetime (Years)	Access and Modul.	Services (*)
						U	D	U	D	R			
Courier (LEO Sat)	LEO 800			72		1.62	2.49						
ECCO (Telebras)	LEO			46									V, D, F, P
E-sat	LEO 1,262		1	6	114	0.149	0.137				10	CDMA	D
FAISAT	LEO 1,000		6	26 (36)	100	0.137	0.149				7–10	FDMA	D, Vm, P, AT
GEMnet	LEO 1,000	50–90	4 + 1	32 + 6	45.2	0.137	0.149	2.4/4.8			5–7	QPSK	D, P, E, MR, AT
Globalstar	LEO 1,414	47–65	8	48 + 8	262 426	1.62 5.13	2.49 6.98	4.8/9.6	4.8/9.6	T	7.5	FDMA, CDMA QPSK	V, D, F, P, GPS
Iridium	LEO 775–785	86.4/90	6 (7)	66 + 6 (77)	689	1.62	1.62	2.4/4.8	2.4/4.8	IS	5–8	FDMA + TDMA QPSK	V, D, F, P
Koskon (Polyot)				32									V, D, F, P
Leo One USA	LEO 950	50	3	48	125	0.149	0.137	9.6	2.4	T	5–7	FDMA	SC, D BB
M-star	LEO	47	12	72	1,150	48.7	39.0						
Odyssey	LEO 10,354	55	3	12	1,130 1,334	29.25	19.45	4.8	4.8		15	FDMA + TDMA DPSK	D, V, F, P SMS
Orbcomm	LEO 775 739	45/70 45 90	3 +1 +1	28 +6 +2	38.5	0.149	0.138	2.4	9.6	T	4–6	FDMA TDMA	D, Vm, AT

System	Orbit / altitude (km)	Incl.	NP	NS	W	f↓	f↑	D	U	R	No.	Access	Services
Skybridge	LEO 1,469	53	20	80	1,250	13.6	11.7	**2 M**	**20.5 M**	T	10	ATDMA + CDMA, SDMA	BB
Stamet	LEO 1,000	53	6	24	330	0.137	0.148–0.150				5	CDMA	Vm, EA, RDSS
Teledesic	LEO 1,350–1,400	98.2	12	288	795	28.9	19.1	**2 M**	**64 M**	IS	10	ATDMA + CDMA, SDMA	BB
VITAsat	LEO 970	88	1	2	45	0.149	0.137	9.6	9.6		5	FDMA	EA
Ellipso (MCHI)	LEO/MEO 7,845/8,040	116.5	2	10 LEO 6 MEO	300	1.61	2.49	9.6	9.6	T	5	CDMA	V, D, P, E
GESN	LEO/MEO 10,354 10,354			4 GEO 15MEO		37.5	50.2						
@contact	MEO			16		28.8	19.1						
Constellation	MEO 2,000	62–0		35 + 11	200	1.62 2.49	5.17 7.0	9.6	9.6	T	7	CDMA	V, D, F, P, RDSS
ICO	MEO 10,355, 10,390	45	2	10 (+2)	1,800	2.18	2.00	2.4/64	2.4/64	T	10–12	TDMA	V, D, F, P
Orblink	MEO 9,000			7		38	48.2						

Systems whose names appear in bold are further described in text. Data rates shown in bold correspond to Mbit/s (M). Abbreviations and acronyms: Incl. = orbit inclination, NP = number of planes, NS = number of satellites, W = satellite weight, U = uplink, D = downlink, M = Megabit/s, R = Routing type, T = terrestrial, IS = inter-satellite, V = voice; for services (* indicative 2003): D = data, Dm = data messaging, Vi = video, Vm = voice messaging, P = paging, E = e-mail, MR = meter reading, EA = emergency assistance, AT = asset tracking, BB = broadband, SC = smart car, SMS = short messaging service, GPS = global positioning system).

with up to 12 units per launch; each satellite has a *phased-array antenna*, able to generate 16 spot beams, covering together a circle of 5,760-km diameter with handling-capability of 2,148 circuits; ensures soft-handover and multi-satellite-path diversity (several SV handling a given active circuit); relies upon *terrestrial routing* through Earth station gateways (similarly to VSAT); individual circuit capacity is optional from a minimum of 1.2 kbit/s to a maximum of 9.6 kbit/s and 4.8 kbit/s for voice and data, respectively; a specific application of Globalstar concerns mobile-telephone communications covering the whole of central Africa; exploitation bankruptcy was declared in November 2001.

Iridium (LEO, alt. 780 km): Launched in 1988 by Motorola and Iridium Inc., at a cost of $3.4 billion; the name of the project comes from the element "iridium" which has the atomic number 77, corresponding to the initially planned number of SV (7 units in 11 orbit planes); the actual number has been reduced to 66 units (6/plane); the 689-kg SV is designed to last between 5 and 8 years; each SV is able to handle 48 spot beams of 600-km diameter; the total number of constellation cells is thus $66 \times 48 = 3,168$ cells, which cover 147% of the Earth surface; this means that only 2,150 spot beams are necessary for operation ($3,168/1.47 = 2,150$); each beam utilizes 80 channels in 12 frequency sub-bands, giving a total number of circuits of $2,150 \times 80 = 172,000$ circuits; the circuit capacity is 4.8 kbit/s and 2.4 kbit/s for voice and data, respectively; interestingly, the system uses the same frequencies (1,610–1,626 MHz) for both uplink and downlink communications; this is made possible by the technique of *time duplexing* (send or receive mode, either one at a time); an innovative feature of the Iridium system is *intersatellite routing* (ISS), whereby calls can be forwarded between satellites to anywhere, without the use of intermediate Earth stations (an advantage over the Globalstar approach); quadruple multisatellite path diversity is ensured through ISS from a given satellite to its fore-and-aft orbit neighbors and between the two nearest ones in the two belt neighbors; system bankruptcy in November 2000 was followed by take-over from the U.S. Department of Defense (DoD), with continuation of the service (operational since November 1998) for both military and commercial exploitation; on the fun side: amateur astronomers enjoy observing the effect of "flares" or "glints" from Iridium satellites, due to the occasional yet fully predictable effect of sun reflection on their flat aluminum antennas.

Skybridge (LEO, alt. 1,460 km): Launched in 1997–1998 by Alcatel and Loral for $6 billion; represents a second generation of constellation systems, with broadband (BB) service capability; traffic capacities for uplink and downlink communications are 2 Mbit/s and 20.5 Mbit/s respectively, compared with 10–100 kbit/s in previous system generations; in enhanced "on-demand" BB services, multiple circuits ($n \times 2$ Mbit/s and $n \times 20.5$ Mbit/s) can be combined for business or residential applications; based upon two interleaved

Walker constellations (see above) each of 10 planes with 4 satellites (80 total); unlike Iridium but like Globalstar, routing is provided by Earth-stations and Earth-based segments; this simplifies the satellite's design and cost, to the advantage of higher on-board electronic speeds with transceiver-only function; the Earth gateways are also connected by a land-based backbone network to ensure more complete service deployment especially in areas inaccessible to traditional land networks; this feature also allows for efficient *frequency reuse* (SDM) between different gateways; another novelty is the sharing of Ku-band (10–18 GHz) already used by FSS/BSS GEO systems; interference is avoided by a technique called *GEO arc avoidance*, which ensures handover when the satellite is within $\pm 10°$ of the protected GEO arc; the satellite position, determined by GPS, does not require tracking; the 1,250-kg satellites have a 400-kg communications payload (initially planned for 800 kg and 300 kg, respectively) and are designed to last 10 years; they are equipped with phased-array antennas able to continuously steer 18 spot beams; in addition to classic voice/data communication (ISDN, X.25, ATM, IP), BB services offered by Skybridge include: multimedia over BB Internet, video-on-demand (VoD), video telephony and video-conferencing, corporate networking, LAN internetworking, and a whole array of novel IP-based applications such as telecommuting, telemedicine, distance-learning; service should be provisioned in 2005.

Teledesic (LEO, alt. 1,375 km): Launched in 1997–1998 on the initiative and funding of Microsoft and McCaw Cellular Communications Inc., now operated by Teledesic, Inc., at a cost of $9 billion; will provide a "fiber-like" broadband access with circuit capacities of 2.048 Mbit/s for uplink and 64 Mbit/s for downlink; on-demand BB services will offer uplink/downlink capacities at $n \times 2.048$ Mbit/s (up to 100 Mbit/s) and $n \times 64$ Mbit/s (up to 720 Mbit/s), respectively; the initial design consisted in using 840 SV at 695–705-km altitudes; the second design considers 288 SV at 1,375-km altitude (the actual number could now be further reduced to 48; the sun-synchronous orbits are close to polar (98.2°), with 12 orbital planes (of 24 SV each $[12 \times 24 = 288]$, as opposed to the initial concept of 21 planes with 40 SV each $[21 \times 40 = 840]$); the satellites, designed for 10-year service, have phased-array antennas which cover a "super cell" of 9 scanning beams, representing a 160 km $\times$ 160 km square area; each scanning beam has a traffic capacity of $1,440 \times 16$-kbit/s circuits; unlike Skybridge, routing is ensured by ISS, where each satellite can exchange/switch traffic with its 8 neighbors at a standard 155.52 Mbit/s (OC-3/STM-1 of SONET/SDH); an interesting novelty, although not confirmed or publicly detailed, is the use of *laser-based ISS links* which, compared to radio, handle higher capacity with improved signal quality (or lower BER); potential BB IP-based services are similar to that previously listed, with first provisioning being scheduled for 2005.

ICO (MEO, alt. 10,355 km): Established in January 1995 from INMARSAT as a new corporation called ICO Global Communications, Ltd. (ICO for *intermediate circular orbit*), representing a total $5 billion investment; it comprises 10 SV (plus 2 spares) distributed into two orbital planes at 45° inclination and node-longitudes spaced by 180°; routing is ensured by 12 Earth station gateways which are interconnected to the PSTN; access to the ICO network is planned with dual-mode mobile terminals that are compatible with both ICO and local standards (GSM, D-AMPS and JDC); voice and data service capacities are limited to 2.4–64 kbit/s; bankruptcy was declared in November 1999, followed by a $1.2 billion reprieve from investors; service should be provisioned in 2003.

Thuraya (GEO): Launched in 2000 by the United Arab Emirates (UAE) for $1 billion; made of two GEO satellites co-located at 44°E longitude (above South Somalia); through 96 dedicated spot beams, coverage is ensured from Morocco to India, encompassing North/Central Africa, Southern Europe, the Middle-East, and central Asia; realizes global GSM-roaming services for the entire area, with services ranging from Internet access, e-mail, SMS, GPS-over-SMS for geologists/explorers, and "prayer-time" information for Muslims, to global Marine and Oil/Gas network communications;

Broadband GEO systems: Consist in $4–6 billion arrays of high-capacity GEO satellites linked together by ISS, examples being Astrolink (Lockheed Martin, United States), Euroskyway (Alenia Aerospazio, Italy), GE*Star (GE American), and Spaceway (Hughes, United States), for instance (see Table 4.7); the circuit capacity in these systems is 2–20 Mbit/s for uplink and 16–155 Mbit/s for downlink; a direct individual comparison of these offered capacities is difficult because of imprecise or evolving definitions concerning "on-demand" service classes; because of their BB capability (however asymmetric and handicapped by $\frac{1}{4}$-s latencies), appear as natural complements, if not competitors, of big-LEO systems for Internet access.

This selection of LEO, MEO, and GEO systems amply illustrates the progress made by satellite telecommunications in both technology and services, from 10–100-kbit/s to 2-Mbit/s uplink capacities, up to 155 Mbit/s downlink. It is worth describing further two major applications of satellite-based technologies which concern personal/mobile networks and the Internet.

Satellite-Based Mobile Networks

The portable/mobile telephone handsets which connect to MSS constellations were initially called *satphones*; marketing considerations however, now tend to avoid differentiating satellite-capable mobile terminals from ground-based ones, rather emphasizing GSM-like interoperability and roaming. Most MSS systems offer voice services similar to GSM with *dual-mode* capability, i.e., the possibility of roaming between Earth-based (SIM card GSM) and satellite-based systems. Whether the data rate be intrinsically limited (10 kbit/s to 100 kbit/s in upgrades) or broadband (2 Mbit/s uplink) is only a question of class and type of service.

TABLE 4.7 A selection of main GEO (2 satellites and over), HEO and GEO + MEO systems and their characteristics

Constellation or System	NS	Frequency (GHz) U	Frequency (GHz) D	Data Rate (kbit/s) U	Data Rate (kbit/s) D	Routing	Lifetime (Years)	Multiple Access.	Services (*)
ACeS	3 GEO			16	16	T			V, D, P, E
Aster	25 GEO								
Astrolink	9 GEO	29.7 28.4	19.9 18.5	20 M	155 M	IS			V, D, Vi
Celsat	3 GEO					T			V, D, F, P
Euro skyway	3 (5) GEO			2 M	32.8 M	IS			V, D, Vi
Expressway	14 GEO								
GE*Star (plus) Ge Starsys	9,11,24 GEO	28.4	19.9			IS			Dm, BB
Inmarsat 3 (M)	5 GEO	1.55 (2.18)	1.64 (1.99)	64/432	64/432		12		V, D, F
PanAmSat	9,12 GEO								
Thuraya	2 GEO	1.54 3.5	1.64 6.57			T	12–15		V, D, P, E
Archimedes	6 HEO 1,000–26,786 km						10	FDMA	V, D, F
Marathon Mayak (Marathon Arcos)	4 HEO (5 GEO)	1.645	1.544				7		V, D, F, E, D
Pentriad	3 HEO	11.9 (NA) 12.6 (Europe)							

(continued)

TABLE 4.7 A selection of main GEO (2 satellites and over), HEO and GEO + MEO systems and their characteristics *Continued*

Constellation or System	NS	Frequency (GHz)		Data Rate (kbit/s)		Routing	Lifetime (Years)	Multiple Access.	Services (*)
		U	D	U	D				
Spaceway	8 GEO	17.7	29.6	**16 M**	**16 M**	IS	12		V, D, Vi, BB
	20 ME0	28.4	19.9			(MEO)			
Starlynx	4 GEO								
	20 ME0								

Data rates shown in bold correspond to Mbit/s (M). Abbreviation legend: NS = number of satellites (number in parentheses indicates original or final target, additional "+ number" indicates spares), W = satellite weight, U = uplink, D = downlink, M = Megabit/s, T = terrestrial routing, IS = inter-satellite routing, V = voice; for services (* indicative 2003): D = data, Dm = data messaging, Vi = video, Vm = voice messaging, P = paging, E = e-mail, BB = broadband.

As a matter of fact, MSS offer the key advantage of being compatible with the different 2G–3G standards, from GSM, GPRS to UMTS (see Section 4.2). In future developments, this could lead to worldwide interoperability and roaming by use of new types of dual-mode, cellular/BTS or cellular/SV capable terminals. Because of the rapid deployment of land-based cellular networks, MSS can be first conceived as inherently limited to a (still important) market niche of developing countries with poor telecommunications infrastructures, and geographical areas/ provinces with difficult land access or under-populated (high-altitude plateaus, mountains, rain forests, deserts, polar zones) or remote regions (island groups, off-shore platforms). MSS are also very well adapted to private/corporate networks which can be deployed without earthbound/frontier constraints and on top of already-existing but congested networks. As for VSAT, the drawback is the need for the network resource to be shared with a sufficiently large number of terminals (>1,000) for the cost-per-user to remain competitive over other Earth-based network solutions. *Broadband access* with both wireless and "wireline" links is indeed rapidly developing as a low-cost commodity (see Chapter 1 in Desurvire, 2004). Such an irresistible evolution could diminish the interest for BB-MSS, for both cost and performance reasons, at least in developed or densely populated areas.

An alternate development perspective for MSS is network *back-haul*. The service can be automatically switched on to compensate accidental network failures (land and submarine cable breaks, congestion shutdown, viruses, conflicts), thus ensuring uninterrupted traffic continuation. Such a back-haul protection is however limited to networks locally carrying relatively small traffic, compared to that of long-haul optical networks (several Tbit/s, see Chapter 2 in Desurvire, 2004). To simplify, the huge capacity of optical fibers does not fit into the limited bandwidth of radio-satellite links. Cables can also contain tens to hundreds of such fibers. Yet, unlike wired networks, MSS can be rapidly deployed and operational for temporary but extremely important uses to service large populations in events such as manmade or natural disasters.

Satellite-Based Internet

With the advent of BB satellite services (FSS or MSS), the Internet has progressed once again into another global network dimension. At the beginning, it was widely thought that the Internet, more exactly the TCP/IP protocol suite, could not work with GEO/GSO satellites and multi-hop LEO systems because of their relatively long latencies ($\frac{1}{4}$ s, $\frac{1}{2}$ s and over). This latency causes TCP to fail, since it is based upon the principle of acknowledgment of datagram reception under a time constraint, typically 5 ms (see Chapter 3). One solution is to use UDP instead of TCP for the Earth-to-satellite round-trip path, but at the expense of decreased communication reliability (connectionless mode without data flow control, acknowledgment and datagram re-ordering, as described in that chapter). One way out from this dilemma is the modification of TCP into TCPSat, allowing longer datagrams and data payloads to be transmitted between two acknowledgments. The loss is greater in case of error, but less likely in a point-to-point satellite link than in IP networks. The other solution was to develop a protocol that sends fake or "spoofed" acknowledgment messages. Since the acknowledgments are meaningless to the transmitting nodes, it is not so much a solution as a way to overcome the problem of failure in TCP implementation because of the high latency. A second problem comes from the way HTML files (Web-site pages and related text/picture/animation media contents) are loaded from the web server to the Web browser. The usual way is that such a loading is made through progressive steps, starting from text then the pictures and so on, which involves some iterations and delays, with a higher risk of failure for media-rich pages. A solution, as already used in land-based IP networks, is to use a *cache memory* (see Chapter 3). Such a memory temporarily stores on a hard-disk drive all the secondary information of websites already visited, or most popular, so that re-loading is made at browser (receiving node) level. The satellite operator must then periodically broadcast the corresponding information (and its updates) to the user's cache memory. This approach may not work so well if home-computer users frequently clean up their cache memory for gaining more hard-disk memory space. Indeed, this space may become rapidly congested or already approaching its limits for a variety of reasons, including having too many applications running simultaneously or having downloaded/stored too many large-size datafiles (high-resolution pictures, music, video . . .). However, secondary disk drives can be hooked up, and some Internet services also offer "archival" memory space for rent.

With uplink speeds generally limited to 150 kbit/s–1.5 Mbit/s, FSS satellite IP faces tough competition with wired BB Internet access (and possibly wireless, soon). Yet, a market niche exists for regions where wired BB access is definitely not available for the foreseeable future, and also if it can be advantageously combined with other services such as *digital TV broadcasting* and other *contents* distribution (downlink speeds of 2–48 Mbit/s). The existing standard called *digital video broadcast by satellite* (DVB-S), based on MPEG image coding, is also able to transmit IP payloads (DVB-IP vs. DVB-TV). In this case, the approach is similar to CATV where the investment and resource are shared by a group of residential customers, at a lower user cost, with *IP multicasting* as an added-value service. Thus access could represent a mix of "bundled" (wholesale, multicast, broadcast, one-way) or "un-bundled" (individualized, unicast, two-way) service types. A question remains whether such a package mix could technically meet and satisfy real expectations from the BB Internet, in particular considering BB interactive access services such as *video on demand* (VoD) and *gaming*. Not to mention low-cost voice/video-telephony, which some may already think of as being a "freebee" to be provided on top of the above services. The key for such a market is to realize an adequate balance in capacity provisioning between the downlink and the uplink (called here "*return channel*," just as for the wired ADSL (*asymmetric digital subscriber line*; see Chapter 1 in Desurvire, 2004). Another potential stumbling block is the extension and integration of such services to mobile with GSM/GPRS compatibility.

A new standard called *DVB-RCS* (DVB-*return-channel over satellite*), also called "*ADSL in the sky*" was created for the above purpose. Such a standard effectively simulates a two-way BB communication link, with bandwidth-allocation flexibility (256 kbit/s–1.5 Mbit/s uplink and 128 kbit/s–8 Mbit/s downlink). The architecture of a satellite-based BB access network, shown in Figure 4.49,

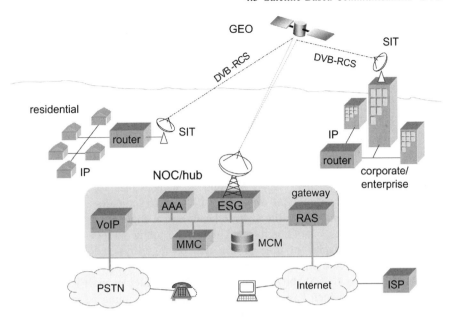

FIGURE 4.49 Basic satellite-based IP network for residential and corporate applications (GEO = geostationary satellite, DVB-RCS = digital video broadcasting with return channel system, NOC = network operations center, ESG = Earth-station gateway, SIT = satellite-interactive terminal, VoIP = voice over IP, PSTN = public-switched telephone network, AAA = authorization and accounting center, MMC = multimedia contents provisioning center, MCM = master cache memory, RAS = remote access server, ISP = Internet service provider).

is not so different from that already seen in Figure 4.47 concerning VSAT networks. The hub with its constituents is called the *network operations center* (NOC). The main difference is the inclusion of *IP gateways* and *routers* at both hub and receiving nodes, with a *remote access server* (RAS) providing IP address allocation and the gateway to the Internet, servers for provisioning multimedia contents and telephony, a "master" IP cache memory, and an *authorization and accounting* (AAA) proxy-server platform. The VSAT dish is also replaced by a more fancy *satellite-interactive terminal* (SIT) with dish sizes of 50–90-cm diameter, but also 1.20 m–1.80 m–3.70 m for locations remote from the NOC. This type of network concerns both residential and corporate/enterprise with different uplink/downlink capacity needs and end-user services (say, essentially entertainment and Web browsing for the first, and multicast/video-conferencing for the second). But the picture becomes not so clear-cut and more uniform if one considers services such as *teleworking* (or *e-working*), and *telelearning* (or distance learning, or *e-learning*). Other FSS examples mixing residential and enterprise areas are teleshopping, telesurveillance and telemedicine, to quote a few. A new paradigm is that in an increasing share of business concerns activities that can be performed from home/personal offices, and where occasional team interactivity can be supported through virtual room/intranet environments, thus saving commute time, energy and costly office space in downtown areas.

Corporate and enterprise applications of BB networking are not limited to Internet video conferencing (to save time, travel costs and speed up decision/process loops). They also concern "layer 2" networking with traditional Ethernet and frame-relay/ X.25 LANs (see Chapter 3), for which IP would not provide any qualitative enhancement (see Chapter 3). As supported by new (or possibly already operable) FSS standards, these layer 2 data networks could be configured while alleviating "wiring" constraints, reliance and dependency with respect to PSTN and other existing infrastructures, and increased flexibility for deployment and dimensioning. Other FSS applications may use the "traditional" PSTN or GSM telephony system for the return channel, because interactivity is not key to the service, the matter being a routine client/server exchange of data. A recently developed FSS protocol is the *unidirectional link routing* (UDLR), which simulates a two-way communication with a simple satellite dish. Related service applications concern much smaller market niches, but of tangible and immediate public interest. Representative examples are the display of time information on highways and bus stops, and restocking of medicine products for pharmacies.

As BB access (>100 kbit/s to 40 Mbit/s) to the Internet via FSS is another technology choice which has clear advantages, a key remaining issue is the end-user cost. By 1999, a 128-kbit/s-1-Mbit/s downlink-capacity, two-way IP access with satellite return channel (DVB-RCS) would represent a $2,000–15,000 monthly charge, in addition to the $1,000–2,500 investment for the SIT dish/transceiver (recalling that this is a common-antenna type service). The monthly fees become 1–2 orders of magnitude lower with telephone as the return channel, but they significantly differ according to the class of service (up/down capacities, guaranteed bit-error rate, percentage service-time availability, etc.). A representative example of DVB-S/DVB-RCS system, to be operational in 2004, is the €56M European-driven *AmerHis* system, which will service BB internet and VoD to both Europe and South America through four GEO spot beams. It could be seen as a valid illustration of the capacity of BB-FSS to be rapidly deployed on continental scales and at minimal cost, *all the way* to the business/residential areas (last mile), without the impediments of classical wired networks which rely upon transcontinental/transoceanic cables and the legacy-network infrastructure.

4.3.4 High-Altitude Platform Systems (HAPS)

High-altitude platform systems (HAPS), also referred to as *HAP technologies*, are flying gateway stations which are positioned in the high atmosphere at altitudes of about 20–25 km (center of stratosphere). This altitude range is defined by a narrow window of relatively mild windspeeds. Indeed, the winds steadily increase from ground to 10–15 km where they reach maximum speeds of 150–250 km/h. The windspeed then decreases towards a minimum near 20–25-km altitudes and increases again. The minimum reached in this high-altitude window is 0–20 km/ h, the value depending upon the location and season. The HAPS is therefore located well above normal air traffic (10 km) but still relatively close to the Earth, in comparison to LEO/constellation satellites (100 km). Unlike LEO, but similarly to GEO/GSO satellites, they occupy a quasi-permanent position in the sky. Unlike

GEO systems, the latency or round-trip time of radio signals is negligible (0.1–0.5 ms). As described further, HAPS can also be designed for lower altitudes in the 3–10-km range.

Although HAPS are not gravitating satellites but flying stations, their network topology and service features are very similar to the ones previously described, i.e., that of VSAT and GEO-based BB-access. For a long time, the military have used the technology of *high-altitude long-endurance [platform]* (HALE, HALEP) communications, from which the HAPS concept is in fact derived. The HALEP could rapidly and safely be deployed over large region/battlefield footprints (10–100-km diameter) with the function of relaying messages between command and mobile ground stations, and radar surveillance. HALEP or HAPS can be either *airships*, another word for lighter-than-air balloons or aerostats, or *aircraft* which are manned or un-manned airplanes. *Unmanned aerial vehicles* (UAV) are also known as "*drones*" in military terminology and use. Unlike balloons, aircraft need to be propelled to fly, which requires some energy reserve and thus defines a maximum flight cycle (see below). Inherently, balloons are voluminous objects which are slow to maneuver and stabilize, unlike aircraft which can fly according to predefined circle patterns.

Both craft types must be designed to be able to resist winds of 150 km/h at least, upon reaching the 10–15-km altitudes while en route to or from their 20–25-km operating point. Besides wind, HAPS must be resistant to two other constraints that are the very-low *atmospheric pressure* and *temperature*. At this altitude, the pressure range is $P = 5,900$–$5,500$ Pascals (1 Pa $= 10^{-5}$ atmosphere, with 1 atm $= 1.033$ kg/cm^2 for normal atmospheric pressure at ground level, corresponding to 760 mm mercury height in *Torricelli* barometers). Thus, the HAPS environmental pressure is about $1/20$ that of ground level. The temperatures prevailing at these altitudes are nominally $T = -56.5°C$ to $T = -51.5°C$ (for 20 km to 25 km, respectively). For balloons, the volume V of gas expands according to the law $P \times V/T = const.$, which means a relative volume increase of $PT'/(P'T) = 13.5$ to 14.5 times when the balloon reaches its high-altitude position. At ground level, the initial volume of lighter-than-air (buoyant) gas must be sufficient to lift the craft hull/envelope and its payload. The HAPS balloon must therefore be designed so as to maximize its hull mechanical resistance and protection against accidental or defective leaks, all this with minimal structural weight and maximal effective payload. This last consideration also applies to HAPS aircraft, with another constraint which is the significantly thinner air (90 g/cm^3 instead of 1,225 g/cm^3 on ground), requiring relatively large wing surfaces and lengths (30–70 m). Since the craft must resist high winds, its hull/wings design must represent an optimum between mechanical resistance, atmospheric drag and weight, with additional constraints differing with the type of propulsion (jet or electrical, see below). Finally, the on-board power resource must be sufficient to ensure propulsion, positioning/maneuvering, powering instruments and communication circuits, and radio-antenna broadcasting. For manned vehicles, this also includes reserves for take-off and landing operations, and inside temperature/pressure conditioning, but under extreme conditions compared to normal private jetliners.

The HAPS *footprint* is defined by both altitude and elevation angle. The minimum elevation angle required for a clear LOS is different for densely-populated areas, suburban areas or country areas, which can present a variety of obstacles such as skyscrapers, buildings, hills, roof tops or even trees. Thus, the minimum elevation can be 15° in the country and 45 to 70–80° in city downtowns, corresponding to very different footprints. It is easily shown (see Exercises at the end of the chapter) that the footprint diameter is approximately given by the following formula:

$$d = 2R_{\mathrm{E}} \left\{ \cos^{-1}\left(\frac{R_{\mathrm{E}}}{R_{\mathrm{E}} + h} \cos\gamma \right) - \gamma \right\} \qquad (4.63)$$

where $R_{\mathrm{E}} = 6{,}378$ km is the Earth's radius, h is the HAPS altitude and γ is the elevation angle, all angles being expressed in radians (1 rad $= \pi/180$ deg). Figure 4.50 shows the different footprint diameters that can be obtained with $h = 20$–25 km and $\gamma = 15°$, 45°, and 80°. It is seen that, as expected, the footprint strongly depends upon the elevation angle, allowing coverages over 150–180 km in country areas (15°) to 7–9 km in dense urban areas (80°). Taking the extreme example of a U.S. city downtown such as Manhattan with 300-m tall buildings and 50-m wide streets (85° minimum elevation for LOS to clear from ground), we find from the above formula that the footprint is still 3.5–4.5 km. In fact the footprint can be arbitrarily increased if antennas are being shared by a business or a residential community, and placed on building rooftops, as one would expect in such areas. Even larger zones can be covered by using several HAPS forming a "constellation." It is an interesting feature that HAPS constellations occupy fixed positions in the sky, in contrast with satellite constellations. Unlike GEO systems, they can be installed at any latitude with any range of elevation angles, namely up to 90°.

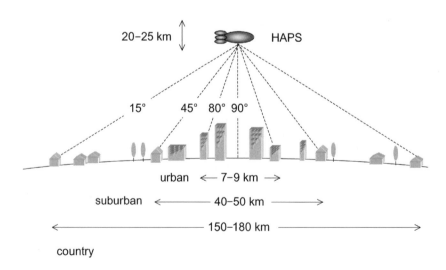

FIGURE 4.50 Footprints corresponding to HAPS as defined by the elevation angle (15° to 90°) and the HAPS altitude range (20–25 km).

We shall then briefly describe the different HAPS technologies concerning airships and aircraft. Some of the main systems features are also summarized in Table 4.8.

Airship/aerostats are designed for keeping a stable position within 1 km³; this is ensured by both longitudinal and transversal motors, which are powered by photovoltaic solar cells; the day/night cycles impose energy-storage capability with regeneration; their semirigid hull is ellipsoidal with typical sizes of 200–250 m (for length) and 30–60 m (for diameter); alternatively, the ARC system (see Table 4.8) uses low-altitude (3–10.5 km), small-size (46-m length) ships attached to the ground or tethered through a 1-inch cable; the cable supplies power and also data through optical fibers; such a cabled configuration requires the local airspace to be restricted as "no-fly zones"; the airship operating lifetimes are estimated to be from 3–5 years to 10 years, depending upon aging characteristics.

Aircraft are designed to follow specified circular trajectories of 4–10-km radius and 2–3-km altitude accuracy; typical operating altitudes are 16–19 km; they can be manned jet-planes as in Halo (*high-altitude long operation* [aircraft] (see Table 4.8) with three rotations of 8 hours per day; unmanned versions like the NASA-sponsored SkyTower (see Table 4.8), entirely powered by solar cells, can operate for up to 6 months without interruption (this is remarkable considering that 50% of this operation period corresponds to night time); jetplanes can carry greater payloads (up to 1,000 kg) than un-manned solar-powered planes (100–250 kg).

As Table 4.8 shows (when information is available) the HAPS services concern both fixed and mobile broadband access (1, 2, 5, 10, to 125 Mbit/s). As with BB-FSS for satellites, the first service type is referred to as B-FWA for *broadband fixed wireless access*. The frequency allocation for B-FWA is 47.9–48.2 GHZ for uplink and 47.2–47.5 GHz for downlink. Because of rain-attenuation constraints in certain areas (e.g. Asia), other bands could be used, near 28 GHz and 31 GHz for instance. The ARC tethered solution is quite interesting from the technology standpoint of using a hybrid fiber-optic cable for uplink and radio for downlink; such a configuration is inherently limited as far as altitudes and footprints are concerned (cable resistance, airspace restriction issues), but the optical link can handle thousand-fold more uplink payloads, thus opening new dimensions for BB access to large number of suburban users.

The HAPS network layout is very similar to that of satellite-based networks such as the early VSAT (Fig. 4.47) and the more elaborate Internet-capable version (in Fig. 4.49). A fully deployed HAPS network must comprise an airborne plant (aircraft, airships), possibly with a fleet of several interconnecting platforms (similar to ISS) for increased coverage and soft handover purposes, and a ground-station hub for service provisioning and connection to PSTN/ISP gateways. The HAPS can provide elevation angles up to 90° at any latitude over 100–200-km footprints with negligible latencies. This is in contrast with GEO systems which have significantly larger footprints, but are also handicapped by latency and

TABLE 4.8 Characteristics of main High-Altitude Platform Systems (early 2003 Internet data)

Project	Altitude	Unit Footprint	Max. Size	Payload	Bit Rate	Power and Cycle
Airship						
Sky Station	21 km	75–150–600 km	200 m	800 kg	2 Mbit/s (U), 10 Mbit/s (D) M: 9.6–16 Kbit/s (Vo), 384 kbit/s (Da)	Solar
Airborne Relay Communications System	3 to 10.5 km	55–225 km	46 m	700 kg	?	Tether fibre-optic cable
Aircraft						
Halo	16–19 km	65–95 km	30 m	900 kg	1–5 Mbit/s (D)	Fuel 3 × 8 h
Sky Tower	17.6 km or 20–21 km	80 km	30–74 m	250 kg	5 Gbit/s (T) 1.5–125 Mbit/s (D)	Solar 6 month
Heliplat	17–20km	60 km with 6 km cells	70 m	100 kg	2–40 Mbit/s (D)	Solar

Bit rate: T = total, U = uplink, D = downlink; services are fixed except when M (mobile) is specified; Vo = voice, Da = data.

Earth-based cellular services, which have much smaller footprint sizes, thus requiring arrays of base stations. Compared to GEO and LEO systems, the power budget of HAPS offers an inherent advantage of 66 dB and 34 dB, respectively, thus enabling significantly higher SNR (or lower signal-power operation).

The possibility for the disruptive HAP technology and services to steal a significant market share from traditional FSS (and possibly MSS), and from conventional Earth-based cellular/GSM services has huge commercial implications. Unlike satellite-based systems (GEO and constellations), HAPS can be rapidly deployed ("overnight service roll-out") and progressively configured at relatively low costs and low "launching" risks. This would be advantageous for many developing countries which do not have access to satellite technologies, or at least remain as yet dependent upon these to implement future BB networks. This is also a solution for rapidly deploying communications in the event of natural or man-made disasters. The exploitation of airship-based HAPS is non-polluting or "environment-friendly," which can gain psychological support for "lasting development" or ecology considerations (yet to be balanced by the uncontrolled multiplication of ugly antennas on rooftops and country landscapes!). Finally, HAPS implementation is relatively free from the network legacy, which opens the wireless telecom market to new entrants and business models for BB Internet services. In addition to the previously described technological features, these are some of the main advantages claimed by HAPS supporters.

However, serious obstacles remain, which concern technology, and legal and commercial aspects. To date, most proposed systems are still at a laboratory or pre-validation stage, and many technology problems remain, in particular in airship powering and optimal design. Indeed, solar exposition is a function of latitude and season, and current solar-cell technology falls short for ensuring other-than-summer coverage of countries located over $\pm 45°$ latitudes (at $45°$ latitudes indeed, the maximum solar flux, as expressed in W/cm^2, drops by more than 50% between June and December). Since atmospheric and weather constraints for airship and aircraft fluctuate over seasons, another issue is the guaranty of service continuity and back-up/back-haul solutions in case of momentary interruption or accidental network failure. Legal issues involve considerations of environment (especially for low-altitude HAPS), safety standards (possibility of flying failing ships to sea for termination) and complex local/government regulations (no-fly zones above major cities, airport traffic control areas). Commercial issues concern the adoption by consumers of new types of fixed (or mobile) terminals operating in the proposed frequency window, which would come in addition to mobile phones and satellite-dish appliances. It is also debatable whether premises' antennas for HAPS should be fixed or steerable, which in any case implies some extra installation cost. While offering potential commercial advantages to operators, the new HAPS services should appear seamless to the end-users, both for fixed and (2.5–3G) mobile applications, preferably with added quality value (e.g., competitive $cost/month, higher uplink/downlink bandwidth options, etc.). Since little information is available concerning these more detailed service offers (and the potential number of users concerned for the highest service class), it is not clear at this stage whether HAPS indeed heralds the "decline" of satellite-based systems, and will succeed some day in gaining a share of wireless access networks, even under the form of a synergetic or niche-market solution. Deployment of the first HAPS systems is scheduled for 2004.

■ 4.4 FIXED WIRELESS NETWORKS

As the name indicates, the concept of *fixed wireless communications* concerns all networks where signals are transmitted in free space through radio waves between nodes that occupy permanent or fixed positions. This is the case of *point-to-point* radio links and BB *multipoint* satellite networks such as illustrated in Figure 4.24 and Figure 4.49, respectively. The main application of multipoint, fixed wireless networks concerns the *local loop*, a term to designate the access to BB services over short distances (1–2 miles) from residential or bu7siness areas. Local loops are either of the *wireless* type (i.e., based upon radio or microwave links) or the *wireline* type (i.e., based upon wired links, which can be RF/microwave or lightwave). Thus, *wireless local loops* (WLL) represent a major alternative to wireline access loops.

An obvious advantage of WLL over wired systems is that they can be deployed anywhere without having to dig tunnels into the ground or pulling wires in the air. But this argument fades if one considers that many cities already have a complete underground network infrastructure, from electricity ducts, to gas, water and sewage pipes. Special technologies have been devised to drive wires through such infrastructures. Other possibilities are offered by train, subway and river networks for the deployment of wired networks in large cities. A drawback of WLL is that it cannot be implemented in dense urban areas where radio waves could be blocked or be subject to strong multipath interference. Apart from such debatable differences between wireline and wireless access, a key consideration is the *available bandwidth* in the uplink and the downlink paths. Clearly, optical fiber networks can carry orders of magnitude more bandwidth than other network types, wired or wireless. But their practical deployment all the way up to the "last mile," at residential or business premises, still remains an expensive option which could take 5–10 years (or even more) to be generalized. In most residential areas, it could be more advantageous and cheaper to resort to wireless rather than fiber access, because of no apparent need for so much bandwidth reserve. But such an argument could prove wrong in view of the progress of the Internet and demand for BB services, which could rapidly saturate the network resource. Overlooking optical-fiber access, WLL also compete with traditional wired networks based upon coaxial-cable and twisted-pair/telephone lines, with some key advantages in bandwidth and transmission-distance potential. It can also be seen as not only a competitor but a back-haul resource in the event of local wired-network failures (e.g., accidental cable cuts due to road works).

In this subsection, we shall consider BB-WLL, more commonly called *broadband wireless access* or BWA, as a mature technology alternative to wireline BB access (which is the full topic of Chapter 1 in Desurvire, 2004). As we shall see, BWA concerns both *multipoint* networks and *point-to-point* links. In the last case, the carrier frequencies can also be optical (*free-space optics*, or FSO). We will also describe other applications and standards which concern *wireless local area networks* (WLAN) and ATM over fixed-wireless networks (WATM).

4.4.1 Broadband Wireless Access (BWA)

The two main WLL standards for wireless, *multipoint distribution service* (MDS) and broadband access are referred to as *multipoint multichannel distribution systems* (MMDS) and *local multipoint distribution systems* (LMDS). Both systems are *licensed*, meaning that one must pay for the right to exploit the corresponding frequency bands, unlike other access systems (see *spread-spectrum* further below).

The earliest system, MMDS, was initially conceived for local TV broadcasting. It consisted in the progressive attribution of 6-MHz channels within a group of 31 slots located at 2.5–2.6-GHz frequencies plus two slots near 2.150 GHz. Referring to Figure 4.2, it is seen that the second spectral region has since been taken over by 3G/UMTS, except in North America where it is currently reserved for fixed WLL. Concerning the first spectral region, only 11 slots were attributed to MMDS, the remnant being used for educational TV, referred to as *instructional fixed television services* (ITFS). These slots could be leased to local TV operators upon the condition that 40 hours/week be dedicated to educational programs. Such subscriber-based TV services have been known as "*wireless cable.*"

Except in certain countries like Mexico or Australia, the worldwide development of MMDS has been slowed down or even blocked because of its incompatibility with local/national TV standards. For instance, the European standard, *called phase alternating line* (PAL), is based on 8-MHz-wide slots with higher image resolution. By 1998, the MMDS was opened to BB wireless internet access (Internet BWA), with an impressive offer of 10 Mbit/s downlink and 256 kbit/s uplink. These numbers become 1 Mbit/s (up) and 36 Mbit/s (down) in shared-capacity mode. Access services with such differences in uplink/downlink capacities are referred to as *asymetric*. Note that in the BWA world, one rather refers to uplink/downlink paths as *upload/download* or *upstream/downstream*, the first appellations being more generic to mobile or satellite systems. Later on, the MMDS consortium renamed their technology as "WDSL" for *wireless digital subscriber line*, in the spirit of evoking an alternative to the wired X-DSL services described in Chapter 1 in Vol. 2. But such a WDSL appellation should be used in full awareness that the XDSL family is based upon a completely different technology. As a matter of fact, the only feature WDSL and XDSL have in common is to be both *digital subscriber loops*. As a key advantage over LMDS (see below), the 2.5-GHz MMDS signals propagate over relatively long distances (100 km) under rain/snow and foliage obstacles conditions, with a single mast antenna being able to cover cells as wide as 50 km (30 miles).

The second system, LMDS, operates at 28–31-GHz frequencies (in three windows of 27.5–28.35 GHz, 29.1–29.25 GHz and 31–31.3 GHz) and is designed for total aggregated capacities of 150 Mbit/s per cell. Because the frequencies are about ten times higher, the signals are highly susceptible to rain or snow conditions and radio-path obstacles, which limit the cell sizes to 8 km, and even down to 4 or

1–2 km in denser urban areas. Because of these limitations, LMDS is referred to as a *line-of-sight* (LOS) access technology, in contrast to MMDS. The LMDS cell is divided into either five sectors of 72° each or four sectors of 90° each. The base station antenna, located in the center of the cell, uses therefore 4 or 5 directive antennas with the same carrier frequency but alternative horizontal and vertical polarizations corresponding to each sector. Figure 4.51 shows in the 90°-sector case how adjacent cells can be arranged with "polarization reuse" in order to form a square-tile pattern free from intercell interference (note that the terminal antennas in each sector must be chosen of the 'vertical' or 'horizontal' polarization type). Each LMDS sector can thus be serviced with downstream capacities of 30–45 Mbit/s, or 155 Mbit/s in shared-capacity mode. Access to this bandwidth is mediated by either TDMA or FDMA (multiple frequency carriers in a given cell), as illustrated in Figure 4.51. A typical configuration uses four carriers in a given cell (a form of *space-division multiplexing* or SDM), which are re-used at 90° angles and only twice. As seen in the figure, such a SDM arrangement makes it possible to form larger square-tile patterns from the same cell type, which are free from frequency overlap. The modulation formats corresponding to TDMA and FDMA are further detailed below. The cell's base stations are connected to a central service hub (*network operations center*, or NOC) and the PSTN/ISP gateways through high-capacity wired links such as fiber-optic cables. The LMDS network layout is illustrated in Figure 4.52.

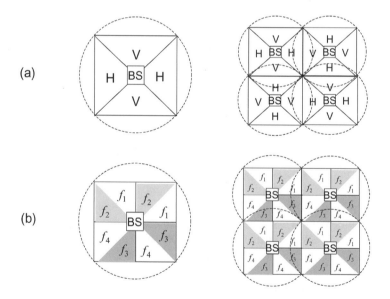

FIGURE 4.51 (a), Left: division of a LMDS cell into four 90° sectors with orthogonal polarizations (H = horizontal, V = vertical, BS = base station); right: arrangement of cells in tile-like pattern showing polarization reuse in adjacent cells and sectors. (b), Left: division of a LMDS cell into eight 45° sectors with four-frequency reuse pattern; right: arrangement of cells in tile-like pattern showing frequency reuse in adjacent cells and sectors.

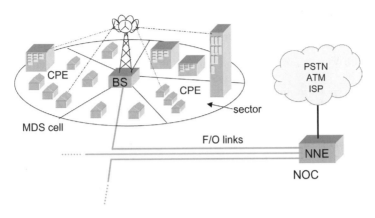

FIGURE 4.52 Basic layout of a local multipoint distribution service (LMDS) for wireless broadband access (WBA), showing unit cell with own sectors and connection to the network (BS = base station, CPE = customer premises equipment, NOC = network operations center, NNE = network node equipment).

In LMDS, the access mode is either TDMA or FDMA, with a coding format of 4-, 16- or 64- QAM (quadrature-amplitude modulation) at 1 to 10-Mbit/s symbol rates. TDMA may also use binary DPSK (differential phase-shift keying). Remember that, 16- and 64-QAM are 2 and 4 times more bandwidth-efficient than 4-QAM (also called QPSK, see Chapter 1). Thus a 64-QAM signal theoretically requires $1/\log_2 64 = 1/\log_2 2^6 = 1/6$ of the bandwidth of a binary signal at the same bit rate (e.g., 0.16 MHz for 1 Mbit/s). Due to receiver limitations, the actual ratio is closer to $1/5$ (e.g., 0.20 MHz for 1 Mbit/s), this estimation not taking into account further decrease due to overhead control data. Considerations of higher bandwidth efficiency would thus indicate a preference for the highest modulation levels (such as 64-QAM) over conventional binary signals. However, signals with higher modulation are much more sensitive to interference than at lower modulation. This requires one to include forward-error-correction (FEC), which increases the overhead at the expense of actual bandwidth-efficiency. Also, higher-modulation signals have more limited transmission range, thus restricting the cell diameter, and for a sectored cell, increasing the number of sectors and frequency re-use. Having more sectors, more frequencies and more cells increases the overall network cost. These different reasons favor the use of the less-efficient QPSK (4-QAM) or self-adapting schemes that locally combine different 4/16/64-QAM coding levels.

The MMDS/LMDS power budget and received signal-to-noise ratio (SNR) are calculated in the same way as previously described (see Section 4.1). Here, we shall provide some useful engineering formulas to help the reader understand how MMDS/LMDS are designed through a certain number of subsystem parameters.

To recall, the relation between received (P_{rec}) and transmit (P_{tran}) powers at antenna levels is simply [see equation (4.20)]:

$$P_{rec}(dBW) = P_{tran}(dBW) + G_T(dBi) + G_R(dBi) - 20\log_{10}\left(\frac{4\pi r}{\lambda}\right) - M(dB) \quad (4.64)$$

where G_T, G_R are the transmit/receive-antenna gains, r is the distance and $\lambda = c/f$ is the wavelength, with $c = 3 \times 10^8$ m/s being the speed of light and f the carrier frequency. The last term, M, is introduced to provide an extra *power margin* accounting for fading effects. Fade margins are typically $M = 15$ dB, but the value depends upon the type of geographical site with associated rainfall, multipath reflections and foliage density.

It is easy to convert this formula using kilometer and GHz units, which gives:

$$P_{rec}(dBW) = P_{tran}(dBW) + G_T(dBi) + G_R(dBi)$$
$$- 20\log_{10}(r_{km}) - 20\log_{10}(f_{GHz}) - M(dB) - 92.44 \quad (4.65)$$

Depending upon the design, transmit powers can vary from 5 W to 20 W (+7 dBW to +13 dBW). The transmit-antenna gain G_T depends upon its directivity, and hence upon the number n of sectors to be serviced within the cell. A rough estimation is $G_T(dBi) = 2.15 + 10\log_{10} n$, where 2.15 dB is the gain of a half-wave dipole antenna. Thus for a 5-sector cell, $G_T(dBi) = 2.15 + 10\log_{10} 5 \approx 9.1$ dBi. The receive antenna gain varies with the antenna design and essentially its aperture; typical values are $G_R = 18–24$ dBi.

The transmit and receive signal powers defined in equation (4.65) are taken at antenna level. Their relation to signal powers at transmitter and receiver levels (Q_{trans}, Q_{rec}) is the following (see Fig. 4.1 for reference):

$$\begin{cases} P_{trans} = Q_{trans}(dBW) + G_{post}(dB) - T_{line}^{trans}(dB) \\ Q_{rec} = P_{rec}(dBW) + G_{pre}(dB) - T_{line}^{rec}(dB) \end{cases} \quad (4.66)$$

where the gains (G_{post}, G_{pre}) are introduced by a *post-amplifier* (transmitter) and a *preamplifier* (receiver), and the losses ($T_{line}^{trans}, T_{line}^{rec}$) are introduced by the microwave transmission lines between the transceivers and the antennas, including the effect of combining/splitting signals. Typical line loss is $T_{line} = 4–6$ dB. Post- and preamplifier gains values are of the order of $G_{port} = +10/15$ dB and $G_{pre} = +20/35$ dB, respectively (post-amplifiers are highly saturated, due to the high input signal levels, to the contrary of preamplifiers).

The second key design parameter is the received SNR. It is given by the following relation:

$$SNR_{rec} = Q_{rec}(dBW) - F(dB) - N(dBW) \quad (4.67)$$

where F is the receiver noise figure (including the preamplifier), $N = k_B T_{eff} \Delta f$ is the thermal noise power associated with the transmitted signal in bandwidth Δf at effective temperature T_{eff} ($k_B = 1.38 \times 10^{-23}$ J/K), see Section 4.1. Noise figures are typically $F = 2.5/3$ dB. The thermal noise power is at minimum $N = k_B T \Delta f$ where $T = 290$ K, which gives $N = 1.38 \times 10^{-23} \times 290 \times 1.10^6 = 4 \times 10^{-15} \equiv -144$ dBW for a channel bandwidth of $\Delta f = 1$ MHz, and $N' = N \times 100 = 4 \times 10^{-13} \equiv -124$ dBW for a channel bandwidth of $\Delta f = 100$ MHz (the channel bandwidth is a function of the modulation format (e.g., PSK or M-ary QAM). The received bit-error-rate (BER) is a function of SNR, as shown in Chapter 1. The BER functional dependence with SNR depends upon the modulation format utilized. The SNR required for BER = 10^{-9} is typically SNR = +25 dB or somewhat higher, depending upon the modulation format. Alternatively, the required SNR can be significantly decreased (e.g., SNR < 15 dB) when error-correction (FEC) is implemented. One then refers to it as "required SNR before correction."

Formulas (4.65), (4.66), and (4.67) thus make possible to optimize the radio-system design according to requirements such as maximum distance and required SNR for nominal BER, given the channel carrier frequency and bandwidth, transceiver/antenna parameters and fade margins.

Notwithstanding its higher capacity offer, a main disadvantage of LMDS concerns the cost of terminal or *customer-premises equipment* (CPE). As per early 2003 quotes, a LMDS CPE would cost about $5,000, which essentially restricts the market to business/enterprise BWA, in contrast to the cheaper MMDS/WDSL. With LMDS, the capital network investment is located on the CPE side rather than the operator's side. From the last viewpoint, this represents an advantage for rapid network deployment and progressive scalability at minimum capital-investment cost, in comparison to wired-loop solutions where the situation is somehow reversed. Finally, MMDS and LMDS networks can be overlaid to adapt the cost and coverage according to the local region density and bandwidth needs.

In addition to MMDS and LMDS which use licensed bands, it is possible to implement BWA in other bands at 2.4–2.483 GHz and 5.725–5.825 GHz, referred to as *industrial, scientific and medical* (ISM, or "ISM-2.4" and "ISM-5.8"), and which are unlicensed. Its access mode is CDMA (spread-spectrum), with a theoretical capacity of 11 Mbit/s and unrepeated transmission distance of 8 km. Because its exploitation is unlicensed, it can be deployed anywhere without specific control but with serious problems of potential overlap and interference. For this reason, applications of unlicensed spread-spectrum systems mostly concern *indoor use*, such as WLANs (see Section 4.4.2). Alternatively, ISM can be used in *point to-point* links with directive antennas and zig-zag repeater paths to avoid interference, as illustrated in Figure 4.24 (see below). Offered BWA capacities for Internet use are 2.048 to 6.144 Mbit/s in shared mode, corresponding to a minimum of 256 kbit/s downstream capacity per user with 8 to 24 users per service cell, respectively.

A last category of fixed-wireless systems consists in *point-to-point microwave links* (in contrast to multipoint systems such as MMDS and LMDS). The available frequencies are of both the licensed or unlicensed types, depending upon the country or continental area. The aforementioned, unlicensed ISM bands (ISM-2.4 and ISM-5.8) are examples in use worldwide. The licensed 28–31 GHz LMDS band can also be used for point-to-point links. There are several other bands at 5.4–5.7 GHz (Europe, un-licensed), at 5.1–5.3 GHz and 5.7–5.8 GHz (*un-licensed national information infrastructure*, or U-NII, in US), and at 18, 23, 26, 38, 60, and 94 GHz (for licensed use in the world, United States, Japan or Europe). Note that the U-NII band overlaps with WLAN (see Section 4.4.2), but without interference since the applications are different (i.e., point-to-point short-haul versus multipoint indoors).

While point-to-point links can be repeated to extend the transmission distance (e.g., $N \times 8$ km), most applications concern single-span links, referred to as *wireless bridges*. A typical example of a wireless bridge is a LOS radio connection between two skyscrapers belonging to the same corporation, commercial trade center or campus. Since point-to-point links only concern two end-user stations, the payload capacity can be significantly higher (i.e., 10–155 Mbit/s and up to 1 Gbit/s) but at a significant user cost, which confines their application to business/enterprise communications. In dense urban areas, such high-capacity microwave links may yet represent a cheaper or more practical alternative than optical fiber. This consideration must be balanced with the higher cost of CPE, especially for 1 Gbit/s capacities for which 60/94-GHz carriers (W band) are required. In contrast, a single optical fiber may provide 622 Mbit/s to 2.5 Gbit/s and over at somewhat lower CPE cost, and tens of fibers can be put into a single cable link. Another application of wireless bridges concerns telephone services in the vicinity of country borders. Indeed, the cost of an international phone call between two "local" foreign cities can be significantly reduced if this call is directly forwarded across the border between adjacent COs from the two countries.

Point-to-point wireless links can form larger microwave networks, each terminal acting as a base-station node for servicing a local cell. The network topology can be a *ring* (nodes forming a polygon), a *star* (ring with central node) or a mesh (all nodes being inter-connected), depending upon the region to be covered and the level of redundancy and network protection required. Different operator owners can unite their local network resource to extend the microwave coverage to much larger areas and share the benefit improvements, a commercial practice called *daisy chaining*. Consistently, the resulting network is referred to as a *daisy-chained wireless local loop*.

As previously mentioned, the highest capacities are currently offered by the W radio band (60 and 94 GHz), which is itself located in the 30–300 GHz EHF region of the EM spectrum, see Figure 4.6. There are no licensing plans for exploiting bands located at EHF frequencies above 100 GHz. As Figure 4.5 shows, these frequencies are located near the *near-infrared* spectrum. Because they are relatively close to the terahertz (1 THz = 1,000 GHz) they are usually called *T-rays*. In this frequency range, the "rays" can traverse flesh but be blocked by metal, like X-rays, but without harmful effect, which can be used for security/screening purposes (e.g. airport passenger control). Apart from their current applications to civilian/military radar systems, T-ray bands above 300 GHz might represent another "last-frontier" Eldorado for future radio-communication systems with multi-gigabit/s capacities.

4.4.2 Free-Space Optics (FSO)

What about even-higher frequencies for point-to-point wireless links? We fall then into the *optical* domain, which opens another dimension for unlicensed wireless services, also known as *free-space laser communications*, or for the technology side, *free-space optics* (FSO). Implementing and exploiting FSO links is indeed free from any regulations other than equipment safety standards and prior path-exploitation rights. While the associated technologies for transmitters and receivers are completely different from microwave ones, FSO benefits from the early works started by NASA in the mid 1960s and advances made since in military infrared-optics and more recently, digital *lightwave* communications. Unlike microwaves and T-rays, optical-infrared waves are blocked by any type of nontransparent materials and rapidly absorbed by the atmosphere, which limits the applications to indoor or short-haul outdoor LOS links. On the other hand, optical beams are immune to any radio and optical interference from other optical carriers. Another advantage is that unlike in radio and wired microwave systems, FSO signals are immune to *eavesdropping* or third-party tapping on the signal path. They are also invisible (being at infrared-frequencies), thus making them artificially "acceptable" to the public, just like radio signals.

In outdoor applications, the FSO beam must stay clear of any *sun rays*, whose position varies with hour and season, including the effect of reflections on glass-covered towers. Since the beam propagates through the atmosphere, it can be subject to refraction bending and to random displacements due to combined hot-air turbulence and wind (an effect known as *scintillation*). It can be momentarily blocked by passing birds, construction cranes, window cleaners (!), and is rapidly absorbed by dust, smoke, fog or snow. Because of the relatively narrowbeam diameters, a precise alignment must be made between the source and the receiver. Such an alignment is threatened by adverse effects such as vibrations, thermal expansion and shocks, and mechanical aging. To counter these effects, terminals can be designed to automatically stay pointed towards each other through *auto-tracking* systems. Finally, the optical parts in the transmitter and receiver must be weather-proof, which is done by glass protection such as provided by an ordinary window. But the window must remain clean and free from pollution dust, frost, moisture, or occasional rain drops! This is in contrast with radio antennas, which can stand extreme weather conditions (except snow for high-elevation satellite dishes) without any material and performance degradation.

An important consideration for FSO deployment is *eye safety*. Signals at infrared wavelengths of 750–980 nm and 1.3 μm, such as available from miniature GaAs and InGaAsP laser sources (see Chapter 2 in Desurvire, 2004), are in fact harmful to the eye, with the risk of permanent retinal damage and even blindness. This is in contrast to 1.5 μm wavelengths, also available from InGaAsP sources, which are blocked by the eye's interior, but under relatively low beam-power conditions (a few mW). Consider indeed that for a 1-mm beam diameter, the relatively low power of 3 mW corresponds to an intensity of 400 mW/cm^2. When focused by the eye onto the retina into a spot of 20 μm size, this intensity becomes ≈ 1 kW/cm^2. If the spurious light rays are absorbed before reaching the retina, then no damage occurs, as long as the exposure is limited in time. For this reason, 1.5 μm communications have been called "eye-safe," despite the fact that their use is subject to power-safety limits. Since hazards are also caused by time exposure, FSO links may include *automatic power reduction* (APR) systems which detect beam interruption and turn off the beam power accordingly. In any case, all laser-based systems must be compliant to safety standards referring to power and time exposure as *accessible emission limit* (AEL) and *maximum permissible exposure* (MPE) parameters, respectively.

Compared to radio rays which diverge in space according to a more or less directive radiation pattern, light signals emitted from lasers are known to form highly parallel beams. The very high directivity of laser beams alleviates some of the above eye hazards (as long as they stay clear of eyes) and also makes it possible to keep the signal power focused over relatively long distances. It can be shown that for an initial beam diameter d at wavelength λ, the beam diverges with an angle

$$\theta = \pi \frac{\lambda}{d} \tag{4.68}$$

Considering for instance a beam of initial width $d = 1$ mm at $\lambda = 1.5$ μm, the divergence is $\theta = \pi \times 1.5 \times 10^{-3}$ mm/1 mm ≈ 0.005 radians (5 mrad) or 0.3°. At an observation distance D, the beam size (called 'spot size') is $d' = d + 2D \tan \theta) \approx d + 2D\theta$ (angle expressed in radians). For a distance of $D = 100$ m, the spot size is thus 1,001 m or 1 m. As a second example, if the initial beam diameter is 10 cm, we have $\theta = \pi \times 1.5 \times 10^{-3}$ mm/100 mm $\approx 5 \times 10^{-5}$ radians, yielding a spot size of $d' = 20$ cm at a distance $D = 1$ km. For a distance $D = 10$ km, the spot size expands to $d' = 1.10$ m. In order to minimize this effect of beam divergence and reach long distances with relatively small spot sizes, the initial beam diameter must be maximized under cost constraints for the required lens optics. Note that such an initial beam expansion is also useful for safety considerations, since the ray intensity decreases as the square of the beam diameter. In addition, expanded beams are not interrupted or significantly perturbed by the occasional crossing of birds. Instead of laser sources, cheaper infrared *light-emitting diodes* (LED) can be used, but their beam divergence is significantly higher, which limits the transmission distances to a few hundred meters (see example below).

Because most of the power is concentrated near the beam center, a major fraction of the signal power can be captured though 10–25 cm-sized optical lenses or telescopes. Note that the beam divergence is proportional to wavelength, thus shorter wavelengths such as 750 nm would correspond to twice-smaller divergence angles compared at 1.5 μm wavelength. Since the lens/telescope apertures are inherently limited for primary reasons of cost, the effect of beam divergence causes significant loss for distances over a few hundred meters. For instance, a spot-size change from 1 cm to 1 m corresponds to an aperture ratio of $(10/100)^2 = 1/100 = -20$ dB. This loss combines with *atmospheric propagation loss*, which (in decibel units) increases in proportion to distance. Under normal atmospheric conditions where LOS visibility is typically 2 km, the attenuation is 8.5 dB/km. But in temporary conditions of heavy fog, such as in coastal areas, this attenuation can be as high as -200 dB/km (85 m visibility)! In contrast, heavy rain corresponds to $-30/40$ dB/km, which shows how much atmospheric attenuation may impact on the link's *power budget* (see Chapter 1 for definition and previous subsection concerning radio links, see also Chapter 1 in Desurvire, 2004 for optical systems). Since relatively high signal powers (10–1,000 mW [+10 dBm to +30 dBm]) can be produced at source level, the achievable transmission distances can be extended in the kilometer range, and even up to 15 kilometers, typically (see examples further below). The actual maximum distance depends upon the source's characteristics and available power, the modulation format used, the corresponding receiver sensitivity and the nominal received bit-error-rate (e.g., BER $= 10^{-9}$), with 99.99% usage guarantee under specified temperature ranges (e.g., -20°C to $+50$°C) and other climatic conditions (rain and fog attenuation). A typical distance reference is 500 m for 1 Gbit/s at receiver BER $= 10^{-12}$ under a 200-dB, worst-case power budget. As in any digital system, the implementation of *forward-error correction* (FEC, see Chapter 1) makes it possible to significantly increase transmission distance (e.g., 5,000 m) at the expense of a relatively small bandwidth overhead. An important reference for "carrier-class" exploitation is the *system availability*, which is rated between

99.9% and 99.99%, corresponding to probabilities of about 8.5 h and 50 mn system-outage times per year. For FSO, carrier grade exploitation seem to limit system hauls to 500 m or less, but such a figure is highly dependent upon geography/region considerations.

As in optical fibers, the FSO bit rate can be increased by use of *wavelength-division multiplexing* (WDM), which combines different optical carriers into a single beam, and to some limited extent by use of *space-division multiplexing* (SDM), which uses closely spaced beams at the same carrier frequency. Commercially available products (see www references in the Bibliography at end of this book) provide duplex FSO systems operating at either 770–800-nm, 1,300-nm or 1,550-nm wavelengths, and covering distances between 100 m and 1,500–4,000 m, with bit rates from 155–622 Mbit/s up to 1.25–2.5 Gbit/s. Typical bit-rate/distance combination offers can be, for instance:

2.5 Gbit/s over 1,000–4,000 m for major carriers:

155–622 Mbit/s over 500–5,000 m for WLAN;

45 Mbit/s over 15 km for any application type;

2.5 Mbit/s over 300 m using cheap LED sources.

These examples illustrate the higher bandwidth capabilities offered by FSO compared to point-to-point microwave/radio systems. At equal service performance, CPE costs could be a few times higher than microwave radio links, but substantially less than point-to-point fiber links. For approximate reference, a 500-m–155-Mbit/s FSO link costs less than $20,000. Ring, mesh, and star FSO networks can be built the same way as in radio systems (e.g., LMDS), now referred to as *wireless optical networks* (WON). Although FSO have high beam directivities with dense FDM potential, the number of WON nodes is more or less limited by the availability of tall-building rooftops with clear LOS. Despite such a limitation, but considering installation cost advantages compared to fiber (the "last-mile" issue) and the higher bit-rate offer, WONs could be deployed at a massive scale on futuristic urban skytops as "*fiber-like bandwidth*."

4.4.3 Wireless LAN (WLAN) and Wi-Fi

The possibility of connecting together computers without using any cables is of considerable impact not only for enhanced networking connectivity, but also for wireless access to the Internet and services from any kind of transportable machine or appliance. In addition to the improvement in mobility and flexibility these reasons have been the key motivation to develop *wireless local area networks* or WLAN. As described in Section 2.5, the acronym LAN is a generic reference to privately owned computer networks having a "local" deployment with 1–10-km maximum station-to-station distance. LANs are governed by specific *access protocols*, which control the way stations communicate. Such access protocols can be either *random* (spontaneous connections at random times) or *deterministic* (orderly connections, for instance in preassigned time slots). While the first mode provides maximum network functionality and connectivity between users, it requires a set of rules to solve a problem known as *contention*. Contention happens when two stations attempt to communicate at the same time to the same destination machine, resulting in data-packet *collision*. A leading random-access protocol that was developed for wired LAN is the well-known *Ethernet*. As described in Chapter 2, Ethernet is based upon a smart protocol called CSMA/CD for *carrier-sense multiple access with collision detection*. This protocol is implemented in a network layer immediately above the physical layer, which is referred to as *data link* (or layer 2). Within layer 2, the CSMA/CD protocol is specifically handled by a sub-layer known as *medium access control* (MAC). Because the functionality of a WLAN is primarily that of a LAN, it is sensible that its access protocol be very similar to Ethernet, with its own CSMA and MAC-sublayer features. But a key difference between wired LAN and WLAN is that in the last case, collisions are being "avoided" rather than "detected." This approach lead to a new protocol called CSMA/CA, for CSMA with *collision avoidance*. Basically, CA is achieved by sending data only when a given frequency channel happens to be free. A second main difference with wired Ethernet is that WLAN intrinsically operates in *broadcast* mode, as opposed to a *connection-oriented* mode where a specific channel is reserved each time two end stations communicate.

The first standardization attempts of WLAN started more than ten years ago (1992) by *ETSI* (European Telecommunications Standards Institute), leading to an early protocol called *HiperLan*. They were followed by IEEE (Institute for Electrical and Electronics Engineers) under the committee known as "802.11," and by another working group called *HomeRF*. Here, we shall only focus on the 802.11 Standard which, with the recent development of portable (laptop) computers and other devices, has gained widespread popularity and seemingly taken the lead. The 802.11 group actually came up with three standards called *802.11a*, *802.11b* and *802.11g*. The first operates the 5-GHz band, while the other two operate in the 2.4-GHz band. In order to differentiate these three standards, it is necessary to describe the possible transmission-access modes that can be implemented in WLAN. One possibility is to code signals on optical/infrared waves, similarly to the technology used in some TV remote-controls (see the following discussion on so-called IrDA networks). This approach was nearly abandoned in favor of radio signaling, as based upon *spread-spectrum* coding (see Section 4.2). The three different and incompatible coding schemes retained for 802.11 WLAN are called FHSS (for *frequency-hopping spread-spectrum*), DSSS (for *direct-sequence spread-spectrum*), and OFDM (for *orthogonal frequency-division multiplexing*). They can be described as follows:

FHSS consists in making the signal frequency-hopping periodically among a cyclic set of 79 frequency tones at 2.4 GHz; these tones are located within the ISM-2.4 band (see previous subsection on BWA); the approach has been abandoned in favor of the next two;

DSSS is the same as in cellular CDMA, where signals are sent through a single-frequency carrier, this carrier being over modulated at a higher bit rate (called chip rate) by a code-word chosen from a set of orthogonal codes; frequency hopping is implemented on top in order to reduce interference; in an enhanced version, further CDMA coding is used, referred to as CCK (*comp-lementary code keying*), which allows bit-rate increase under optimal or inter-ference-free conditions;

OFDM simultaneously transmits on all frequency tones available.

The Standard 802.11a (5 GHz) is based upon OFDM, with a throughput capacity of 54 Mbit/s. Because of a sizeable overhead, the actual payload capacity of 802.11a is 31 Mbit/s. The standard 802.11b (2.4 GHz) is based upon DSSS and CCK enhance-ment. Its throughput capacity is 11 Mbit/s, corresponding to an actual payload capacity of 6 Mbit/s. The standard 802.11g (2.4 GHz) is based upon OFDM, like 802.11a, with 54/31 Mbit/s throughput/payload capacity. It is important to note that these capacities correspond to optimal, interference-free conditions. Interfer-ence effects correspond to spurious packet collisions and deletion as "packet errors." With increased packet-error rate, the effective data rate decreases between two-thirds (for 50% rates) to one order of magnitude (for 80–90% rates).

The popular (nonprofit) trademark *Wi-Fi*, a short for *Wireless Fidelity*, corre-sponds to the 802.11b standard. It also known as *wireless Ethernet*, not only because of conceptual similarity but for commercial appeal through a matching comparison with the first-generation of 10 Mbit/s wired Ethernet (as called 10base2, 10base5 and 10baseT; see Chapter 2). As previously stated, Wi-Fi thus operates at 2.4 GHz with 11 Mbit/s throughput capacity or 6 Mbit/s actual payload capacity per frequency channel. Three such channels are available within ISM-2.4 (2.40–2.83 GHz), corresponding to a maximum of $3 \times 11 = 33$ Mbit/s. The communication range for Wi-Fi is approximately 100 m indoors and up to 300 m outdoors or inside an empty warehouse or airport hall. Apart from using inter-mediary access points, the transmission range can be improved by more powerful directive antennas, although this approach is not in the true "spirit" of light-weight and serendipitous Wi-Fi networks!

The name Wi-Fi5 was given to the 802.11a. Unlike the previous Wi-Fi version, it is based on OFDM and operates at 54 Mbit/s (31 Mbit/s actual). For indoors applications, it can use any of 8 channels located within the U-NII (*unlicensed national information infrastructure*) band at 5.15–5.35 GHz. For outdoors appli-cations, it can use 4 channels in the second ISM (*industrial, scientific and medical*) band at 5.8 GHz (ISM-5.8, covering 5.725–5.825 GHz). Two Wi-Fi5 channels can be combined to double the payload (i.e., 72 Mbit/s), referred to as *turbo-mode* by Atheros Corp. The maximum throughput capacity of Wi-Fi5 is $8 \times 54 = 432$ Mbit/s for indoors and $4 \times 54 = 216$ Mbit/s for outdoors. The com-munication range for Wi-Fi5 is approximately 70 m indoors. As with 802.11b, the range becomes a few times greater inside empty building volumes. The range can also be improved by use of directive and more powerful antennas.

New generations of computers will be able to support both Wi-Fi and Wi-Fi5 standards (802.11a and 802.11b) as *dual-mode* machines. In Europe, Wi-Fi5 is not used since the 5-GHz band has been reserved by ETSI for the future develop-ment of its own *HiperLan* standard (referred to as *HiperLan2*). By the end of 2002, however, the ban for 802.11a was lifted by some of the EU countries. The specifications of HiperLan2 are practically identical (OFDM, 54 Mbit/s), except that it is not based on CSMA Ethernet. Rather, it is based upon TDMA, which is a collision-free access scheme and can include priority hierarchies between network stations. Progress is currently made towards reaching a common worldwide standard for both 802.11a (Wi-Fi5) and HiperLan2.

There are different ways to configure a WLAN. A set of stations put together in close vicinity can form a direct meshed network, each one being able to freely access any other. Alternatively, the network access can be centralized through an *access point*. The access points are transceivers which can have several functions. They can act as a simple relaying device between two stations, thus extending the WLAN range. A mobile station can also be connected from one access point to another. In this case, connection is ensured via *soft handover* (802.11) or *hard hand-over* (HiperLan2), as in GSM-cell roaming (see Section 4.2). Finally, the access point can act as a bridge between a WLAN and a LAN, or as a gateway to the Internet. These different configurations are illustrated in Figure 4.53. In any of these uses, the access point acts as a relay to broadcast the signals it detects, thus playing the role of a hub for the WLAN. Alternatively, the access point can act as a switch, with frequency selection for multicasting signals only to specific user subgroups called WLAN *segments*. The switch is advantageous over the hub since different subgroups can communicate independently of each other, unlike in the hub case. For further reference, the protocols for multiple access to the Internet via a single Wi-Fi access point (or gateways) are called NAT (for *network access translation*) and DHCP (for *dynamic host control protocol*), which roles are to dynamically assign temporary IP addresses to the individual computer clients from the connected group.

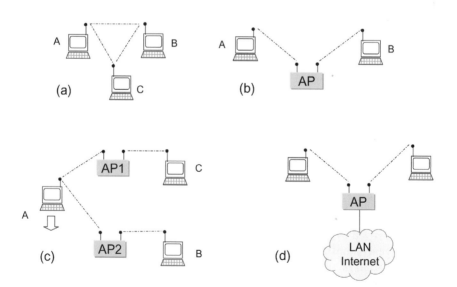

FIGURE 4.53 Possible configurations of wireless LAN: (a) meshed network, (b) connection through single access point (AP), (c) soft handoff connection via two access points, (d) use of access point as bridge or gateway to LAN or Internet.

A disadvantage of WLAN is that, by default of any secured protection, the broadcast signals can be intercepted by any party or eavesdropper who could be "driving by." This is also true in wired LAN since coaxial wires can be tapped into without traffic interruption, but this operation requires some level of building intrusion and manipulation. The 802.11 protocol allows for a built-in protection system called *wired equivalent privacy* (WEP). This WEP can be activated from the unprotected default mode to provide signal encryption and password access. However, some users may not activate this protection, which exposes their corporate or private networks to the hazards of eavesdroppers, attackers and other *hackers* (see last subsection of Chapter 3 on Internet jargon). The WEP protection itself has some drawbacks. Indeed, while the encryption is based upon a 40-bit encryption algorithm called *RC4* (see Chapter 3 in Desurvire, 2004), all WLAN users must share the same encryption key, which is not convenient to change. Like any encryption system, the key can be eventually found (as called "code cracking") by use of computation algorithms, a key length of 40 bits not requiring really powerful machines to process. Since WEP keys are not conveniently changeable, they remain in place for sufficiently long times for such programs to operate and succeed. But the fact that automatic WEP-cracking programs have been made available on the Internet is of more immediate and serious consequence! A new WEP-like encryption system was therefore developed, called *temporary-key integrity protocol* (TKIP), which is 128 bits long. It is however subject to the same weaknesses as its predecessor, save the somewhat longer times needed to crack it. The end solution to this security problem is to place network *firewalls* (see Section 3.3) between the Internet and each of the WLAN access point. Firewalls can indeed deny access to intruders. But they are expensive and make the network infrastructure more complex to develop and upscale. Besides, the problem of eavesdroppers being able to freely roam inside the WLAN zone (even as protected by WEP or TKIP), while standing "behind" the firewalls, say somewhere next door or driving by on the street, still remains to be solved. They may not be able to connect and read in clear, but they can store the data for future and more advanced cracking/processing. In addition, the fact that the same key must be shared by the different WLAN users is a generic concern for security. By 2002, there was no user authentication mechanism to prevent "legitimate session hijacking" by hackers. The future security issues of Wi-Fi and possible evolutions from WEP/TKIP are now under discussion by the *802.11i* standard group, including a new (yet contested) protocol referred to as *802.1X*. See more on this issue in Chapter 3 of Desurvire, 2004.

Apart from the safety considerations for business and privacy concerns, Wi-Fi has become extremely popular as a means to connect to the Internet without any wire, anywhere a public access point can be found: in cyber-cafes, cafeterias/restaurants, hotels, shopping malls, airport lounges, convention centers, university campuses, country resorts, and highway rest areas, to provide a few examples in an open list. These access points, called "hot-spots," make it possible for business and private people to access their e-mail or browse the Internet while traveling or being simply away from home or office. Some companies already use this feature to save office space, allowing employees to work not only while on the move or during lunchtime, but also on the premises, preferably in quiet or sunny outdoor places!

As Figure 4.53 indicates, and as previously mentioned, a single Internet connection (e.g., via any modem, including DSL) can be shared by multiple users, for instance by family members at home or on vacation. Moreover, peripheral resources like printers can be shared as well by user groups, now free from the hassle of wires and other rear-panel plugs. Another application of Wi-Fi is public broadcasting, for instance of useful service and urban/local information for tourists and travelers equipped with laptop computers. The rapid deployment of Wi-Fi (single and dual-band products) is illustrated by the sales of the computer-interface cards, which in 2002 are already estimated to exceed 1 million per month. While such a growth comes in competition with BWA from 3G cellular systems, it could be viewed as a complementary option. The coexistence between mobile cellular (3G and up) and WLAN "802.11" technologies, as they are now called, may in fact prove synergetic and successful. The idea is to combine high-speed access from WiFi-WLAN and migrate to 3G access from wide-area networks (WAN) according to instant bandwidth needs. For this reason, provisions for seamless GPRS/3G/802.11 roaming are under way. Considering still the relatively slow progress of 3G systems (see Section 4.2), such a transparent and universal communication means could take a few more years to come to reality.

4.4.4 Personal-Area Networks (PAN): IrDA and Bluetooth

As we have described Wi-Fi/Wi-Fi5 are currently leading standards for WLAN, which makes it possible for practically any laptop computers to "talk together" and connect to the Internet near dedicated access points. But such wireless connections are not limited to computers and in fact concern a vast array of devices and appliances. A nonexhaustive list includes printers, photocopiers, modems, mobile phones, pagers, headsets, electronic notebooks, *personal digital assistants* (PDA), digital photo/video cameras, portable music/DVD players, and pace-makers, for instance. Short-range wireless connectivity also concerns internet-controlled home systems (e.g., surveillance, fire alarms), medical and industrial instruments, telematics (short-range remote data acquisition) and various private/public services such as virtual ticketing, rentals, house meter-reading, highway tolling and city-parking control. Two leading standards for these applications are IrDA (*Infrared Data Association*) and *BlueTooth* ™.

Infrared data links (IrDA) were initially developed as an inexpensive solution (≤$2) for removing wires between computers and printers and also allowing wireless computer file transfers. The IrDA technology is based upon cheap/miniature infrared LED and photo-detectors which provide LOS wireless connectivity within a 30° angle, 1-meter range and with bit rates from 115 kBit/s to 4–16 Mbit/s. The bit rates are grouped into four classes: SIR (*serial infrared*) for 115 kbit/s, MIR (*medium infrared*) for 1.152 Mbit/s, FIR (*fast infrared*) for up to 4 Mbit/s and VFIR (*very fast infrared*) for 16 Mbit/s and possibly higher rates. Such a family of "wireless optical cables" is particularly adapted to applications where radio interference must be avoided. The IrDA protocol enabling fixed-wireless (stationary) connection to wired LAN is called *IrLAN*. Potential and future eye-safety issues of IrDA systems are under the control of the IEC-825-1 standard (*International Engineering consortium*).

The intriguing name *Bluetooth* comes from the legendary Denmark Viking Harald Blåtand ("bluetooth") who unified Scandinavia around 920A.D. and was known for his love of blueberries which stained his teeth. Based upon the 2.4-GHz ISM band, the Bluetooth technology offers low-cost ($5 chip), omni-directional and short-range connectivity (10 cm to 10 m) without LOS requirement (rays traverse walls and briefcases). The signal format is spread-spectrum frequency hopping, similar to the 802.11 standard but with fast, 3.2-KHz frequency-hop cycles. The maximum bit rate is currently 720 kbit/s, but a new IEEE standard (under the 802.15 working group) could bring it to 20 Mbit/s. Up to eight Bluetooth units can be interconnected, forming what is called a *piconet*. A larger network including more than eight Bluetooth connection devices is called a *scatternet*. One also generally refers to this as a *personal area network* (PAN), the human-oriented version of a LAN. The applications supported by Bluetooth are called *profiles*, and a single device can so far include up to nine such profiles. The profile list includes no less than 13 items, the first one called "core profile" being mandatory. The profile family covers functions as varied as cordless telephony (up to three simultaneous duplex channels), wireless headset, two-way intercom, fax-modem emulation (via mobile phone), serial-port emulation (such as *universal serial bus* or USB), device-to-device file transfer (e.g., digital camera and music, PDA to laptop/desktop computer), and LAN/Internet access. When considering data exchange, Bluetooth and IrDA have a significant overlap. Both use a client/server protocol called OBEX (*object exchange*) which allows direct data transfer between applications. Wireless data exchange has its own problems of security and confidentiality. In this respect, the omni-directional and long-range features of Bluetooth can be a handicap in comparison with the short-range, high-directivity IrDA. The fact of being able to point-and-shoot a given device (or a physical person) removes the need to search for the intended recipient among a possibly exhaustive list of surrounding users and appliances. However, both IrDA and Bluetooth include security, authentication and optional encryption mechanisms. In contrast with IrDA, which is a LOS point-to-point connection technology, the key advantage of Bluetooth piconets is the possibility of forming omni-directional piconets that traverse walls, remaining connected and synchronized with moving appliances (e.g., cellular phone).

4.4.5 Wireless Internet Access: WAP and i-Mode

Direct access to the Internet via mobile/cellular telephones is possible through two leading standards called WAP (*wireless application protocol*), and *i-mode*, which we shall describe hereafter. The concept of a "*mobile Web service*" is not to download full web pages on a 2″ × 1″ handset display, but rather to provide convenient cellular access to the Internet with low bandwidth requirements, in the form of simplified, miniature web pages. Examples of such services are flight/train schedules, weather and traffic reports, account balances, stock quotes, yellow pages, live sports results, interactive gaming, in addition to the usual e-mail and www browsing uses.

For the purpose of this *mobile internet access*, two specific programming languages were developed in replacement and simplification of HTML, the language which defines text, style, format, graphics and hyper-links in Internet Web-site pages (see Chapter 3). For the US-driven WAP and the Japan-driven (NTT's DoCoMo) i-mode, these are called *WML (*wireless markup language*), and* C-HTML (compact HTML, also written cHTML), respectively. In contrast to C-HTML, WAP also comes in as a new and complex stack of protocols which are adapted to the cellular-communication environment, as characterized by relatively high bit-error-rates, latencies and timing jitter. While the WAP pages travel through the Internet via HTTP and TCPIP, they are encapsulated under different payloads through the OSI network levels (see Figures 2.18 and 3.15). The equivalent TCPIP datagrams corresponding to WAP messages are handled under WDP (wireless datagram protocol) on both "transportnetwork" layers 43. Encryption, session and application follow a series of protocols similar to, but somewhat more complex than, with IP. In contrast, the C-HTMLi-mode approach can be viewed as a direct, no-frills simplification of HTML web pages, making it possible to implement it via the same high-level Internet layers, i.e., from HTTP (layer 5) and to the top (layer 7). Note that the definition of a backward-compatible standard common to WAP and C-HTML is under way, as based upon a new version of WML, called WML2 or XHTML (extended HTML).

Beside being inherently slow and sometimes unreliable, a drawback of WAP is that Internet browsing is limited to a restricted number of web sites which offer WML pages, in contrast to the unlimited www based on the more complex HTML pages. In contrast to WAP, i-mode has paid close attention to user-friendly terminals, with fancy color displays and one-click preset Internet access. Another difference between the two access systems is the end-user focus, i.e. businessman versus teenage. The initial success of i-mode, to be eventually followed by WAP, was to offer services such as fun cartoons and competitive games (e.g., "Catch a Fish" in Finland). Understandably, web-sites that contain too much text, pictures and animated banner adds cannot be available to either WAP or i-mode access, which limits the commercial interest of further developing websites towards wireless "internauts." Also, expectations from confirmed internauts might be disappointed by the observed difference in browsing quality and access times. But as we have seen, other solutions such as Wi-Fi more fully address the issue of *wireless Internet access*, although they are dependent upon the local availability of "hot-spot" or wireless-modem access points, unlike for cellular networks which are deployed nearly everywhere. A potential upgrade for WAP is its eventual integration into the GSM-based, broadband GPRS. The developments of Internet access through mobile phones are however the key point and appeal of future 3.5G–4G systems (see Section 4.2).

✓ EXERCISES

Section 4.1

4.1.1 What is the E-field amplitude associated to an EM wave with intensity of $I = 1$ W/cm^2 in vacuum? Same question in a glass medium with index $n = 1.45$?

4.1.2 An EM wave with power $P = 1$ mW is captured into a hollow cylindrical guide having $d = 1$ cm diameter and being filled with vacuum. What are the EM intensity and the corresponding E-field amplitude?

4.1.3 What is the length of a $\lambda/2$ dipole antenna emitting at $f = 2$ GHz?

4.1.4 A radio link operating at 5-GHz frequency has a transmitting antenna with absolute gain $G_A = +70$ dB, and a receiving antenna with effective aperture $A_R = 1$ m². What is the transmit power required to receive $P_R = 1$ μW (-60 dBW) at a distance of $r = 500$ km? Express the result in dBW and dBm.

4.1.5 A satellite is located at 36,000 km above the Earth. What is the dB loss introduced by free-space transmission from the shortest path to/from the satellite? Provide the power budget assuming a satellite-antenna gain of 30 dB and a receiver-antenna aperture of 1 m².

4.1.6 Calculate the power absorption loss (dB) caused by a rainfall of rate $R = 10$ mm/h on radio waves at $f = 5$ GHz and 10 GHz, assuming vertical polarization and horizontal propagation over 250-km distance.

4.1.7 Calculate the power absorption loss (dB) caused by a rainfall of rate $R = 100$ mm/h on a Ka-band (30-GHz) Earth-to-satellite radio system, assuming horizontal polarization, propagation at $\theta = 45°$ angle, and effective rain height of $h = 3$ km.

4.1.8 Show that the maximum incident angles for reflecting radio waves onto the ionosphere layers (heights $h = 50-250$ km) are in the range $\phi = 74-83°$. Calculate the maximum possible transmission distance for a single Earth-to-Earth round-trip path (Earth radius $R_E = 6,378$ km).

4.1.9 Calculate the ionospheric reflection coefficient during daytime (plasma density $N = 2 \times 10^{11}$ m⁻³) of a $f = 14$ MHz radio-wave for a $\phi = 74°$ incidence with horizontal polarization.

4.1.10 Prove the parabolic approximation formula for the Fresnel-zone radii:

$$H \approx \sqrt{n\lambda \frac{d'(d - d')}{d}}$$

where d is the LOS path length, d' the distance from the origin and n is an odd integer.

4.1.11 Calculate the beam clearance for a $f = 2$ GHz radio link with $d = 1$-mile antenna-to-antenna distance (express the result as maximum Fresnel radius and safety cone angle).

4.1.12 What is the noise frequency of multipath fading, assuming a 2-GHz radio carrier and a mobile-antenna speed of 75 mph?

4.1.13 What is the effective temperature of an UHF antenna emitting at $f = 1$ GHz in outer space and in the Earth's atmosphere ($T = -270°C$ and 20°C)? Conclusions?

4.1.14 An Earth-based radio system ($T = 290$ K) utilizes an antenna with noise temperature $T_{ra} = 50$ K, a transmission line with power loss of $L = -0.2$ dB and a preamplifier with noise temperature $T_{amp} = 2610$ K. Calculate the effective system temperature and the corresponding noise figure. To how many decibels corresponds the end-to-end SNR degradation?

4.1.15 An Earth-to-satellite communication link at carrier frequency $f = 2$ GHz has a system temperature of $T_S = 30$ dBK, a satellite antenna gain $G_T = 30$ dB, a satellite transmission power of $+3$ dBW, a mobile receiving antenna gain of $G = 15$ dB. The satellite is orbiting at an altitude of $r = 36,000$ km. Calculate the receiver CNR and SNR, assuming a bandwidth of $\Delta f = 100$ kHz.

4.1.16 A satellite-based communication system is characterized by an uplink CNR of $CNR_U = +75$ dBHz and a downlink CNR of $CNR_D = +50$ dBHz. What is the net CNR corresponding to the round-trip path? Conclusion?

Section 4.2

4.2.1 Prove that in a given network, the traffic intensity A is equal to the number of calls Q that can be expected during the average holding time H.

4.2.2 Prove that in a network with traffic intensity A, the occupation rate of a single channel is equal to $B = A/(1 + A)$. [Use result of exercise 4.2.1.]

4.2.3 A mobile-network cell is 10 km in diameter and the area has a density of 10 mobile users per km². During 1 peak hour of monitoring time, it is observed that 5% of these users place calls with an average call duration of 10 mn. What is the traffic offer, and how many channels are required to provide a grade service of 0.01?

4.2.4 Show that for a network having an infinite number of channels which is filled on first-channel-available basis, the occupancy for channel N, as expressed in erlangs, is given by

$$O(N) = [B(N - 1) - B(N)]A$$

where A is the traffic offer, and $B(N)$ the Erlang formula for the blocking rate in an N-channel network.

Section 4.3

4.3.1 What is the altitude of geosynchronous satellites with respect to sea level?

4.3.2 Taking $r_0 = 35,785$ km as the reference distance for geostationary satellites, what are the distances and corresponding altitudes for which satellites can complete exactly $N = 2, 4,$ 12 or 24 revolutions per 24-h day (corresponding to periods of 12 h, 6 h, 2 h, or 1 h, respectively)? Conclusion?

4.3.3 What is the maximum inter-city distance covered by a GEO satellite? (Use $r_0 = 42,000$ km and $r_E = 6,400$ km for the Earth-to-satellite distance and Earth radius, respectively.)

4.3.4 What is the intercity distance covered by LEO satellites? How many LEO satellites are required for full coverage of the Earth's perimeter, assuming that one should use two times the theoretical minimum? (Use $h = 100–1,000$ km and $r_E = 6,378$ km for the satellite altitude and Earth radius, respectively.)

4.3.5 Assume a satellite having highly-elliptic orbit (HEO) with major axis $2a = 53,260$ km and eccentricity of $e = 0.74$ (Molnya orbit type). Calculate first the revolution period (in hours) and perigee altitude (in kilometers). Calculate then the nodal regression and perigee precession for polar and equatorial inclinations ($i = 0$ or $90°$, respectively), as expressed in degrees/day at the perigee point. Also, calculate the nodal regression for the critical inclinations i giving $\cos(i) = \pm 1/\sqrt{5}$ (use $J_2 = 1.0823 \times 10^{-3}$ and $R_E = 6,378$ km).

4.3.6 Provide the inclination i required for a Sun-synchronous circular orbit at $h = 500$ km altitude? (Use $R_E = 6,378$ km, $\mu = 3.986 \times 10^{14}$ m³/s² and $J_2 = 1.0823 \times 10^{-3}$.)

4.3.7 Over what amount of time does a geostationary satellite drift by $0.1°$ from its nominal position, due to the precession of perigee effect? (Use $J_2 = 1.0823 \times 10^{-3}$ and $R_E = 6,378$ km.)

4.3.8 The International Space Station (ISS) has a near-circular orbit with cruising altitude of 388 km. Calculate (a) its revolution period in hours and minutes, (b) its angular speed in degrees per minute, (c) its cruising speed in meters per second and relative to the speed of sound (300 m/s), (d) its speed relative to a point on Earth at latitude λ, assuming an orbital inclination of $i = 51.63°$ and a $90°$ observation angle (zenith), and (e) its angular speed relative to the Equator, in degrees per minute. Concerning the last result, how does it compare with the apparent motion of stars in the sky at the Equator's zenith?

4.3.9 How many MEO satellites are necessary to cover the Earth's circumference, assuming 10% coverage overlap and an altitude range of 5,000–15,000 km? (Clue: Calculate for both altitudes the maximum intercity distance, as defined by the satellite's horizon, and use $r_E = 6,378$ km for the Earth's radius.)

4.3.10 How many LEO satellites are necessary to cover the Earth's circumference, assuming 10% coverage overlap and an altitude range of 100–1,000–1,500 km? (Clue: Calculate for the three altitudes the maximum intercity distance, as defined by the satellite's horizon, and use $r_E = 6,378$ km for the Earth's radius.)

4.3.11 Evaluate the minimal cost of two typical constellation systems made of either $N = 48$ or 66 LEO satellites, assuming satellite (SV) masses of $M = 250$ kg or 700 kg, respectively, assuming for both a unit cost of $U = 75$ k$/kg and a launching cost of $L = 15$ k$/kg.

4.3.12 Show that for a HAPS at altitude h and seen above the minimum elevation angle γ, the footprint diameter d is approximately given by the formula ($R_E = 6,378$ km, $h = 20$–25 km):

$$d = 2R_E \left\{ \cos^{-1} \left(\frac{R_E}{R_E + h} \cos \gamma \right) - \gamma \right\}$$

a MY VOCABULARY

Can you briefly explain each of these words or acronyms in the wireless, mobile, and satellite network context?

Absolute gain (antenna)	Ascending node ATM	C-HTML (wireless) Carried traffic	Coulomb force CPE
Absolute temperature	Attenuation AuC (GSM)	Carrier-to-noise ratio	Cryptography CSMA/CA
ADC (mobile)	B3G	CCK (WLAN)	(Ethernet)
ADSL	Back-fire	CCS	CSMA/CD
AEL (laser safety)	Back-haul	cdma2000 (mobile)	(Ethernet)
Alternative current	(network)	CDMA	CSPDN
Amplitude	Base station	cdmaOne (mobile)	Current (electrical)
AM	Beam steering	cdmaTwo (mobile)	Cycles per second
AMPS (mobile)	Belt (satellite,	Cell	CUG (satellite)
AMS (ATM)	service)	Cellular radio	D-AMPS (mobile)
Anomaly (true,	Big LEO	Chip rate	Daisy chaining
mean eccentric)	Big Bang	Ciphertext	(wireless)
Antenna	Blue shift	Clarke's belt	dB
Aperture	Bluetooth	Clearance	dBHz
Apogee	Boltzmann's	CLI (GSM)	dBi
APR (laser safety)	constant	Cluster size	dBK
Apsidal rotation	Booster	CN (UMTS)	dBm
rate	Broadside	CNR	dBW
Apsides	BSC (GSM)	CO (PSTN)	DCS-1800
Argument of	BSS (GSM)	Codec	DQPSK
perigee	BTS (GSM)	Comsat	Delay spread
Array (antenna)	BWA (wireless)	Constellation	Deregulation
Array factor	C-band	(satellite)	Descending node

DGPS
DHCP (WLAN)
Dipole
Directive gain
    (antenna)
Directivity
    (antenna)
Director (element)
Dish antenna
Doppler frequency
    shift
Downlink (path)
Dowload (path)
Downstream (path)
DSL
DSSS (WLAN)
EDGE
E-GPRS
EHF
e-learning
E-wave
e-working
Eccentric anomaly
Eccentricity
Eclipse (Earth,
    Moon)
Effective rain height
Effective
    temperature
Efficiency (antenna)
E-HSCSD (mobile)
EIR (GSM)
Electric charge
Electrical field
Electromagnetic
    (field/wave)
Electron
Electron-volt
Element (orbital)
Elevation angle
ELF
Ellipse
EM
EMSS (satellite)
Encryption
End-fire
Epoch (time)
Erlang (unit)
Erlang's formula
ETACS (mobile)

ETSI
EV
Fade margin
Far-field
Frequency hopping
FDD
FDM
FDMA
FEC
Feed (antenna)
FHSS (WLAN)
Field line
Field strength
FIR (wireless)
Firewall (WLAN)
FM
F-NMT (mobile)
Focus
FOMA (mobile)
Footer
Footprint
FPLMTS (mobile)
Frequency
Frequency-division
    duplexing
Frequency-scanning
    (antenna)
Fresnel zone
Friis transmission
    formula
FSO (wireless)
FWA, B-FWA
Gain (antenna)
Gamma rays
GEO
Geo-stationary
    (orbit)
Geo-synchronous
    (orbit)
GGSN (UMTS)
GMSC (UMTS)
GMSK
GPRS (mobile)
Grade (GPRS)
Gravitation
    (constant)
Gold codes
    (CDMA)
GOS
GPRS (mobile)

GPS
GSM
GSM slot
GSM frame
GSM superframe
HALE
HALEP
HALO (aircraft)
Handoff
Handover
HAP
HAPS
Hard handover
HBE (satellite)
Header
HEO
Hertz
HF
HLR (GSM)
Hot spot (WLAN)
HPBW
HSCSD (mobile)
HTML (Internet)
Hub (satellite)
Hyperframe (GSM)
HiperLan (WLAN)
HiperLan2
    (WLAN)
i-mode
IEC
IEEE
Impedance
IMT-2000 (mobile)
Inclination (orbit)
Incumbent
INMARSAT
INTELSAT
Intensity
    (EM wave)
Intensity (traffic)
Interference
Intersymbol
    interference
Ion
Ionization
Ionosphere
IR
IrDA (WLAN)
IrLAN (WLAN)
IS-95x (mobile)

ISDN
ISI
ISM (wireless)
ISM-2.4
ISM-5.8
Isotropic antenna
ISS
J-TACS (mobile)
JDC (mobile)
Ka-band
Képler
Ku-band
L-band
LAN
Laser
Last mile
Latitude
LED (optics)
LEO
LF
LH
Lightwave
Line of sight (LOS)
LMDS
Long-haul
Longitude
Longwave radio
Loss (transmission)
Lunisolar attraction
LW
MAC (Ethernet)
MAN
MCS-L1/L2
    (mobile)
Mean
Mean anomaly
Medium-haul
MEO
Meteor trail
Microcell
Microstrip antenna
Microwave
    Microwaves
    (domain)
MF
MH
MIR (wireless)
MMS (mobile)
MMDS
Mobile (radio)

Mobile commerce
Molnya orbit
MPE (laser safety)
MSC (cellular)
MTSO (cellular)
MUF
Multiframe (GSM)
Multimedia (3G)
Multipath fading
Multipath
    interference
MW
n^{th}-resonant orbit
Network C (mobile)
NADC (mobile)
NAMTS (mobile)
NAT (WLAN)
NCC (satellite)
NCN (ATM)
NE
Network element
Newton (unit)
NF
NMT (mobile)
NNE (LMDS)
NOC (MDS)
Node (ascending/
    descending)
Node B (UMTS)
Node regression
Noise
Noise figure
Noise temperature
NSS (GSM)
OBEX (wireless)
Octogonal PSK
OFDM
Offered traffic
Offset feed
    (antenna)
OMC (GSM)
Optical
    (communications)
Orbit
Orbital elements
OSS (GSM)
Over-dense trail
Overhead
Overhearing (GSM)
Overshoot

Ozone
Ozonosphere
Paging
PAL
PAN
Parabolic antenna
Parasitic element
    (antenna)
Patch antenna
Payload
PCM
PCN (mobile)
PCS (mobile)
PDA (mobile)
PDC (mobile)
Perigee
Perihelion
Phase (orbit)
Phased-array
    antenna
Photon
Picocell
Piconet
PIN
Plaintext
Planck's constant
Planck's law
Plasma
Potential
    (difference of)
Pound
Power budget
Precession of
    Perigee
Primary radiation
    pattern
Profile (Bluetooth)
PSPDN
PSK
PSTN
PTT
Q-band
Quanta
Quantum noise
R2000 (mobile)
Radar
Radiation pattern
Radio (wave)
Radiotelephone
Rainbow

Rainfall
    attenuation/loss
Rayleigh fading
RC4
Red shift
Reflection
    (coefficient)
Reflector (antenna,
    element)
Refraction
Refractive index
Regression (node)
Repeater station
Return channel
    (satellite)
Reuse (frequency)
Right ascension
RMTS (mobile)
RNC (UMTS)
RNS (UMTS)
Roaming
Room temperature
S-band
Satellite
Satphone
Scattering
Scatternet
Secondary lobe
Secondary radiation
    pattern
Semi-synchronous
    satellite
SDMA
SDR (mobile)
SGSM (UMTS)
SH
Short haul
SHF
Side-lobe
Sidereal day
Signal-to-noise
    ratio
SIM (GSM)
SIR (wireless)
Skywave
Slot class (GPRS)
Smart antenna
SMS (GSM)
SNCC (satellite)
SNR

Soft handover
Spectral efficiency
Spectrum
Speed of light
Spoofing (satellite)
Spot area (satellite)
Spot beam
    (satellite)
Spread spectrum
Spreading factor
Standard deviation
Station keeping
Stratosphere
Sun outage
    (satellite)
Super-synchronous
    (satellite)
Surface wave
SW
T-rays
TACS (mobile)
TCH/Fx (GSM)
TCPSat
TDM
TDMA
Telelearning
Teleshopping
Telesurveillance
Teleworking
TEM
Temperature
    (effective
    antenna)
Thermal noise
Thuster
Time duplexing
TKIP (WLAN)
TLE (satellite)
Traffic intensity
Traffic offer
Transverse
    EM-wave
Triangulation
Troposphere
Tropospheric
    scattering
True anomaly
TSF
Two-line element
    format

U-NII (wireless)
UAV (craft)
UDLR (satellite)
UE (UMTS)
UHF
UMTS
UMTS Forum
Under-dense trail
Uniform field
Uplink (path)
Upload (path)
Upstream (path)
USB
USDC (mobile)
UTC
UTRAN (UMTS)
UV

V-band
Van Allen (belts)
Variance
VFIR (wireless)
Videophone
Visible light
VHF
VLF
VLR (GSM)
Volt
VoD
VSAT
W-band
Walker orbit
Walsh codes
  (CDMA)
WAN

WAP
Wavelength
WDP (wireless)
WDSL
Webphone
WEP (WLAN)
Wi-Fi
Wi-Fi5
Wireless access
Wireless bridge
Wireless Ethernet
Wireline
WLAN
WLL
WML
  (wireless
  Internet)

WML2
  (wireless
  Internet)
WON
X-band
X-ray
XHTML (internet)
XOR
Yagi−Uda array
  antenna
ZM
802.11
802.11a
802.11b
802.11g
802.11i
802.1X

Solutions to Exercises

Section 1.1

1.1.1 Calculate how much energy is contained in the wave generated by the pebble's fall into the pond, assuming that the pebble mass is $m = 100$ g and that it is launched from height of $h = 1$ m (elementary physics background required).

The energy in the wave is equal to the kinetic energy of the pebble at the time of impact (assuming full conversion). The kinetic energy is equal to the potential difference from launch to impact, i.e., $\Delta U = mg(h_2 - h_1) \equiv mgh$, where $g = 9.81$ m/s^2 is the gravitational acceleration coefficient. So $\Delta U = 0.1$ kg $\times$ 9.81 ms^{-2} $\times$ 1 m $= 0.98 \approx 1$ joule.

1.1.2 Assume the wave defined by equation (1.1) to be a light wave with $\lambda = 1.5$ μm wavelength. What are the frequency and the period of this wave? (Speed of light $c = 3 \times 10^8$ m/s.)

The frequency is given by $\nu = c/\lambda = 3 \times 10^8$ m/s$/(1.5 \times 10^{-6}$ m$) = 2.0 \times 10^{14}$ Hz $\equiv$ 200 THz (terahertz). The period is $T = 1/\nu = 1/(2 \times 10^{14}$ s$^{-1}) = 5.0 \times 10^{-15}$ s $\equiv$ 5×10^{-3} ps (picosecond).

1.1.3 Calculate the power density in an optical fiber whose core diameter is 4 μm, assuming that the light signal power is 1 mW. Convert the result in W/cm^2. What can be concluded?

The core cross section or surface is $S = \pi \times (4 \times 10^{-6}$ m$)^2 = 50 \times 10^{-12}$ m^2. The power density or intensity is then $I = P/S = 1 \times 10^{-3}$ W$/(50 \times 10^{-12}$ m$^2) = 2 \times 10^7$ W/m^2. Since 1 m$^2 = 1 \times (10^2$ cm$)^2$, we have 2×10^7 W/m$^2 = 2 \times 10^7$ W$/(10^2$ cm$)^2 = 2 \times 10^3$ W/cm^2, or 2 kW/cm^2. Conclusion: in an optical fiber, a power as small as 1 mW corresponds to an enormous power density, like a kW lamp whose light would be focused onto a square centimeter surface.

1.1.4 Assume a waveform defined by equation (1.10). What is the modulation index m for which the total power in the side-bands is equals to that in the carrier tone?

By definition, the total side-band power is $2 \times (mA_M/2)^2 = m^2 A_M^2/2$. The carrier tone power is A_M^2. Equating the two gives $m^2 A_M^2/2 = A_M^2$ or $m = \sqrt{2}$.

1.1.5 Without (!) a calculator, convert in dBm the following powers: $P_1 = 10$ nW, $P_2 = 4$ μW, $P_3 = 500$ μW, and $P_4 = 25$ mW.

A nanowatt is 10^{-6} mW or -60 dBm. So $P_1 = 10$ nW is -50 dBm. A microwatt is 10^{-3} mW or -30 dBm. A factor of four is $+6$ dB. So $P_2 = 4$ μW is $-30 + 6 = -24$ dBm. Then 500 μW is half of 1000 μW $= 1$ mW or 0 dBm. A factor of $\frac{1}{2}$ is -3 dB. Thus $P_3 = (1000/2)$ μW $= -3$ dB $+ 0$ dBm $= -3$ dBm. Finally, 25 mW is one-quarter of 100 mW or $+20$ dBm. One-quarter is -6 dB. Thus $P_4 = (100/4)$ mW $= -6$ dB $+ 20$ dBm $= +14$ dBm.

Section 1.2

1.2.1 Make the table of binary numbers ranging from 0 to 15.

Wiley Survival Guide in Global Telecommunications: Signaling Principles, Network Protocols, and Wireless Systems, by E. Desurvire
ISBN 0-471-44608-4 © 2004 John Wiley & Sons, Inc.

$15 = 2^4 - 1$, so numbers fit into 4 bits:

1 = 0001	6 = 0110	11 = 1011
2 = 0010	7 = 0111	12 = 1100
3 = 0011	8 = 1000	13 = 1101
4 = 0100	9 = 1001	14 = 1110
5 = 0101	10 = 1010	15 = 1111

1.2.2 Convert your year of birth into the binary system, and show the result in two octets. (Clue: Decompose the number in powers of two, starting from the highest.)

Two octets is 16 bits. Since $2^{11} = 2048$ and $2^{10} = 1024$, the bits #11 and above are zero and the bit #10 is one. Assume you are born in year 1955, so my birth date is of the form 000001XX_XXXXXXXX (underscore to separate the two octets). Remains then to convert $1955 - 1024 = 931 = 2^9 + 419 = 2^9 + 2^8 + 163 = 2^9 + 2^8 + 2^7 + 35 = 2^9 + 2^8 + 2^7 + 0 + 2^5 + 3 = 11\_10100011$. Combining the results yields $1955 = 00000111\_10100011$

1.2.3 Draw the voltage diagram of the bit string 1001 1000 1101, first in Manchester and then in differential Manchester code, assuming in the second case that the bit preceding the string is 1 in low state.

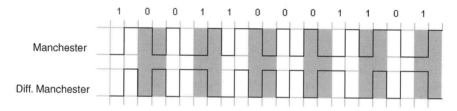

1.2.4 Draw the voltage diagram of the bit string 1001 1000 1101, first in AMI and then in HDB2 code, assuming that the bit preceding the string is 1 in low state.

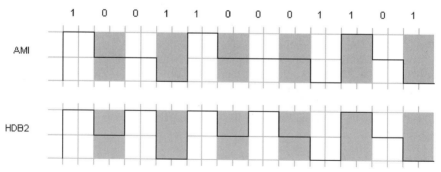

1.2.5 Generate a Gray-code 4-bit symbol correspondence for 8-ary PAM with the same coding rule as in the quaternary space diagram shown in Figure 1.12.

Eight-ary signals are coded with three bits, starting from 000. The rule is to modify a single bit from one symbol to the next, chosing the one having the lowest weight and giving a different code word. Thus we get 000, 001, 011, then 010 (according to the lowest-weight rule). The next symbol must be 110 since the first two bits have gone through all changes possible, etc. As a result, we obtain the following 8-ary space diagram:

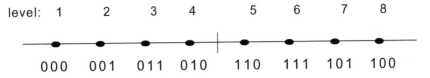

level: 1   2   3   4   5   6   7   8

000   001   011   010   110   111   101   100

Section 1.3

1.3.1 Calculate the nonlinear quantization increments in the case of μ-law with $\mu = 255$ and 1-byte sample coding, assuming a waveform with amplitude range $[-1, +1]$. Calculate the quantization noise power, and compare it with the noise obtained in the case of uniform quantization.

The 1-byte coding requires 8 quantization levels. The compressed (normalized) waveform sample has amplitudes between -1 and $+1$, which are mapped by equal increments of $\frac{1}{4}$. Thus the 8 intervals, which we label "k," and their corresponding binary values can be tabulated as follows:

Negative Amplitudes	Positive Amplitudes
$k = 1, 000 = [-1, -0.75]$	$k = 5, 100 = [0, 0.25]$
$k = 2, 001 = [-0.75, -0.5]$	$k = 6, 101 = [0.25, 0.5]$
$k = 3, 010 = [-0.5, -0.25]$	$k = 7, 110 = [0.5, 0.75]$
$k = 4, 011 = [-0.25, 0]$	$k = 8, 111 = [0.75, 1]$

On the other hand, the μ-law is defined through

$$(a)\ f_{\text{comp}} = -\frac{\log(1 + \mu|f|)}{\log(1 + \mu)}$$

for negative values of the waveform f, and

$$(b)\ f_{\text{comp}} = \frac{\log(1 + \mu f)}{\log(1 + \mu)}$$

for positive values of the waveform. Consider only the positive-amplitude region, as defined by (a). Two adjacent quantization levels i and $i-1$ with separation $\frac{1}{4}$ in the compressed waveform correspond to two ordinates Y_i and $Y_{i-1} (Y_{i-1} < Y_i)$ in the uncompressed waveform. The relation between these two ordinates is then

$$\frac{1}{4} = \frac{\log(1 + \mu Y_i)}{\log(1 + \mu)} - \frac{\log(1 + \mu Y_{i-1})}{\log(1 + \mu)} \equiv \frac{1}{\log(1 + \mu)} \log \frac{1 + \mu Y_i}{1 + \mu Y_{i-1}}$$

which converts into ($\mu = 255$):

$$\frac{1 + \mu Y_i}{1 + \mu Y_{i-1}} = \exp\left(\frac{\log(1 + \mu)}{4}\right) = (1 + \mu)^{1/4} = 4$$

and finally

$$Y_{i-1} = \frac{Y_i}{4} - \frac{3}{4\mu} = \frac{Y_i}{4} - 0.0029$$

Starting from the maximum $Y_i = 1$, we thus obtain $Y_{i-1} = 1/4 - 0.0029 = 0.24705$. The corresponding increment is $\Delta_{NL}(i, i-1) = Y_i - Y_{i-1} = 1 - 0.24705 = 0.7529$. The same operation with levels $i-1, i-2$ yields the next increment value $\Delta_{NL}(i-1, i-2)$, etc. The same calculations can be done with the negative amplitudes (going up from level j to level $j+1$, etc.), but using formula (b). The results are summarized in the table below:

Negative Amplitudes	$\Delta_{NL}(k)$	Positive Amplitudes	$\Delta_{NL}(k)$
$k = 1$ (j to $j+1$)	0.7529	$k = 5$ ($i-4$ to i-3)	0.0029
$k = 2$ ($j+1$ to $j+2$)	0.1882	$k = 6$ ($i-3$ to $i-2$)	0.0470
$k = 3$ ($j+2$ to $j+3$)	0.0470	$k = 7$ ($i-2$ to $i-1$)	0.1882
$k = 4$ ($j+3$ to $j+4$)	0.0029	$k = 8$ ($i-1$ to i)	0.7529

It can be checked that within truncation errors, the sum of increments in each column is unity. It is seen that the increments shrink then expand as we move up the quantization scale ($k = 1$ to $k = 8$).

Section 1.4

1.4.1 Determine the mean and standard deviation of the following number set:
$\{-1, 9, 5, 2, -8, 10, -2, -3, 7, 1, 2\}$
The mean is given by

$$\langle n \rangle = (-1 + 9 + 5 + 2 + (-8) + 10 + (-2) + (-3) + 7 + 1 + 2)/10 = 2$$

The mean-square is given by

$$\langle n^2 \rangle = [(-1)^2 + 9^2 + 5^2 + 2^2 + (-8)^2 + 10]^2 + (-2)^2 + (-3)^2 + 7^2 + 1^2 + 2^2]/10 = 33.8$$

The standard deviation is then

$$\sigma = \sqrt{\langle n^2 \rangle - \langle n \rangle^2} = \sqrt{33.8 - 2^2} = 5.45$$

1.4.2 Assume a normal distribution $p(x)$ with zero mean and standard deviation σ. Determine the abscissa x for which the distribution is equal to $\frac{3}{4}, \frac{1}{2}, \frac{1}{4}$, and $\frac{1}{8}$ of its peak value. Make an approximate plot of the normal distribution, as based upon these values, while assuming $\sigma = 1$.
The normal PDF with zero mean is defined as

$$p(x) = p_{peak} \exp\left(-\frac{x^2}{2\sigma^2}\right), \qquad \text{where } p_{peak} = 1/(\sigma\sqrt{2\pi})$$

We search the value of x for which $p(x) = p_{peak}/N$, where $N = 4/3$, 2, 4, and 8. Solving the equation

$$\exp\left(-\frac{x^2}{2\sigma^2}\right) = \frac{1}{N}$$

yields (taking the natural logarithm of both sides: $x^2 = 2\sigma^2 \log N$, or $x = \sqrt{2 \log N}\sigma$. Thus, the solutions are $x = \sqrt{2\log(4/3)}\sigma$, $\sqrt{2 \log 2}\sigma$, $\sqrt{2 \log 4}\sigma$, $\sqrt{2 \log 8}\sigma$, or $x = 0.75\sigma$, 1.17σ, 1.66σ, 2.03σ.

Assume next $\sigma = 1$. The peak value $(x = 0)$ is then $p_{peak} = 1/\sqrt{2\pi} = 0.4$. We then can make an approximate plot of the normal distribution, according to the following table values:

$X =$	$p(x) =$
0	0.4
± 0.75	$0.4/(4/3) = 0.3$
± 1.17	$0.4/2 = 0.2$
± 1.66	$0.4/4 = 0.1$
± 2.03	$0.4/8 = 0.05$

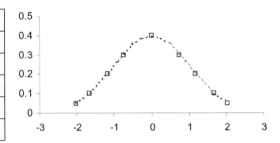

Section 1.5

1.5.1 Assume a transmission system which is $L = 100$ km long. The power received at the system end represents 1% of the input. What is the trunk loss rate, in dB/km? What is the % of received power in a system having the same trunk loss but a length of $L = 150$ km? By definition,

$$\text{Loss}_{dB/km} = \frac{1}{L_{km}} 10 \log_{10}\left(\frac{P_{out}}{P_{in}}\right) \quad \text{or with } P_{out}/P_{in} = 1\%$$

$$\text{Loss}_{dB/km} = \frac{1}{100\,km} 10 \log_{10}(0.01) = \frac{-20\,dB}{100\,km} = -0.2\,dB/km$$

Reversing the formula, we get

$$\frac{P_{out}}{P_{in}} = 10^{loss \times L/10}.$$

For a system of 150-km length, we thus have

$$\frac{P_{out}}{P_{in}} = 10^{-0.2 \times 150/10} = 1 \times 10^{-3} = 0.1\%$$

1.5.2 Assume a light wave at carrier frequency $f_c = 200$ THz, whose spectrum is 120 GHz wide. What is the time broadening experienced by the waveform after 100-km propagation in a medium having a dispersion coefficient of 17 ps/(nm·km) and carrier velocity of $c = 3 \times 10^8$ m/s?

The minimum and maximum frequency components are $f_1 = f_c - 60\,GHz = 199,940\,GHz$ and $f_2 = f_c + 60\,GHz = 200,060\,GHz$. The corresponding wavelengths are given by $\lambda = c/f$ or $\lambda_1 = 3 \times 10^8/(199,940 \times 10^9) = 1.50045 \times 10^{-6}$ m and $\lambda_2 = 3 \times 10^8/(200,060 \times 10^9) = 1.49955 \times 10^{-6}$ m. The wavelength separation is $\Delta\lambda = \lambda_1 - \lambda_2 = $

$(1.50045 - 1.49955) \times 10^{-6} = 0.9 \times 10^{-9}$ m $= 0.9$ nm. The waveform time broadening is $\Delta t = DL\Delta\lambda = 17_{\text{ps/nm-km}} \times 100_{\text{km}} \times 0.9_{\text{nm}} = 1,530$ ps or 1.5 ns.

1.5.3 A 10-Gbit/s system continuously transmits data with a bit error rate of BER $= 10^{-12}$. How many errors are made over the course of a single day?

A BER of 10^{-12} corresponds to one error every 10^{12} received bits. The rate of 10 Gbit/s corresponds to $10 \times 10^9 = 10^{10}$ bits received every second. Therefore, it takes $10^{12}/10^{10} = 100$ seconds (in average) to get a single error. A day is made of 24 h$\times$ 60 mn $\times$ 60 s $= 86,400$ s. The number of error per day is therefore $86,400/100 = 864$ errors.

1.5.4 What is the Q-factor requirement for achieving a bit-error rate very near 10^{-12}?

The bit-error rate is defined by BER$\approx 1/(Q\sqrt{2\pi}) \exp(-(Q^2/2))$. We must find the value Q_0 such that BER $< 10^{-12}$, which requires a bit of guessing since the inverse function $Q = f^{-1}(\text{BER})$ is not known *a priori*. We know, however, that Q_0 is greater than $Q_1 = 6$ (BER $< 10^{-9}$). Taking the ratio of the two BERs yields

$$\frac{Q_0}{Q_1} \exp\left(-\frac{Q_0^2 - Q_1^2}{2}\right) = \frac{10^{-12}}{10^{-9}} = 10^{-3}$$

Since the exponential term has more weight than the ratio Q_0/Q_1, we can make the approximation

$$\exp\left(-\frac{Q_0^2 - Q_1^2}{2}\right) \approx 10^{-3}$$

which gives $Q_0^2 - Q_1^2 \approx -2\log(10^{-3}) = 9.21$, or $Q_0 \approx \sqrt{Q_1^2 + 9.21} = 6.72$. Replacing this value of Q_0 into the BER definition, one finds BER $= 9.2 \times 10^{-12}$, which is too high. Increasing Q_0 by 5% ($Q_0 = 7.03$) gives BER $= 1.05 \times 10^{-12}$, which is acceptably close to BER $= 10^{-12}$.

1.5.5 How long would it take to make a measurement of BER $= 10^{-14}$ in a 10-Gbit/s system, assuming that at least 10 error counts would be required?

A count of 10 errors at BER $= 10^{-14}$ would require detecting $10 \times 10^{14} = 10^{15}$ bits. At a bit rate of 10 Gbit/s, every second correspond to $10 \times 10^9 = 10^{10}$ received bits. The time required is therefore $10^{15}/10^{10} = 10^5$ seconds, or $10^5/3600 = 278$ hours, or $278/24 = 11\frac{1}{2}$ days. Such an impractical measurement time justifies the Q-factor extrapolation method.

1.5.6 What are the ideal receiver sensitivities in photons/bit for homodyne and heretodyne ASK signals? Express the figures in dBm, assuming a 2.5-Gbit/s bit rate and $\lambda = 1.55$ μm wavelength.

From Table 1.3, we see that for ASK signals the homodyne/heterodyne power sensitivities are $2P$ and $4P$, respectively, where P is the sensitivity for homodyne PSK. Expressed in photons/bit, the ideal PSK sensitivity is known to be 9 photons/bit. For ASK signals, therefore, the figures are $\langle n \rangle = 18$ photons/bit (homodyne) and $\langle n \rangle = 36$ photons/bit (heterodyne). The corresponding powers are given by $\langle P \rangle = \langle n \rangle h\nu B = \langle n \rangle h(c/\lambda)B$. Using $h = 6.62 \times 10^{-34}$ J.s and $c = 3 \times 10^8$ m/s, we get for the photon energy $h\nu$ (joules) and power at $B = 2.5$ Gbit/s (watt):

$$h(c/\lambda) = (6.62 \times 10^{-34} \text{ Js})(3 \times 10^8 \text{ m/s}/(1.55 \times 10^{-6} \text{ m})) = 1.28 \times 10^{-19} \text{ J}$$

$$h\nu B = (1.28 \times 10^{-19} \text{ J})(2.5 \times 10^9 \text{ s}^{-1}) = 3.2 \times 10^{-10} \text{ W} = 3.2 \times 10^{-7} \text{ mW}$$

Hence, the sensitivities of $\langle P \rangle = 18 \times 3.2 \times 10^{-7}$ mW $= 5.7 \times 10^{-6}$ mW $= -52.3$ dBm (homodyne) and $\langle P \rangle = 36 \times 3.2 \times 10^{-7}$ mW $= 1.1 \times 10^{-5} = -49.4$ dBm (heterodyne).

1.5.7 A lightwave receiver has a sensitivity of $P^{\text{sens}} = -40\,\text{dBm}$. The transmission line has a loss coefficient of $-0.2\,\text{dB/km}$. The available transmitter power is $P^{\text{trans}} = +15\,\text{dBm}$. Assuming a 3-dB system margin, what is the maximum system length permitting error-free transmission?

The power budget writes

$$P^{\text{trans}}_{\text{dBm}} + T_{\text{dB}} = M_{\text{dB}} + P^{\text{sens}}_{\text{dBm}}, \quad \text{or}$$

$$T_{\text{dB}} = -P^{\text{trans}}_{\text{dBm}} + M_{\text{dB}} + P^{\text{sens}}_{\text{dBm}} = +15\,\text{dBm} + 3\,\text{dB} - 40\,\text{dBm} = -22\,\text{dB}$$

The maximum system loss for error-free (BER $= 10^{-9}$) transmission is $-22\,\text{dB}$. The corresponding length is $L = -22\,\text{dB}/(-0.2\,\text{dB/km}) = 110\,\text{km}$.

1.5.8 Assume for a transmission line a loss coefficient of $-0.3\,\text{dB/km}$. What would be the transmitter powers necessary to obtain a signal power of 1 mW at distances $L = 100$ km, 250 km and 500 km? How would these figures change with in-line amplifiers placed every 50 km? What can be concluded from the result?

The line loss corresponding to each of the three distances is

$$T_1 = -0.3\,\text{dB/km} \times 100\,\text{km} = -30\,\text{dB}$$
$$T_2 = -0.3\,\text{dB/km} \times 250\,\text{km} = -75\,\text{dB}$$
$$T_3 = -0.3\,\text{dB/km} \times 500\,\text{km} = -150\,\text{dB}$$

A signal power of 1 mW (0 dBm) measured at these distances would then require transmitter powers of

$$P_1 - 30\,\text{dB} = 0\,\text{dBm} \rightarrow P_1 = +30\,\text{dBm}, \quad \text{or 1 W}$$
$$P_2 - 75\,\text{dB} = 0\,\text{dBm} \rightarrow P_1 = +75\,\text{dBm}, \quad \text{or 31.6 kW(!)}$$
$$P_3 - 150\,\text{dB} = 0\,\text{dBm} \rightarrow P_1 = +150\,\text{dBm} \quad \text{or 1 TW(!!)}$$

Consider now the use of line amplifiers every 50 km. The signal loss experienced over a 50km trunk section is $T = -0.3\,\text{dB/km} \times 50\,\text{km} = -15\,\text{dB}$. Since the line amplification periodically restores the signal power, the transmitter power required for any distance multiple of 50 km is therefore $P - 15\,\text{dB} = 0\,\text{dBm} \rightarrow P = +15\,\text{dBm}$, or 31.6 mW. The conclusion is that line amplification makes possible to overcome the tremendous loss effect in long transmission system, reducing the power budget requirement to that of a single inter-amplifier segment.

1.5.9 Calculate the noise figure of a 100-km-long microwave transmission line with loss of 1 dB/km and amplifiers placed every 10 km (amplification preceding loss), assuming an amplifier temperature of $T_A = 626.5$ K.

The line loss between two amplifiers is $10\,\text{km} \times 1\,\text{dB/km} = 10\,\text{dB}$, corresponding to the linear transmission loss $T = 10^{-1} = 0.1$. The lossy line element thus has a noise figure of $F_{\text{loss}} = -10\,\text{dB}$. Since the line is 100 km and the amplifier spacing is 10 km, the total number of amplified segments if $k = 10$. To achieve transparency, the microwave amplifier gain is set to $G = 10$ or $+10$ dB. The amplifier noise figure is then defined by

$$F_{\text{amp}} = 1 + \frac{T_A}{T_0} \quad \text{with} \quad T_0 = 290\,\text{K} \quad \text{and} \quad T_A = 626.5\,\text{K}$$

Thus $F_{\text{amp}} = 1 + 626.5/290 = 3.15$ or $+5.0$ dB.

According to formula in equation (1.54), the system noise figure is given by

$$F = k\left(F_{amp} + 1 - \frac{1}{G}\right) - 1, \quad \text{or}$$

$$F = 10\left(3.15 + 1 - \frac{1}{10}\right) - 1 = 39.5 \quad \text{or} \quad -15.9\,\text{dB}$$

1.5.10 Assume you want to realize high-gain amplifier made by cascading two smaller amplifiers of gains G_1 and G_2 with corresponding noise figures F_1 and F_2. Consider the following cases:

(a) $G_1 > G_2$, $F_1 = F_2 = F$
(b) $G_1 = G_2 = G$, $F_1 > F_2$

In which order would you place the amplifiers in each case to obtain the smallest possible noise figure for the twin-amplifier system? What do you conclude from the results?

The two possible configurations for the twin-amplifier system are amplifier "1" (G_1,F_1) followed by amplifier "2" (G_2,F_2), or the reverse, as referred to by configurations A and B, respectively.

The cascading formula gives the corresponding system noise figures:

$$F(A) = F_1 + \frac{F_2 - 1}{G_1}$$

$$F(B) = F_2 + \frac{F_1 - 1}{G_2}$$

For each of the two cases (a) and (b), we obtain

$$F(A,a) = F + \frac{F-1}{G_1}, \quad F(B,a) = F + \frac{F-1}{G_2}, \quad G_1 > G_2$$

$$F(A,b) = F_1 + \frac{F_2 - 1}{G}, \quad F(B,b) = F_2 + \frac{F_1 - 1}{G}, \quad F_1 > F_2$$

We see in the case (a) that since $G_1 > G_2$ we have $F(A,a) < F(B,a)$, thus $F(A,a)$ represents the minimum-NF configuration of case (a).

In case (b), we calculate the NF difference

$$F(A,b) - F(B,b) = F_1 + \frac{F_2 - 1}{G} - F_2 - \frac{F_1 - 1}{G} = (F_1 - F_2)(1 - 1/G)$$

We see that the NF difference is positive, since $F_1 - F_2 > 0$ and $1 - 1/G > 0$. Thus $F(B,b)$ represents the minimum-NF configuration of case (b).

In conclusion, the minimum NF is achieved by placing first in the sequence either

- the amplifier of highest gain (if they have equal individual noise figures)
- the amplifier of smallest noise figure (if they have equal individual gains)

1.5.11 An amplified system with $k = 100$ transparent segments of identical noise figures $F_{seg} = 3.16$ ($F_{seg}^{dB} = +5\,\text{dB}$) is operating error-free. What is the decibel SNR used at transmitter level?

The NF cascading formula is $F = F_1 + \dfrac{F_2 - 1}{H_1} + \dfrac{F_3 - 1}{H_1 H_2} + \cdots$.

Here, all NFs are equal to F_{seg} and all transmissions H_i are equal to unity (transparency). Thus, we have for k segments:

$$F = F_{\text{seg}} + \frac{F_{\text{seg}} - 1}{1} + \frac{F_{\text{seg}_3} - 1}{1 \times 1} + \cdots \equiv k F_{\text{seg}} - (k - 1)$$

Using $k = 100$ and $F_{\text{seg}} = 3.16$, we get $F = 100 \times 3.16 - (100 - 1) = 217$ or $F_{\text{dB}} + 23.3$ dB. Error-free transmission is achieved with $Q = 6$, or with a received/output SNR of $\text{SNR}_{\text{out}} = Q(Q + \sqrt{2}) = 6(6 + 1.414) = 44.5$ or $\text{SNR}_{\text{out}}^{\text{dB}} = +16.5$ dB. The transmitted/input SNR is thus given by $\text{SNR}_{\text{in}} = F \times \text{SNR}_{\text{out}}$, or in decibels $\text{SNR}_{\text{in}}^{\text{dB}} = \text{SNR}_{\text{out}}^{\text{dB}} + F_{\text{dB}} = +16.5$ dB $+ 23.3$ dB $= 39.8$ dB, or $\text{SNR}_{\text{in}}^{\text{dB}} \approx +40$ dB, approximately.

Section 1.6

1.6.1 What is the bit error rate of the message HAMPY_HXLIDEYS, assuming that each letter or symbol takes two bytes and that the message is communicated within a single second?

The 1-second message comprises 14 letters/symbols or byte pair, i.e., $14 \times 2 \times 8$ bits $= 224$ bits. The bit rate is therefore 224 bit/s. Three letters or symbols, M, X, and E, are wrong, which in each time is caused by a single bit error. The bit error rate is therefore $3/224 = 1.3 \times 10^{-2}$ bit/s.

1.6.2 Define a syndrome table corresponding to the linear block code (6,3), assuming a parity matrix equal to

$$P = \begin{pmatrix} 1 & 1 & 0 \\ 0 & 1 & 1 \\ 1 & 1 & 1 \end{pmatrix}$$

Then detect possible single-bit error in the three received code words $Y_1 = (010100)$, $Y_2 = (001111)$ and $Y_3 = (110001)$ and identify the original/corrected message words.

This block code of $n = 6$ bits corresponds to $k = 3$ message bits and $m = 3$ parity bits. With the parity matrix P given as an assumption, the generator matrix takes the form:

$$\tilde{G} = [I_k | P] = \begin{pmatrix} 1 & 0 & 0 & 1 & 1 & 0 \\ 0 & 1 & 0 & 0 & 1 & 1 \\ 0 & 0 & 1 & 1 & 1 & 1 \end{pmatrix}$$

The parity-check matrix $\tilde{H}$ and its transposed matrix $\tilde{H}^T$ are, by definition,

$$\tilde{H} = [P^T | I_m] = \begin{pmatrix} 1 & 0 & 1 & 1 & 0 & 0 \\ 1 & 1 & 1 & 0 & 1 & 0 \\ 0 & 1 & 1 & 0 & 0 & 1 \end{pmatrix} \quad \text{and}$$

$$\tilde{H}^T = \left[\frac{(P^T)^T = P}{I_m} \right] = \begin{pmatrix} 1 & 1 & 0 \\ 0 & 1 & 1 \\ 1 & 1 & 1 \\ 1 & 0 & 0 \\ 0 & 1 & 0 \\ 0 & 0 & 1 \end{pmatrix}$$

We can now generate the block codes $Y = (y_1 \cdots y_6)$ corresponding to the $2^k = 8$ possible $X = (x_1, x_2, x_3)$ 3-bit messages. We know that the first bits of the block code are the same as the message bits. So we just have to calculate the y_4, y_5, y_6 bits and complete the following table:

x_1	x_2	x_3	y_1	y_2	y_3	y_4	y_5	y_6
0	0	0	0	0	0			
0	0	1	0	0	1			
0	1	0	0	1	0			
0	1	1	0	1	1			
1	0	0	1	0	0			
1	0	1	1	0	1			
1	1	0	1	1	0			
1	1	1	1	1	1			

For this, we use the relation $Y = X\tilde{G}$ for the last three bits, i.e.,

$$y_4 = x_1 \times 1 + x_2 \times 0 + x_3 \times 1$$
$$y_5 = x_1 \times 1 + x_2 \times 1 + x_3 \times 1$$
$$y_4 = x_1 \times 0 + x_2 \times 1 + x_3 \times 1$$

(remembering that in the binary system $1 + 1 = 0$), and we can complete the table as follows

x_1	x_2	x_3	y_1	y_2	y_3	y_4	y_5	y_6
0	0	0	0	0	0	0	0	0
0	**0**	**1**	**0**	**0**	**1**	**1**	**1**	**1**
0	1	0	0	1	0	0	1	1
0	**1**	**1**	**0**	**1**	**1**	**1**	**0**	**0**
1	0	0	1	0	0	1	1	0
1	0	1	1	0	1	0	0	1
1	**1**	**0**	**1**	**1**	**0**	**1**	**0**	**1**
1	1	1	1	1	1	0	1	0

(the three messages and code words highlighted in bold are used later as examples)
There are six possible (single) bit errors in the block code, corresponding to the error patterns $E = (100000), (010000) \ldots$ to (000001). We can then calculate the 3-bit syndrome vector S, using the relation $S = E\tilde{H}^T$ (3 bits = 6 bits $\times$ [6 $\times$ 3] matrix), the result of which being listed in the following table:

Error Pattern	Syndrome
000000	000
100000	110
010000	011
001000	111
000100	100
000010	010
000001	001

We shall now analyze the received code words

$$Z_1 = (010100), \qquad Z_2 = (001111), \qquad \text{and} \qquad Z_3 = (110001)$$

by calculating their syndromes $S = Z\tilde{H}^T$. This operation gives $S_1 = (111)$, $S_2 = (000)$, and $S_3 = (100)$. Since S_1 and S_3 are nonzero, this means that the corresponding code words contain an error, while the second is error-free. Looking at the previous table, we observe that $S_1 = (111)$ corresponds to the error pattern $E_1 = (001000)$, or the third bit being erroneous. So we make the correction by changing the third bit, i.e., the corrected word is $Z_1' = Z_1 + E_1 = (010100) + (001000) = (011100)$. The same comparison with $S_3 = (100)$ yields $E_3 = (000100)$, or the fourth bit being erroneous. So we make the correction by changing the fourth bit, i.e., the corrected word is $Z_3' = Z_3 + E_3 = (110001) + (000100) = (110101)$. Finally, we decode the corrected blocks by identifying Z_1', $Z_2' = Z_2$, Z_3' with the block codes Y_1, Y_2, Y_3 and corresponding messages X_1, X_2, X_3 using the next-to-last table. The result is

$$Y_1 = (011100) \rightarrow X_1 = (011)$$
$$Y_2 = (001111) \rightarrow X_2 = (001)$$
$$Y_3 = (110101) \rightarrow X_3 = (110)$$

(note that since the code is systematic, we did not need to look at the table to identify the original message words; these are given by the first three bits of the corrected block codes).

1.6.3 An error-correcting code makes possible to bring the bit-error-rate from BER $= 5 \times 10^{-4}$ to BER $= 1.0 \times 10^{-12}$. What is the corresponding coding gain? (Clues: (a) Use $Q = 7.0$ as the Q-factor for BER $= 10^{-12}$ and (b) find the uncorrected Q-factor by progressive approximations.)

By definition, the coding gain is $\gamma = 10\log_{10}(Q_c/Q_{unc})$. The corrected and uncorrected Q-factors, Q_c, Q_{unc} correspond to BER $= 1.0 \times 10^{-12}$ and BER $= 5 \times 10^{-4}$, respectively. Thus $Q_c = 7.0$.

The bit-error rate is defined by

$$\text{BER} \approx \frac{1}{Q\sqrt{2\pi}} \exp\left(-\frac{Q^2}{2}\right)$$

The ratio of the two BERs yields the identity

$$\frac{Q_{unc}}{Q_c} \exp\left(-\frac{Q_c^2 - Q_{unc}^2}{2}\right) = \frac{10^{-12}}{5 \times 10^{-4}} = 2 \times 10^{-9}$$

Taking the decimal logarithm (times 10) of the expression, we get

$$10\log_{10}\left[\frac{Q_{unc}}{Q_c}\exp\left(-\frac{Q_c^2-Q_{unc}^2}{2}\right)\right] = -86.9 \quad \text{or equivalently:}$$

$$10\log_{10}\left(\frac{Q_{unc}}{7}\right) - \left(\frac{49-Q_{unc}^2}{2}\right) \times 10\log_{10}(e) = -86.9$$

$$10\log_{10}(Q_{unc}) - 10\log_{10}(7) + 2.1 \times Q_{unc}^2 - 53.2 = -86.9$$

$$f(Q_{unc}) = 2.1 \times Q_{unc}^2 + 10\log_{10}(Q_{unc}) = 26.9$$

This is a transcendental equation in Q_{unc} which must be solved by progressive approximations.

We know that $Q_{unc} < 6$. The values $Q_{unc} = 3$ and $Q_{unc} = 4$ yield $f(3) = 23.6$ and $f(4) = 39.6$, showing that $3 < Q_{unc} < 4$. Increasing the lower bound by 10%, we get $f(3.3) = 28.0$, which is too high. Decreasing the value by 5%, we get $f(3.13) = 25.5$. Increasing by 2% gives $f(3.19) = 26.3$, then by another 1%: $f(3.22) = 26.8$, which is better than 0.5% of the target $f(Q_{unc}) = 26.9$. Thus $Q_{unc} \approx 3.22$, which yields the coding gain $\gamma = 10\log_{10}(Q_c/Q_{unc}) = 10\log_{10}(7/3.22) = 3.3$ dB.

Section 1.7

1.7.1 A movie has 99% chances of being played in a theater on a certain date. What is the information of its being canceled, compared to its being played?

The probability for the movie being played is $p_1 = 0.99$, corresponding to the information $I_1 = -\log(p_1) = 0.01$. The contrary event has the probability $p_2 = 1 - 0.99 = 0.01$, corresponding to the information $I_2 = -\log(p_2) = 4.6$. The information in the movie being cancelled is 450 times higher than its being played.

1.7.2 Preparing his/her vacations, an individual knows that the probability that

    (a) the car will have a mechanical problem is 0.1%

    (b) one of the participants be sick is 10%

    (c) serious business calls will happen are 40%

Give the entropy associated with the vacations event, assuming that all three cases are uncorrelated possibilities. How far is the result from maximum entropy?

The entropy associated with the vacations is given by $H = -\sum_n p_n \log_2 p_n$, where p_n represent the probability associated with all possible events and their unfortunate possible combinations! Indeed, we can have events (a), (b), or (c) occurring alone or together, by two or by three. The different possibilities are defined by the following probabilities:

Event (a) alone: $p(a)[1 - p(b)][1 - p(c)] = 0.01 \times 0.9 \times 0.6 = 0.0054 = p_1$

Event (b) alone: $[1 - p(a)]p(b)[1 - p(c)] = 0.99 \times 0.1 \times 0.6 = 0.0594 = p_2$

Event (c) alone: $[1 - p(a)][1 - p(b)]p(c) = 0.99 \times 0.9 \times 0.4 = 0.3564 = p_3$

Event (a) and (c) alone: $p(a)[1 - p(b)]p(c) = 0.01 \times 0.9 \times 0.4 = 0.0036 = p_5$

Event (b) and (c) alone: $[1 - p(a)]p(b)p(c) = 0.99 \times 0.1 \times 0.4 = 0.0396 = p_6$

All events together: $p(a)p(b)p(c) = 0.01 \times 0.1 \times 0.4 = 0.0004 = p_7$

None of the events: $[1 - p(a)][1 - p(b)][1 - p(c)] = 0.99 \times 0.9 \times 0.6 = 0.5346 = p_8$

It can be checked that the sum of all the above probabilities, $\sum_{n=1}^{8} p_n$ equals unity.

From the entropy definition, we find $H = 1.52$ bits. The maximum-entropy situation would correspond to the catastrophic situation where all eight possible events would have equal likelihood, namely $p_n = 1/8 = 0.125$. Thus $H_{max} = -8 \times (1/8) \log_2 (1/8) = 3$ bits. The conclusion is that the first scenario, as defined in the problem, has much less entropy than the catastrophic situation. This is because the successful outcome (unperturbed vacations, corresponding to $p_8 = 0.5346$) is much more certain, and has unevenly distributed risk, in opposition to the other case.

1.7.3 An 8-letter alphabet (A, B, C, D, E, F, G, H) has the following symbol probabilities: $p(A) = 8u$, $p(B) = 7u$, $p(C) = 6u$, $p(D) = 5u$, $p(E) = 4u$, $p(F) = 3u$, $p(G) = 2u$ and $p(H) = u$, where u is a strictly positive real number. Determine a possible Huffman code for this alphabet and provide the corresponding coding efficiency.

We first determine the value of u using the property that the sum of all probabilities is unity, i.e.,

$$1 = 8u + 7u + 6u + 5u + 4u + 3u + 2u + u = 36u, \text{ thus } u = 1/36, \text{ and}$$

$$p(A) = 8/36 = 0.2222 \quad p(B) = 7/36 = 0.1944$$
$$p(C) = 6/36 = 0.1666 \quad p(D) = 5/36 = 0.1388$$
$$p(E) = 4/36 = 0.1111 \quad p(F) = 3/36 = 0.0833$$
$$p(G) = 2/36 = 0.0555 \quad p(H) = 1/36 = 0.0277$$

We put the probabilities in decreasing order in an EXCEL spreadsheet. We then draw the following diagram by proceeding through the Huffman algorithm, starting from the two lowest probabilities, assigning 0 and 1 bits, summing the two probabilities, and reordering for the next stage.

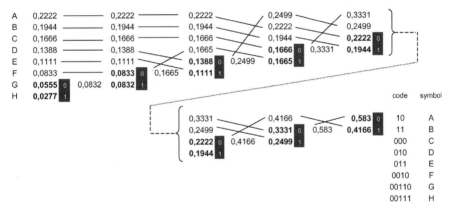

We calculate the average code-word length and the alphabet entropy, which gives

average word length				entropy	
0,2222	×	2	0,4444	$-p \log 2(p)$	0,48216
0,1944	×	2	0,3888		0,45932
0,1666	×	3	0,4998		0,43072
0,1388	×	3	0,4164		0,3954
0,1111	×	3	0,3333		0,35217
0,0833	×	4	0,3332		0,29866
0,0555	×	5	0,2775		0,2315
0,0277	×	5	0,1385		0,14331

$$\langle I \rangle = 2,8319 \text{ bits} \qquad H = 2,79323$$

It is seen that the coding efficiency is $\eta = H/\langle l \rangle = 2.79323/2.8319 = 98.6\%$. If we had coded this 8-symbol alphabet with equal lengths of 4 bits, the efficiency would be $\eta = 2.79323/4 = 69.8\%$

Section 2.1

2.1.1 Draw permanent connection diagrams corresponding to $N = 2$ to $N = 6$ subscribers. By inference, demonstrate the formula $M_N = N(N - 1)/2$ giving the number of required connections for any N subscribers.

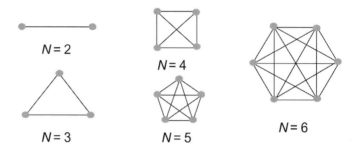

The diagrams for $N = 2$ to $N = 6$ show that the formula is correct. Assume then that the formula is correct up to order N, so that $M_N = N(N - 1)/2$. Is the formula correct to order $N + 1$? We would have $M_{N+1} = (N + 1)N/2$, or identically $M_{N+1} = M_N + N$. The answer is yes, since an extra user would need N new connections. This inference reasoning proves the formula to all orders.

2.1.2 Describe the sequence of events during a call from person A to person B within a local loop, corresponding to the following situations:

- B is absent (no answering machine).
- B is busy with another call.
- Same as above, but the CO offers the possibility of automatic call-back of B to A.

B is absent: same steps as 1–3 (see Section 2.1); (4) after several ringing signals, CO senses the idle state of B; (5) CO sends a voice message to A indicating unavailability of person B; (6) person A hangs up and CO drops the case.

B is busy with another call: same steps 1–3 as above; (4) CO identifies the number of person B as being already switched or busy; (5) CO sends a signal to A indicating that person B's line is busy; (6) person A hangs up and CO drops case; (7) if A remains unhooked, CO will send a voice signal for person A to hang up and call again later.

Automatic call-back: same steps 1–4 as previously; (5) CO sends a signal to A indicating that person B's line is busy, and that if person A presses a touch (say "5"), automatic callback will be made at person A's expense; (6) person A hangs up; (7) person B terminates call and hangs up; (8) CO senses that B line is available and then sends a ringing signal to B; (9) person B picks up handset; (10) CO sends ringing signal to A; (11) person A picks up handset; (12) CO establishes circuit connection, etc. [complete from (11) assuming that person A is not available].

2.1.3 In reference to the analog repeater shown in Fig. 2.3, calculate the net gain experienced by the signals leaking through the hybrids after one then N circulations though the loop.

Assume $G = 11$ dB, an excess branch loss of $E = 4$ dB, and a hybrid leakage loss of $F = 6$ dB.

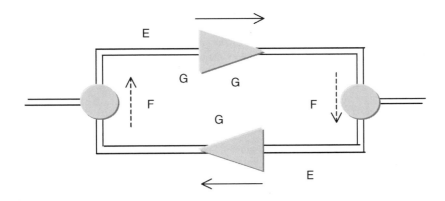

The total loop loss is seen to be $L = 2E + 2F$ or 20 dB. The total loop gain is $G' = 2G$ or 22 dB. Therefore, an excess loop gain of $G' = G - L = 2$ dB is experienced by the leakage signal. At the second re-circulation, the net cumulated gain for the leakage signal is $2 + 2 = 4$ dB, and so on. After a certain number of recirculations N, the cumulated leakage power $[P_{leak} + G(1 + 2 + \cdots + N)]$ is sufficiently high to saturate the amplifier, leading to loop instability (gain drops and circulation stops, then gain restarts) or steady-state breakdown.

2.1.4 The loss of optical fibers is 0.2 dB/km at 1.55-μm wavelength. How many amplifiers are required in an inter-office optical trunk of length 500 km where the signal loss would not exceed 15 dB between each repeater stage?

Since the maximum allowable signal loss is 15 dB, the amplifiers must be spaced not farther apart than $Z_a = 15$ dB/0.2 dB/km $= 75$ km. Since 500 km/75 km $= 6.6$, one only needs 6 amplifiers at 15 dB gain (the remaining distance of $500 - 450 = 50$ km will bring down the signal to $50_{km} \times 0.2_{dB/km} = 10$ dB, which is in the margin.

Section 2.3

2.3.1 What is the time duration of a 1-byte voice sample? Show it in two different ways.

Voice sampling is make at a rate of 8,000 times per second. The corresponding sample length, as coded into 8 bits (1 byte), is therefore $1/(8000 \text{ s}^{-1}) = 125$ μs. Another way to show this: the digital voice rate is 64 kbit/s; thus a single bit duration is $1/(64 \text{ kbit/s}) = 15.625$ μs; the byte length is therefore 8×15.625 μs $= 125$ μs.

2.3.2 How many framing and stuffing bits are included inside a T2 frame (refer to Figure 2.9)?

A T2 frame has a bit rate of 6.312 Mbit/s. There are 8,000 TDM frames per econd, so the corresponding number of bits is 6.312 Mbit/s/$(8,000 \text{ s}^{-1}) = 789$ bits. T2 is made of four T1 frames. T1 has 24 voice-channels, representing a total of $24 \times 8 + 1 = 193$ bits. Without extra bits, T2 $= 4 \times$ T1 would have $4 \times 193 = 772$ bits. The number of extra bits included in the T1 $\rightarrow$ T2 conversion is therefore $789 - 772 = 17$ bits.

Section 2.4

2.4.1 Using Table 1.2 in Chapter 1, write your own initials in ASCII.

This is to check if you can correctly read the ASCII table! Take initials ED. The table gives $E = 1000101$ and $D = 1000100$, with the lowest-weight bit shown at right. The ASCII version of ED is therefore 1000100 1000101, starting with E!

2.4.2 Calculate the byte size of a numerical photo having 640 × 480 pixels with 255 color-intensity levels for each of the red, blue, and green tones. Calculate then the byte size of a 30-s movie clip based on this picture resolution, assuming 18 pictures per second. What is the bit rate required to transmit the video?

Each pixel has three colors with $255 = 2^{8-1}$ color intensities, corresponding to 8 bits (1 byte) coding each—that is, 3 bytes. The photo has $640 \times 480 = 307{,}200$ pixels, corresponding to a total of $307{,}200 \times 3 = 921.6$ Kbytes. A 30-s movie corresponds to 30 s × 18 pictures/s or 540 pictures. The total movie information represents 540×921.6 Kbyte $= 497.61$ Mbyte, about half of a gigabyte, which fills up well a good CD-ROM! The required bit rate is 18 pictures/s of 921.6 Kbyte, or $18 \times 921.6 \times 8 = 132.7$ Mbit/s.

2.4.3 In X.25 networks, not more than 7 payload packets can be transmitted without receiving acknowledgment from the receiving terminal. Assume 1 Mbit/s, 128-bytes packets travelling at the speed 3×10^8 m/s and a point-to point distance of 600 km. How many packets approximately are in the line during transmission? How many complete transmission cycles (7 bytes plus acknowledgment received) can be made every second?

The duration of a single bit is $1/(1 \times 10^6/\text{s}) = 1 \times 10^{-6}$ s. The space occupied by the bit is given by the duration × speed product, $(1 \times 10^{-6}$ s$) \times (3 \times 10^8$ m/s$) = 300$ m. The number of bits in the line is thus 600 km/300 m $= 2{,}000$. The corresponding number of packets (128×8 bits) is $2{,}000/(128 \times 8) = 1.95$ or about two packets. A full transmission cycle requires transmitting seven packets, plus receiving a 1-packet acknowledgment. This represents 8 packet transmissions. The $128 \times 8 = 1{,}204$ bits packet are transmitted at 1 Mbit/s, representing a duration of $1{,}024/(1 \times 10^6/\text{s}) = 1.024 \times 10^{-3}$ s ≈ 1 ms. The complete transmission cycle is thus 8 ms. The number of cycles per second is therefore 1 s$/(8 \times 10^{-3}$ s$) = 125$ cycles.

Section 2.5

2.5.1 Assume a network with four stations located on the four corners of a square with 1-km side. Provide the minimum cable-length requirement for a bus, ring, star and mesh network topologies and class them in order or length. (Advanced): There exists a star-like connection configuration taking the shape of a letter H being squeezed in the middle. Find its minimum length and conclude.

With the four stations forming a square pattern of unity side, the four network types can be drawn as follows:

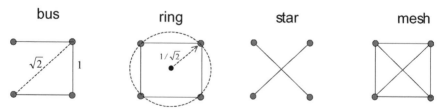

The bus has two possibilities, i.e., using three sides (length $L = 3$) or two sides and a diagonal (length $L = 2 + \sqrt{2} = 3.414$). The first solution is the shortest one.

The ring has two possibilities, i.e., a square (length $L = 4$) or a circle (length $L = 2\pi R = 2\pi/\sqrt{2} = 4.44$). The first solution is the shortest one.

The star has a unique solution of length $L = 2\sqrt{2} = 2.82$.

The mesh has a unique solution of length $L = 2\sqrt{2} + 4 = 6.82$.

Thus the classification in order of increasing length requirement is star/bus/ring/mesh with lengths 2.82/3/4/6.82.

Advanced: Let's draw a letter H squeezed in the middle, as follows:

We see that this star-like configuration has a potentially shorter cabling requirement, compared to the star, where the horizontal bar of the H has zero length ($L = 2\sqrt{2} = 2.82$) and the straight H-letter where the horizontal bar has unity length ($L = 3$). To find what should be the horizontal bar length (x_1) giving the minimum cable length requirement (L_{min}), we need to define the parameters x_2 and y as follows:

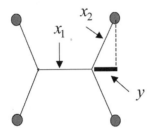

The total length is given by $L = x_1 + 4x_2$. We have $x_1 = 1 - 2y$, and using Pythagoras's theorem, $x_2 = \sqrt{y^2 + 1/4}$. Thus the total length is $L(y) = 1 - 2y + 4\sqrt{y^2 + 1/4}$.

Let's try $y = 1/4$ as a possibility. We get $L = 1 - 2/4 + 4\sqrt{(1/4)^2 + 1/4} = 2.736$. We immediately see that this solution beats the star topology ($L = 2.82$)!

Decrease y by 5%, i.e., $y = 0.24$. We get $L = 2.738$, which is higher. Therefore, $L_{min} \leq 2.736$. Let's go the other way and increase y. Assume, for instance, $y = 1/3 = 0.33$. We get $L = 2.737$, which is higher than the previous solution. Thus we know our minimum length L_{min} is given by a value of y between 0.25 and 0.33, which gives $L_{min} \approx 2.73$. This will suffice for an answer to this exercise (more advanced: taking the derivative of $L(y)$ and solving for the minimum yields $y_{opt} = 1/\sqrt{12}$, which gives $L_{min} \leq 2.732$).

Conclusion: The optimum topology linking with minimal length an array of $N \geq 4$ points located in a plane is not given by the star. With four points, the minimum length is given by an optimal "squeezed H-like" pattern.

Readers really liked this exercise: Verify the above conclusion by putting soapy water between two closely spaced microscope slides. The bubble's surface tension seeks a minimum which gives the squeezed H-like pattern. How would this work in three dimensions, for an array of points located at the eight corners of a cube? Here is the solution (not that intuitive, is it?).

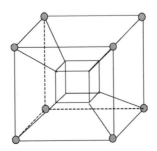

2.5.2 An SMDS network has a double bus of 75-km, node-to-node length. How many frames and time slots are present at a given time in each of the buses? What is the maximum number of time slots that the nodes located at bus ends could simultaneously use before transmission interruption? (Assume 155-Mbit/s rate and frame velocity of $c = 3 \times 10^8$ m/s.)

The SMDS frame is $T = 125$ μs. The frame length is therefore $cT = (3 \times 10^8$ m/s$) \times (125 \times 10^{-6}$ s$) = 37.5$ km, corresponding to two frames per 75-km bus. The time slot is 57 bytes long, corresponding to $57 \times 8 = 456$ bits. At 155-Mbit/s rate, each bit has a duration of $1/(155 \times 10^6 s^{-1}) = 6.45 \times 10^{-9}s \equiv 6.45$ ns, giving a time-slot duration of $T = 456 \times 6.45$ ns $= 2.94$ μs. The time-slot length is therefore $cT = (3 \times 10^8$ m/s$) \times (2.94 \times 10^{-6}$ s$) = 0.882$ km, corresponding to about 85 slots per 75-km bus.

The maximum number of time slots that could be simultaneously used by end nodes is provided by the following situation: (a) The bus has been idle for at least two frame durations (250 μs); (b) the two end bus nodes detecting request counters set at 00 start transmitting, while incrementing the frame counters. Before the first time slot reaches the other end, 85 slots can be transmitted. At this point the 01 counter of the first time slot sets the nodes to interrupt transmission.

Section 3.1

3.1.1 Show that after the conventions of Figure 3.1 (1) the tributary unit group TUG-2 cannot be loaded by a container of C-22 type, and (2) the tributary unit group TUG-3 can be loaded by any of the C-3 container types.

1. After Table 3.1, the C-22 container represents a payload of 8.448 Mbit/s. Loading TUG-2 with a single C-22 and multiplexing 3×7 times into a VC-4 would give a payload of $3 \times 7 \times 8.448 = 177.408$ Mbit/s. But this is above the maximum payload of a VC-4, which is determined by C-4, i.e., 149.76 Mbit/s.

2. For C-3 containers, the maximum payload is that of C-33 (48.384 Mbit/s). Multiplexing 3 times into a VC-4 corresponds to a payload of $3 \times 48.384 = 145.152$ Mbit/s, which is under the maximum VC-4 payload, i.e., 149.76 Mbit/s.

3.1.2 A town of 15,000 homes (95%) and businesses (5%) must be serviced with a new SONET telephone system. Assume that homes are equipped with a single telephone line, while businesses have two lines. Also assume that the daily average traffic pattern is 5 incoming/outcoming calls for homes and 50 for businesses. What is the required OC-n frame capability for handling the full average traffic? Same question if a margin of 50% must be provided to handle network congestion.

Combining homes and businesses, the average daily traffic for all possible terminals represents 0.95×5 calls/day plus 0.05×50 calls/day, or 7.25 calls/day. This number should be multiplied by 10,000 line users, which gives 72,500 calls in average per day. Looking at Table 3.3, we find that OC-192 framing (129,000 to 131,000 voice circuits) is adequate. A 50% margin in case of network congestion would correspond to $1.5 \times 72,500 = 108,750$ circuits. The OC-192 level is therefore sufficient.

Section 3.2

3.2.1 How many ATM cells can be embedded into either a STM-1 frame, using VC-4 containers?

Referring to Figure 3.4, a VC-4 container is made of 261 columns of 9 bytes, representing 2,349 bytes. Removing the 9-byte POH field, the actual payload is thus 2,340 bytes. Since

ATM cells are 53-bytes long, the number of ATM cells that can be encapsulated is therefore $2,340/53 = 44.15$ or 44 cells.

Section 3.3

3.3.1 An IP datagram with a data payload of 2,800 bytes must traverse two networks (A, B) which have MTUs of 1,488 bytes (A), and 648 bytes (B), respectively. Describe how the datagram must be fragmented from source to destination, keeping the number of fragments minimum and assuming a header size of 48 bytes. How many fragments are received at destination, with what payload sizes?

We recall that all fragments payloads should be multiples of 8 bytes, except for the last one. The first network (A) has a maximum transmission unit (MTU) of 1,500 bytes which, removing the 48-byte header, allows payloads of $1,488-48 = 1,440$ bytes. Since $2,800 = 1,440 + 1,360 = 180 \times 8 + 170 \times 8$, the original datagram (D) can be split into two fragments of payloads 1,440 bytes (D1) and 1,360 bytes (D2), respectively.

The second network (B) has an MTU of 648 bytes. This allows payloads of $648-48 = 600 = 75 \times 8$ bytes. For the first fragment D1, we have $1,440 = 600 + 600 + 240 = 75 \times 8 + 75 \times 8 + 30 \times 8$, corresponding to two fragments of 600 bytes and one fragment of 240 bytes. For the second fragment D2, we have $1,360 = 600 + 600 + 160 = 75 \times 8 + 75 \times 8 + 20 \times 8$, corresponding to two fragments of 600 bytes and one fragment of 160 bytes.

The end result of the fragmentation is six fragments, four of which with 600-byte payloads and two with 240 and 160 bytes, respectively.

3.3.2 If all 128-bit IPv6 addresses would be assigned to each square meter on Earth, how many addresses would a square meter contain? Same question with a square micron.

The number of IPv6 addresses is $2^{128} - 1 \approx 3 \times 10^{38}$. What is the surface of the Earth in square meters? Take the earth radius as being $R = 6,400$ km. The Earth's surface is then $S = 4\pi R^2 = 4\pi \times (6,400 \times 10^3 \text{ m})^2 \approx 5 \times 10^{14} \text{ m}^2$. The number of IP addresses per square meter on Earth is thus $3 \times 10^{38}/(5 \times 10^{14}) \approx 6 \times 10^{23}$. One square meter equals $(10^6 \text{ } \mu m)^2 = 10^{12}$ square microns. The number of IP addresses per square microns on Earth is thus $\approx 6 \times 10^{23}/10^{12} = 6 \times 10^{11}$ or 600 billions. This calculation illustrates that a number such as 10^{38} does not have meaningful reference value at a human scale.

3.3.3 Convert the Ipv6 address 801::AAAF:0:DD1E:5A6C in binary format.

First we rewrite 801::AAAF:0:DD1E:5A6C into the exact hexadecimal form:

0801:0000:0000:0000:AAAF:0000:DD1E:5A6C

To recall, hexadecimal numbers range from 0 to 9 and from A to F (0 to F or 0 to 15), corresponding to 4-bits binary numbers. We have A (hexa) = 1010 (binary), B(hexa) = 1011, etc., and F(hexa) = 1111. Thus the first of the IP address group 0801 is equal to the binary number

0000 1000 0000 0001

The other groups are, in binary:

```
0000 = 0000 0000 0000 0000
AAAF = 1010 1010 1010 1111
DD1E = 1101 1101 0001 1110
5A6C = 0101 1010 0110 1100
```

which gives the following complete binary address correspondence (as arranged from high-weight to low-weight bit order with space and underlining introduced for reading clarity):

0000 1000 0000 0001 <u>0000 0000 0000 0000</u> 0000 0000 0000 0000 <u>0000</u>
<u>0000 0000 0000</u>
1010 1010 1010 1111 <u>0000 0000 0000 0000</u>1101 1101 0001 1110 <u>0101</u>
1010 0110 1100

3.3.4 Show that <u>class D</u> addresses for IP multicast applications range from 224.0.0.0 to 239.255.255.255.

The first four bits in the first byte of <u>Class D</u> addresses are 1110. In decimal notation (X.Y.Z.W), the first IP-address number (X) thus corresponds to the binary number 1110 xxxx. The minimum and maximum values are therefore 1110 0000 = 224 and 1110 1111 = 224 + 15 = 239, respectively. The three other values (Y, Z, W) range from 0000 0000 = 0 to 1111 1111 = 255. Thus the smallest and biggest IP multicast addresses are 224.0.0.0 and 239.255.255.255, respectively.

Section 4.1

4.1.1 What is the E-field amplitude associated to an EM wave with intensity of $I = 1\,\text{W}/\text{cm}^2$ in vacuum? Same question in a glass medium with index $n = 1.45$?

According to definition in equation (4.1), the intensity is $I = E^2/(2\eta)$, corresponding to $E = \sqrt{2\eta I}$. In vacuum, we have $\eta = \eta_0 = 377\Omega$, thus $E = \sqrt{2 \times 377\Omega \times 1\text{W}/\text{m}^2} \approx 27.5$ Volt/m. In glass, we have $\eta = \eta_0/n = 377/1.45 = 260\Omega$, thus $E = \sqrt{2 \times 260\Omega \times 1\text{W}/\text{m}^2} \approx 23$ volt/m.

4.1.2 An EM wave with power $P = 1$ mW is captured into a hollow cylindrical guide having $d = 1$ cm diameter and being filled with vacuum. What are the EM intensity and the corresponding E-field amplitude?

The EM intensity is given by the ratio $I = P/S$, where $S = \pi r^2 = \pi d^2/4$ is the guide cross section. We thus have $S = \pi(1\text{cm})^2/4 = 0.78\,\text{cm}^2$, and $I = 1.10^{-3}\text{W}/0.78\,\text{cm}^2 = 1.27 \times 10^{-3}\,\text{W}/\text{cm}^2$. Considering vacuum (impedance $\eta = \eta_0 = 377\Omega$), the E-field amplitude is then $E = \sqrt{2\eta_0 I} = 0.98 \approx 1$ volt/cm.

4.1.3 What is the length of a $\lambda/2$ dipole antenna emitting at $f = 2$ GHz?

The corresponding E-field wavelength is $\lambda = c/f = 3 \times 10^8\,\text{ms}^{-1}/(2 \times 10^9\,\text{s}^{-1}) = 0.15\,\text{m} = 15\,\text{cm}$. The $\lambda/2$ antenna length is thus $L = 7.5\,\text{cm}$.

4.1.4 A radio link operating at 5-GHz frequency has a transmitting antenna with absolute gain $G_A = +70$ dB and has a receiving antenna with effective aperture $A_R = 1\,\text{m}^2$. What is the transmit power required to receive $P_R = 1\,\mu\text{W}$ (-60 dBW) at a distance of $r = 500$ km? Express the result in dBW and dBm.

The received power is given by the relation

$$P_{\text{rec}} = G_A \frac{A_R}{4\pi r^2} P_{\text{trans}}, \quad \text{or} \quad P_{\text{trans}} = \frac{4\pi r^2}{G_A A_R} P_{\text{rec}}?$$

Replacing the parameters with their values in conventional units yields

$$P_{\text{trans}} = \frac{4\pi(500 \times 10^3 m)^2}{10^7 \times 1\,\text{m}^2} 1 \times 10^{-6}\,\text{W} = 0.31\,\text{W} = -5.0\,\text{dBW} = +24.9\,\text{dBm}$$

4.1.5 A satellite is located at 36,000 km above the Earth. What is the dB loss introduced by free-space transmission from the shortest path to/from the satellite? Provide the power budget assuming a satellite-antenna gain of 30 dB and a receiver-antenna aperture of 1 m^2.

The free-space transmission loss is $L_{FS} = 1/(4\pi r^2)$ with $r = 36,000$ km being the shortest-path distance. Thus $L_{FS} = 1/[4\pi(36,000 \times 10^3$ m$)^2] = 6.14 \times 10^{-17}$ m^{-2}, or -162.1 dB with reference to a 1 m^2 aperture.

Considering the transmitter-antenna gain G_T and the receiver-antenna aperture A_R, the actual transmission loss is $L = G_T A_R L_{FS} = 10^3 \times 1$ m$^2 \times 6.14 \times 10^{-17}$ m$^{-2} = 6.14 \times 10^{-14}$, or -132 dB. The power budget is then $P_{rec}(dBW) = P_{trans}(dBW) - 132$ dB.

4.1.6 Calculate the power absorption loss (dB) caused by a rainfall of rate $R = 10$ mm/h on a radio waves at $f = 5$ GHz and 10 GHz, assuming vertical polarization and horizontal propagation over 250 km distance.

In these conditions, we have $\alpha = \alpha_V$ and $\beta = \beta_V$. From Table 4.2, we have $\alpha = 1.17$ and $\beta = 0.001$ for $f = 5$ GHz and have $\alpha = 1.264$ and $\beta = 0.008$ for $f = 10$ GHz. The corresponding dB/km attenuation coefficient is $\gamma = \beta R^\alpha$, or

$$\gamma(5\,\text{GHz}) = 0.001 \times (10)^{1.17} = 0.0147\,\text{dB/km}$$

$$\gamma(10\,\text{GHz}) = 0.008 \times (10)^{1.264} = 0.146\,\text{dB/km}$$

Over a 250-km distance, the corresponding absorption losses are

$$L(5\,\text{GHz}) = -0.0147\,\text{dB/km} \times 250\,\text{km} = -3.6\,\text{dB}$$
$$L(10\,\text{GHz}) = -0.146\,\text{dB/km} \times 250\,\text{km} = -36.5\,\text{dB}$$

4.1.7 Calculate the power absorption loss (dB) caused by a rainfall of rate $R = 100$ mm/h on a Ka-band (30-GHz) Earth-to-satellite radio system, assuming horizontal polarization, propagation at $\theta = 45°$ angle, and effective rain height of $h = 3$ km.

The total distance traversed through the rainy atmosphere is given by $d = h/\sin\theta = 3$ km$/\sin 45° = 4.2$ km. According to the formula, the attenuation coefficient is $\gamma = \beta R^\alpha$ with

$$\alpha = \frac{\beta_H \alpha_H + \beta_V \alpha_V + (\beta_H \alpha_H - \beta_V \alpha_V)\cos^2 45°}{2\beta} = \frac{3\beta_H \alpha_H + \beta_V \alpha_V}{4\beta}$$

$$\beta = \frac{\beta_H + \beta_V + (\beta_H - \beta_V)\cos^2 45°}{2} = \frac{3\beta_H + \beta_V}{4}$$

The values of α_H, α_V, β_H, β_V at $f = 30$ GHz can be approximated by interpolation from the data given in Table 4.2 for 20–40 GHz. We find: $\alpha_H \approx 1.019$, $\alpha_V \approx 0.997$, $\beta_H \approx 0.212$, and $\beta_V \approx 0.189$, which gives $\alpha = 1.013$ and $\beta = 0.206$. The attenuation coefficient is then $\gamma = 0.206(100)^{1.013} = 21.8$ dB/km. The total absorption loss is therefore $L = \gamma \times d = 21.8$ dB/km $\times 4.2$ km $= 91.5$ dB.

4.1.8 Show that the maximum incident angles for reflecting radio waves onto the ionosphere layers (heights $h = 50$–250 km) are in the range $\phi = 74$–$83°$. Calculate the maximum possible transmission distance for a single Earth-to-Earth round-trip path (Earth radius $R_E = 6,378$ km).

The Earth horizon and the different parameters are represented in the figure below.

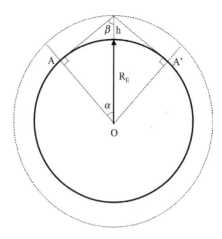

From this figure, the following relations are easily established:

$$\cos \alpha = \frac{R_E}{R_E + h} = \cos(\pi - \pi/2 - \beta) = \sin \beta, \quad \text{or} \quad \beta = \sin^{-1}\left(\frac{R_E}{R_E + h}\right)$$

Using $h = 50$ km or 250 km and $R_E = 6{,}378$ km, one finds the maximum possible incidences $\beta = 82.8° \approx 83°$ and $\beta = 74.2° \approx 74°$, respectively.

The maximum distance $A–A'$ covered (see figure) is given by $d = R_E \times (2\alpha_{rad})$, where α is expressed in radians. We have $\alpha_{rad} = (\pi/180)\alpha_{deg}$ and $\alpha_{deg} = 90° - \beta_{deg}$. Thus, we find $d = 6{,}378$ km $\times (\pi/180) \times 2 \times (90° - \beta_{deg})$, which gives $d = 1{,}603$ km for the low-altitude case ($\beta = 83°$) and 3,562 km for the high-altitude case ($\beta = 74°$).

4.1.9 Calculate the ionospheric reflection coefficient during daytime (plasma density $N = 2 \times 10^{11}$ m^{-3}) of a $f = 20$ MHz radio-wave for a $\phi = 74°$ incidence with horizontal polarization.

The refractive-index perturbation is $\Delta = 80.5 \times N/f^2 = 80.5 \times 2 \times 10^{11}$ m$^{-3}/(20 \times 10^6$ Hz$)^2 = 0.04$, giving $1 - \Delta = 0.96$. For horizontal polarization, the reflection coefficient is given by equation (4.25):

$$R_H = \left| \frac{\cos 74° - \sqrt{0.96 - \sin^2 74°}}{\cos 74° + \sqrt{0.96 - \sin^2 74°}} \right|$$

which gives $R_H \approx 0.157$, or about 16%.

4.1.10 Prove the parabolic approximation formula for the Fresnel-zone radii:

$$H \approx \sqrt{n\lambda \frac{d'(d - d')}{d}}$$

where d is the LOS path length, d' the distance from the origin and n is an odd integer.

The figure below shows the LOS path $(A–B)$, the approximated parabolic curve for the Fresnel boundary, and the corresponding radius $H(x)$.

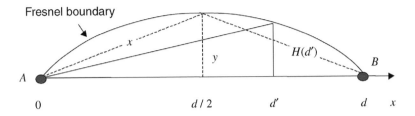

By assumption, the radius $H(x)$ of the Fresnel boundary is approximated by the parabolic function $H^2(x) = ax^2 + bx + c$, with the condition $H(0) = H(d) = 0$. These two conditions give $H^2(x) = -ax(d - x)$. To find the parameter a we consider the mid-point at $x = d/2$. The corresponding (square) radius is $y^2 = H^2(d/2) = -ad^2/4$. The total path length for the reflected beam is $2x$, corresponding to an excess distance of $n\lambda/2(n = 1, 3, 5, \ldots)$ with the LOS path, or

$$\text{(i)} \qquad 2x = d + n\lambda/2$$

By construction, we also have

$$\text{(ii)} \qquad x^2 = y^2 + d^2/4$$

Combining (i) and (ii) yields

$$\frac{1}{4}\left(d + \frac{n\lambda}{2}\right)^2 - a\frac{d^2}{4} = \frac{d^2}{4} \quad \text{or} \quad a = -\frac{n\lambda}{d} - \left(\frac{n\lambda}{2d}\right)^2 \approx -\frac{n\lambda}{d} \quad \text{(assuming } \lambda \ll d)$$

This result yields $H^2(d') \approx n\lambda x(d - d')/d$ or $H(d') \approx \sqrt{n\lambda d'(d - d')/d}$, which is the definition that had to be proven.

4.1.11 Calculate the required beam clearance for a $f = 2$ GHz radio link with $d = 1$-mile antenna-to-antenna distance (express the result as minimum obstacle distance and safety cone angle).

The maximum Fresnel radius is $H \approx \sqrt{\lambda d/2}$. Using $\lambda = c/f = 3 \times 10^8/(2 \times 10^9) = 0.15$ m and $d = 1600$ m, we get $H \approx \sqrt{0.15 \times 1600/2} = 10.9$ m. The minimum obstacle distance is $0.6H$ or 6.5 m. The safety cone angle is $\beta = 2\alpha \times 0.6 = 1.2 \tan^{-1}(\sqrt{2\lambda/d})$ or $\beta = 1.2 \tan^{-1}[\sqrt{2 \times 0.15/1600}] = 0.94° \approx 1°$.

4.1.12 What is the noise frequency of multipath fading, assuming a 2-GHz radio carrier and a mobile-antenna speed of 75 mph?

In practical units, the speed of 75 mph corresponds to $75 \times 1.6 = 120$ km/hour, or 120×10^3 m/3600 $= 33.3$ m/s. For 2 GHz signals, the wavelength is 15 cm, corresponding to a number of wavelengths traversed per second of $33.3/0.15 = 222.2$/s. This rate corresponds to a noise frequency of 222 Hz or 0.2 KHz, which falls into the audition-frequency range.

4.1.13 What is the effective temperature of a UHF antenna emitting at $f = 1$ GHz in outer space the and in the Earth atmosphere ($T = -270°$C and 20°C)? Conclusions?

We apply the formula

$$T_{\text{eff}} = \frac{hf}{k_B} \left(\frac{1}{\exp\left(\dfrac{hf}{k_B T}\right) - 1} + \frac{1}{2} \right)$$

with $k_B = 1.38 \times 10^{-23}$ J/K and $h = 6.62 \times 10^{-34}$ J-s. Using $f = 10^9$ Hz, $T = -270 + 273 = 3$ K (outer space), and $T = 273 + 20 = 293$K (atmosphere), we get

(Outer space): $hf/k_B T = 1.599 \times 10^{-2}$ and $T_{\text{eff}} = 2.97$ K.

(Atmosphere): $hf/k_B T = 1.637 \times 10^{-4}$ and $T_{\text{eff}} = 292.97$ K

Conclusions: (1) The effective antenna temperature is sligthly below the actual physical temperature, and (2) at radio frequencies and over the whole temperature range considered, the quantum noise correction is negligible.

4.1.14 An Earth-based radio system ($T = 290$ K) utilizes an antenna with noise temperature $T_{\text{ra}} = 50$ K, a transmission line with power loss of $L = -0.2$ dB, and a preamplifier with noise temperature $T_{\text{amp}} = 2610$ K. Calculate the effective system temperature and the corresponding noise figure. To how many decibels does the end-to-end SNR degradation correspond?

We convert first the loss into a linear-scale number, i.e., $L = 10^{L_{\text{dB}}/10} = 10^{-0.2/10} = 0.95 \equiv \varepsilon$. The system temperature is given by the formula

$$T_S = T_{\text{ra}} + T(1/\varepsilon - 1) + T_{\text{amp}}/\varepsilon$$

giving $T_S = 50 + 290(1/0.95 - 1) + 2610/0.95 = 2812.6$ K. The system noise figure is $F = 1 + T_S/T$ or $F = 1 + 2812.6/290 = 10.69$ or 10.3 dB. The end-to end SNR degradation is therefore 10.3 db.

4.1.15 An Earth-to-satellite communication link at carrier frequency $f = 2$ GHz has a system temperature of $T_S = 30$ dBK, a satellite antenna gain $G_T = 30$ dB, a satellite transmission power of $+3$ dBW, a mobile receiving antenna gain of $G = 15$ dB. The satellite is orbiting at an altitude of $r = 36{,}000$ km. Calculate the receiver CNR and SNR, assuming a bandwidth of $\Delta f = 100$ kHz.

The carrier wavelength is $\lambda = c/f = (3 \times 10^8 \text{ m/s})/(2 \times 10^9 \text{ Hz}) = 0.15$ m. Next, we calculate the isotropic loss according to the definition $L = (4\pi r/\lambda)^2$, or $L = (4\pi \times 36 \times 10^6 \text{ m}/0.15 \text{ m})^2 = 9.04 \times 10^{18}$, or 189.5 dB. The receiver CNR is then given by the formula

$$\text{CNR(dBHz)} = P_{\text{trans}}(\text{dBW}) + G_T(\text{dB}) + G_R(\text{dB}) - L(\text{dB}) - T_S(\text{dBK}) + 228.6$$

or

$$\text{CNR(dBHz)} = 3 \text{ dBW} + 30 \text{ dB} + 15 \text{ dB} - 189.5 - 30 \text{ dBK} + 228.6 = +57.1 \text{ dBHz}$$

For a bandwidth of $\Delta f = 100$ kHz, this CNR corresponds to a SNR of

$$\text{SNR(dB)} = \text{CNR(dBHz)} - 10 \log_{10}(\Delta f_{\text{Hz}}) = 57.1 \text{ dBHz} - 10 \log_{10}(100 \times 10^3 \text{ Hz})$$

$$= +7.1 \text{ dB}.$$

4.1.16 A satellite-based communication system is characterized by an uplink CNR of $\text{CNR}_U = +75$ dBHz and a downlink CNR of $\text{CNR}_D = +50$ dBHz. What is the net CNR corresponding to the round-trip path? Conclusion?

We first convert the CNRs in linear scale, according to $\text{CNR}_{\text{linear}} = 10^{\text{CNR}_{\text{dBHz}}/10}$, which gives $\text{CNR}_U = 3.16 \times 10^7$ Hz and $\text{CNR}_D = 1.0 \times 10^5$ Hz. The net CNR is given by

$1/CNR = 1/CNR_U + 1/CNR_D = 1/(3.16 \times 10^7) + 1/(1.0 \times 10^5) = 1.00316 \times 10^{-5}$, or $CNR = 9.96 \times 10^4$, or $+49.9$ dB. The CNR degradation is essentially due to the downlink noise.

Section 4.2

4.2.1 Prove that in a given network, the traffic intensity A is equal to the number of calls Q that can be expected during the average holding time H.

Proof:

1. Assume first that, during the monitoring time T, there are M incoming calls of duration $h_1, h_2, \ldots, h_M$.

2. Thus, the total holding time is $T' = \sum_{i=1\ldots M} h_i$ and the average holding time per call is $H = T'/M$.

3. By definition, the traffic intensity is the ratio of total holding time to the monitoring time, or $A = T'/T$.

4. The number of calls during the monitoring time is M; therefore, the number of calls to expect during the average holding time is reduced in the proportion $Q = M \times H/T$, or equivalently, $Q = T'/T \equiv A$.

4.2.2 Prove that in a network with traffic intensity A, the occupation rate of a single channel is equal to $B = A/(1 + A)$. [Use result of exercise 4.15.]

Proof:

1. The traffic intensity A corresponds to the total number of incoming calls during the average call duration, or holding-time, H.

2. Calls come therefore to the network at equivalent regular intervals H/A.

3. One can view a given channel as being busy during the call duration, and being idle until a new call comes in.

4. The idle duration is H/A, so the full cycle to handle one call to the next has the duration $H' = H + H/A = H(1 + A)/A$.

5. The relative fraction of time where the channel is busy (or occupancy rate) is therefore $H/H' = A/(1 + A)$.

4.2.3 A mobile-network cell is 10 km in diameter and the area has a density of 10 mobile users per km². During a 1 peak hour of monitoring time, it is observed that 5% of these users place calls with an average call duration of 10 mn. What is the traffic offer, and how many channels are required to provide a grade service of 0.01?

First we need to calculate the number of calling users. The cell area is $S = (10 \text{ km})^2/4 = 78.5$ km². The number of mobile users is therefore $10 \times 78.5 = 785$ users. The fraction of users placing a call during the 1-h observation time is thus $785 \times 5\% = 39.25$ or 40. Given the average call duration of 10mn, the traffic offer is $A = 40 \times 10 \text{ mn}/60 \text{ mn} = 6.6$ Erlangs. To calculate the number of channels N required for a grade of service of $B = 0.01$, we must solve the equation

$$B(N) = \frac{A^N}{N!} \left(\frac{1}{1 + A + \dfrac{A^2}{2!} + \cdots + \dfrac{A^N}{N!}} \right) \leq 0.01 \qquad \text{with } A = 6.6.$$

As a start, we calculate the blocking rate for $N = 12$, which is about twice the traffic intensity. The result is $B(N = 12) = 1.98 \times 10^{-2}$, which is fairly close to the result. We could proceed with $N = 13$, and so on, in order to obtain the right N. But a quicker way is to observe that

$$B(N + 1) < \frac{A}{N + 1} B(N)$$

since

$$B(N + 1) = \frac{A^{N+1}}{(N + 1)!} \left(\frac{1}{1 + A + \dfrac{A^2}{2!} + \cdots + \dfrac{A^{N+1}}{(N + 1)!}} \right)$$

$$= \frac{A}{N + 1} \frac{A^N}{N!} \left(\frac{1}{1 + A + \dfrac{A^2}{2!} + \cdots + \dfrac{A^N}{N!}} \right) \left(\frac{1 + A + \dfrac{A^2}{2!} + \cdots + \dfrac{A^N}{N!}}{1 + A + \dfrac{A^2}{2!} + \cdots + \dfrac{A^{N+1}}{(N + 1)!}} \right)$$

$$= \frac{A}{N + 1} B(N) \left(\frac{1 + A + \dfrac{A^2}{2!} + \cdots + \dfrac{A^N}{N!}}{1 + A + \dfrac{A^2}{2!} + \cdots + \dfrac{A^{N+1}}{(N + 1)!}} \right) < \frac{A}{N + 1} B(N)$$

Therefore, we have

$$B(13) < \frac{6.6}{13} B(12) = 1.005 \times 10^{-2} \quad \text{and}$$

$$B(14) < \frac{6.6}{14} B(13) < \frac{6.6}{14} 1.005 \times 10^{-2} = 4.7 \times 10^{-3}$$

Since $4.7 \times 10^{-3} < B(13) < 1.005 \times 10^{-2}$, the exact solution is $N = 13$.

4.2.4 Shows that for a network having an infinite number of channels which is filled on first-channel-available basis, the occupancy for channel #N, as expressed in erlangs, is given by

$$O(N) = [B(N - 1) - B(N)]A$$

where A is the traffic offer, and $B(N)$ is the Erlang formula for the blocking rate in a N-channel network.

For a N-channel network, the blocking rate is given by Erlang's formula:

$$B(N) = \frac{A^N}{N!} \left(\frac{1}{1 + A + \dfrac{A^2}{2!} + \cdots + \dfrac{A^N}{N!}} \right)$$

Considering a network with an infinite number of channels, we just need to analyze what happens to the traffic in the first few channel slots. For this, we make the following drawing:

O_1			P_1	
O_1	O_2		P_2	
O_1	O_2	O_3	P_3	
O_1	O_2	O_3	O_4	P_4

In each line of the drawing, O_1 is the traffic handled by channel 1, O_2 is the traffic handled by channel 2, etc., and the same for $P_1, P_2, \ldots$, etc., which correspond to the traffic "lost" or "not handled" by channels 1, 2, $\ldots$, etc. If A is the traffic offer, the conservation of traffic dictates that

$$Q_1 + P_1 = A$$
$$O_2 + P_2 = P_1$$
$$\ldots$$
$$O_k + P_k = P_{K-1}$$

If we sum up all the above, we get

$$\sum_{i=1\ldots K} O_i + \sum_{i=1\ldots K} P_i = A + \sum_{i=1\ldots K-1} P_i$$

or, noting that the sums on P_i only differ by the last term:

$$\sum_{i=1\ldots K} O_i + P_K = A$$

This leads to a set of iterative relations:

$$O_1 + P_1 = A \leftrightarrow O_1 = A - P_1$$
$$O_1 + O_2 + P_2 = A \leftrightarrow O_2 = A - P_2 - O_1 = P_1 - P_2$$
$$\ldots$$
$$O_1 + \cdots + O_K + P_K = A \leftrightarrow O_K = A - P_K - (O_1 + \cdots + O_K) = P_{K-1} - P_K$$

We have thus obtained the channel occupancies (O_i) and we just need to specify now what are the quantities P_i. Consider the above drawing. The first line corresponds to the situation of O_1 erlangs handled by Channel 1, and P_1 erlangs lost by that channel. Since the loss rate for a 1-channel network is $B(1)$, we have $P_1 = A \times B(1)$, meaning that P_1 is the traffic blocked by the first channel, as expressed in erlangs. The second line shows two channels handling traffic. Consistently, the corresponding lost traffic is $P_2 = A \times B(2)$. It is clear then that $P_K = A \times B(K)$, which with the previous result defining the channel occupation O_K proves the theorem. As a fool proof, we can sum up all the channel contributions. The result is the following:

$$S = O_1 + O_2 + \cdots + O_K = A + (P_1 - P_2) + (P_2 - P_3) + \cdots + (P_{K-1} - P_K) \equiv A - P_K$$

When the number of channels is infinite ($P_K \to \infty$), we obtain $S \to A$, meaning that the sum of all channel occupancies matches the traffic offer A.

Section 4.3

4.3.1 What is the altitude of geosynchronous satellites with respect to sea level?

The sideral day has duration of $T_s = 23$ h, 56 min, 4.1 s, corresponding to 23 h $\times$ 3,600 s $+ 56$ mn $\times$ 60 s $+ 4.1$ s $= 86,164.1$ s. Equating the sideral day with the circular-orbit period, i.e.,

$$T_0 = 2\pi \sqrt{\frac{r_0^3}{\mu}} = T_S$$

yields

$$r_0 = \left[\mu \left(\frac{T_S}{2\pi} \right)^2 \right]^{1/3} \quad \text{or} \quad r_0 = \left[3.986 \times 10^{14} \left(\frac{86,1641}{2\pi} \right)^2 \right]^{1/3}$$

$$= 4.2163 \times 107 \quad \text{or} \quad 42,163 \, \text{km}$$

The satellite's altitude with respect to sea level is given by subtracting the Earth radius $r_E = 6{,}378$ km, which gives the altitude of 35,785 km.

4.3.2 Taking $r_0 = 35{,}785$ km as the reference distance for geostationary satellites, what are the distances and corresponding altitudes for which satellites can complete exactly $N = 2$, 4, 12, or 24 revolutions per 24-h day (corresponding to periods of 12 h, 6 h, 2 h, or 1 h, respectively)? Conclusion?

Completing N revolutions per day for a satellite means that its period is $T_S = T_0/N$, where $T_0 = 2\pi\sqrt{r_0^3/\mu}$. If r_1 is this satellite's distance, we thus have $r_1 = r_0/N^{2/3}$. This gives for the distances and corresponding altitudes ($a = r_1 - 6{,}378$ km):

$$r_1 = 35{,}785 \, \text{km}/2^{2/3} = 22{,}543 \, \text{km}, \quad a = 16{,}165 \, \text{km} \ (N = 2)$$

$$r_1 = 35{,}785 \, \text{km}/4^{2/3} = 14{,}201 \, \text{km}, \quad a = 7{,}823 \, \text{km} \ (N = 4)$$

$$r_1 = 35{,}785 \, \text{km}/12^{2/3} = 6{,}827 \, \text{km}, \quad a = 449 \, \text{km} \ (N = 12)$$

$$r_1 = 35{,}785 \, \text{km}/24^{2/3} = 4{,}300 \, \text{km}, \quad a = -2{,}077 \, \text{km} \ (N = 24)$$

Conclusion: The last result shows that it is not possible to realize satellites having 1-h revolution periods, since the theoretical altitude would be less than the Earth radius. However, a 2-h period is possible since the corresponding altitude is about 450 km.

4.3.3 What is the maximum intercity distance covered by a GEO satellite? (Use $r_0 = 42{,}000$ km and $r_E = 6{,}400$ km for the Earth-to-satellite distance and Earth radius, respectively.)

We need fist to draw a picture of the situation in order to visualize what is meant by "intercity-distance covered":

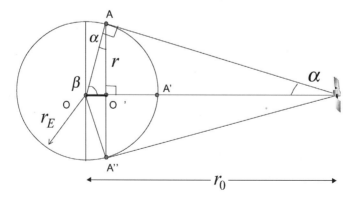

From the above figure, we see that the maximum intercity distance covered by the GEO satellite is the length L of the arc $A - A' - A''$ intercepted by the cone of half-angle α. This angle is defined by $\sin \alpha = r_E/r_0 = 6{,}400 \, \text{km}/42{,}000 \, \text{km} = 0.152$, or $\alpha = 8.75°$ or 0.152 radians. The angle β is equal to $\pi/2 - \alpha$, or $\pi/2 - 0.152 = 1.42$ radians. The length of the arc

$A - A' - A''$ is $L = 2r_E\beta = 2 \times 6,400\,\text{km} \times 1.42 = 18,176\,\text{km}$, which is the maximum intercity distance.

4.3.4 What is the intercity distance covered by LEO satellites? How many LEO satellites are required for full coverage of the Earth's perimeter, assuming that one should use two times the theoretical minimum? (Use $h = 100–1,000$ km and $r_E = 6,378$ km for the satellite altitude and Earth radius, respectively.)

Referring to the figure of previous exercise, we have $r_0 = h + r_E$, giving the range $r_0 = 6,478$ (low orbit case) to $r'_0 = 7,378$ km (high-orbit case). These two distances define the range of angles:

$\sin \alpha = r_E/r_0 = 6,378\,\text{km}/6,478\,\text{km} = 0.984$, or $\alpha = 79.9°$ or 1.39 radians (low orbit)
$\sin \alpha' = r_E/r'_0 = 6,378\,\text{km}/7,378\,\text{km} = 0.864$, or $\alpha' = 59.8°$ or 1.044 radians (high orbit)

The corresponding angles β (see figure) are then

$$\beta = \pi/2 - \alpha, \quad \text{or} \quad \pi/2 - 1.139 = 0.431 \text{ radians (low orbit)}$$
$$\beta' = \pi/2 - \alpha', \quad \text{or} \quad \pi/2 - 1.044 = 0.526 \text{ radians (high orbit)}$$

The corresponding lengths of arcs $A - A' - A''$, defining the maximum intercity distance, are thus

$$L = 2r_E\beta = 2 \times 6,400\,\text{km} \times 0.431 = 5,516\,\text{km (low orbit)}$$
$$L' = 2r_E\beta' = 2 \times 6,400\,\text{km} \times 0.526 = 6,732\,\text{km (high orbit)}$$

The Earth perimeter is $P = 2\pi r_E = 2\pi \times 6,378\,\text{km} = 40,074\,\text{km}$. The number of required LEO satellites is therefore

$$N = P/L = 40,074\,\text{km}/5,516\,\text{km} = 4.1 \quad \text{or} \quad 5 \text{ (low orbit)}$$
$$N' = P/L' = 40,074\,\text{km}/6,732\,\text{km} = 5.96 \quad \text{or} \quad 6 \text{ (high orbit)}$$

These numbers correspond to the theoretical limit defined by the Earth's horizon. A better coverage is obtained by doubling the number of required satellites, which comes to $N'' = 10–12$, or 11 to define an average requirement.

4.3.5 Assume a satellite having highly elliptic orbit (HEO) with major axis $2a = 53,260$ km and eccentricity of $e = 0.74$ (Molnya orbit type). Calculate first the revolution period (in hours) and perigee altitude (in kilometers). Calculate then the nodal regression and perigee precession for polar and equatorial inclinations ($i = 0$ or $90°$, respectively), as expressed in degree/day at the perigee point. Also, calculate the nodal regression for the critical inclinations i giving $\cos(i) = \pm 1/\sqrt{5}$ (use $J_2 = 1.0823 \times 10^{-3}$ and $R_E = 6,378$ km). The revolution period is given by $T = 2\pi\sqrt{a^3/\mu}$, with $\mu = 3.986 \times 10^{14}\,\text{m}^3/\text{s}^2$. Replacing the value of $a = 53,260\,\text{km}/2 = 2.663 \times 10^7$ m yields $T = 4.325 \times 10^4$ s (or $4.325 \times 10^4\,\text{s}/(3,600\,\text{s}) = 12.0$ h). The perigee radius is given by $R_P = a(1 - e) = 2.66 \times 10^7(1 - 0.74) = 6.916 \times 10^6$ m or 6,916 km. The perigee altitude is thus $6,916 – 6,378 = 538$ km.

The nodal regression and perigee precession rates (as expressed in radians/day) are given by

$$\delta\varphi/\text{day} = -3\pi J_2 \left(\frac{R_E}{R_P}\right)^2 \frac{\cos(i)}{T} \times 86,400$$

$$= -3\pi \times 1.0823 \times 10^{-3} \left(\frac{6,378\,\text{km}}{6,916\,\text{km}}\right)^2 \frac{\cos(i)}{4.325 \times 10^4 s} \times 86,400$$

$$= -1.73 \times 10^{-2} \cos(i)$$

and

$$\delta\psi/\text{day} = \frac{3\pi J_2}{2}\left(\frac{R_E}{R_P}\right)^2 \frac{5\cos^2(i)-1}{T(s)} \times 86{,}400$$

$$= \frac{3\pi \times 1.0823 \times 10^{-3}}{2}\left(\frac{6{,}378\,\text{km}}{6{,}916\,\text{km}}\right)^2 \frac{5\cos^2(i)-1}{4.325 \times 10^4 s} \times 86{,}400$$

$$= 8.66 \times 10^{-3}\left[5\cos^2(i)-1\right]$$

respectively. The conversion of these angle rates in degrees/day is given by multiplying the expressions by $180/\pi$.

For polar orbits ($i = 0$), we have

$$\delta\varphi = -1.73 \times 10^{-2} \times 180/\pi = -0.99\,\text{deg/day}$$

$$\delta\psi = 8.66 \times 10^{-3} \times 4 \times 180/\pi = 1.98\,\text{deg/day}$$

For equatorial orbits ($i = 90°$), we have

$$\delta\varphi = 0$$

$$\delta\psi = 8.66 \times 10^{-3} \times (-1) \times 180/\pi = -0.49\,\text{deg/day}$$

For the critical inclinations ($\cos(i) = \pm 1/\sqrt{5}$), we have

$$\delta\varphi\,(\text{deg/day}) = -0.99\cos(i) = -0.99(\pm 1/\sqrt{5}) = -(\pm 0.442)\,\text{deg/day}$$

4.3.6 Provide the inclination i required for a Sun-synchronous circular orbit at $h = 500\,\text{km}$ altitude. (Use $R_E = 6{,}378\,\text{km}$, $\mu = 3.986 \times 10^{14}\,\text{m}^3/\text{s}^2$, and $J_2 = 1.0823 \times 10^{-3}$.)

The condition for Sun-synchronicity is that the nodal regression rate per day, $\delta\varphi/\text{day}$, be equal over a full year to $360°$. Using the definition for $\delta\varphi/\text{day}$ as converted in degrees ($\times 180/\pi$) and years ($\times 365.25$),the above condition is equivalent to

$$\delta\varphi/\text{day} = -3\pi J_2\left(\frac{R_E}{r}\right)^2 \frac{\cos(i)}{T} \times 86{,}400 \times \frac{180}{\pi} \times 365.25 \equiv 360°$$

where $T = 2\pi\sqrt{r^3/\mu}$ is the orbit period, and $r = R_E + h$ is the orbit radius. We then get $r = 6{,}378 + 500 = 6{,}878\,\text{km}$ and $T = 2\pi\sqrt{(6.878 \times 10^6)^3/(3.986 \times 10^{14})} = 5{,}676\,\text{s}$. Thus

$$\delta\varphi(\text{deg/year}) = -3\pi \times 1.0823 \times 10^{-3}\left(\frac{6.378\,\text{km}}{6.878\,\text{km}}\right)^2 \frac{\cos(i)}{5{,}676 \times 10^3\,\text{s}} \times 86{,}400 \times \frac{180}{\pi} \times 365.25$$

$$= -2{,}792 \times \cos(i)$$

which must be equal to $365°$, or $\cos(i) = -365/2{,}792 = -0.1307$ or $i = 97.5°$, which is over a polar-orbit inclination by a $7.5°$ angle.

4.3.7 Over what amount of time does a geostationary satellite drift by $0.1°$ from its nominal position, due to the precession of perigee effect? (Use $J_2 = 1.0823 \times 10^{-3}$ and $R_E = 6{,}378\,\text{km}$.)

The daily rate of precession of perigee, as defined in degrees per day, is given by the formula

$$\delta\psi \ (\text{deg/day}) = \frac{3\pi J_2}{2}\left(\frac{R_E}{r}\right)^2 [5\cos^2(i) - 1] \times \frac{180}{\pi}$$

where $r = 42{,}000$ km is the GEO radius and $i = 0°$ the GEO inclination. We thus get

$$\delta\psi \ (\text{deg/day}) = -\frac{3 \times 1.0823 \times 10^{-3}}{2}\left(\frac{6{,}378 \text{ km}}{42{,}000 \text{ km}}\right)^2 (5 - 1) \times 180$$

$$= -2.69 \times 10^{-2} \text{ deg/day}$$

A precession angle of $\delta\psi = -0.1°$ is then obtained in $0.1/(2.69 \times 10^{-2}) = 3.7$ days or about $\frac{1}{2}$ week.

4.3.8 The International Space Station (ISS) has a near-circular orbit with cruising altitude of 388 km.

Calculate (a) its revolution period in hours and minutes, (b) its angular speed in degrees per minute, (c) its cruising speed in meters per second and relative to the speed of sound (300 m/s), (d) its speed relative to a point on Earth at latitude λ, assuming an orbital inclination of $i = 51.63°$ and a $90°$ observation angle (zenith), and (e) its angular speed relative to the Equator, in degrees per minute. Concerning the last result, how does it compare with the apparent motion of stars in the sky at the Equator's zenith?

- The ISS distance to the Earth center is $a = 388$ km $+ 6{,}378$ km $= 6{,}766$ km.
- The ISS revolution period is $T = 2\pi\sqrt{a^3/\mu} = 2\pi\sqrt{\frac{6{,}766 \times 10^3 \text{m}}{3.986 \times 10^{14} \text{m}^3/\text{s}^2}}$ giving $V_R = 7.675 -$ $0.4 \times \cos(51.63°) = 7.426$ km/s. Calculate now the ISS angular speed relative to this point, i.e., $\omega = V_R/a = (7.426 \text{ km/s})/6{,}766 \text{ km} = 1.09 \times 10^{-3}$ radians/s or 3.7 degree/min or $45°$ in 12 min. Note that this angular speed is relatively high to allow one tracking the ISS with an ordinary telescope (to compare with $\omega' = 360°/24h = 0.25°/$ min or $45°$ in 3 h for the maximum angular speed of stars at Equator's zenith). 5,539 s or 1 h, 32 min, 19 s.
- The ISS cruising speed is $V = \sqrt{\mu/a} = [3.986 \times 10^{14} \text{ m}^3/\text{s}^2)/(6.766 \times 10^6 \text{ m})]^{1/2} = 7{,}675$ m/s or 7.675 km/s. This corresponds to the (amazing) speed of *Mach 25.6*, or 25.6 times the speed of sound (300 m/s).
- The ISS orbital inclination is $i = 51.63°$. The relative speed, as viewed from a point at latitude λ, is given by $V_R = V - V_E(\lambda) \times \cos(51.63°)$. At the equator, $V_E(0) = 0.4$ km/s, giving $V_R = 7.675 - 0.4 \times \cos(51.63°) = 7.426$ km/s.
- The ISS angular speed relative to this Equator is $\omega = V_R/a = (7.426 \text{ km/s})/6{,}766 \text{ km} = 1.09 \times 10^{-3}$ radians/s or 3.7 degrees/min or $45°$ in 12 min. This angular speed is relatively high to allow tracking the ISS with an ordinary telescope (compare with $\omega' = 360°/24$ h $= 0.25°/$min, or $45°$ in 3 h, for the maximum apparent motion of stars in the sky, as viewed at Equator's zenith.)

4.3.9 How many MEO satellites are necessary to cover the Earth's circumference, assuming 10% coverage overlap and an altitude range of 5,000–15,000 km? (Clue: Calculate for both altitudes the maximum intercity distance, as defined by the satellite's horizon, and use $r_E = 6{,}378$ km for the Earth's radius.)

We need to refer to the figure of exercise 4.3.3 to visualize the "intercity distance." Using the definitions shown in the figure, we have for the two extreme altitudes $h_{min} = 5,000$ km and $h_{max} = 15,000$ km:

1. $\sin \alpha_{max} = r_E/(r_E + h_{min}) = 6,378$ km$/(6,378$ km$ + 5,000$ km$) = 0.56$, or $\alpha = 34.1°$ or 0.595 radians. The angle β is equal to $\pi/2 - \alpha$, or $\pi/2 - 0.595 = 0.97$ radians (or $55.5°$). The length of the arc $A - A' - A''$ is $L = 2r_E\beta = 2 \times 6,378$ km $\times 0.97 = 12,375$ km, which is the maximum intercity distance for a low-orbit MEO satellite.

2. $\sin \alpha_{min} = r_E/(r_E + h_{max}) = 6,378$ km$/(6,378$ km$ + 15,000$ km$) = 0.29$, or $\alpha = 17.3°$ or 0.302 radians. The angle β is equal to $\pi/2 - \alpha$, or $\pi/2 - 0.302 = 1.26$ radians (or $72.2°$). The length of the arc $A - A' - A''$ is $L = 2r_E\beta = 2 \times 6,378$ km $\times 1.26 = 15,946$ km, which is the maximum intercity distance for a high-orbit MEO satellite.

The Earth's circumference is $C = 2\pi r_E = 40,074$ km. From the above results, we see that with 10% overlap, it takes $N = 1.1 \times C/L$ satellites (N being the closest higher integer) to cover the circumference, i.e.,

$$N = 1.1 \times 40,074/12,375 = 4 \quad \text{for low-orbit MEO}$$
$$N = 1.1 \times 40,074/15,945 = 3 \quad \text{for high-orbit MEO}$$

How many satellites would be required to cover the entire Earth's surface? Since $3\beta = 166 - 216°$, three orbits with node-longitude differences of $60°$ should be sufficient. This gives a required number of $M = 3N = 9-12$ satellites.

4.3.10 How many LEO satellites are necessary to cover the Earth's circumference, assuming 10% overlap and an altitude range of $100-1,000-1,500$ km? (Clue: Calculate for the three altitudes the maximum intercity distance, as defined by the satellite's horizon, and use $r_E = 6,378$ km for the Earth's radius.)

This is the same exercise as the previous one for MEO systems, but with a different scale factor. We first need to refer to the figure of exercise 4.3.3 in order to visualize the "inter-city distance." Using the definitions shown in the figure, we have for the three altitudes $h_{min} = 100$ km, $h_{max} = 1,000$ km, and $h'_{max} = 1,500$ km:

(a) $\sin \alpha_{max} = r_E/(r_E + h_{min}) = 6,378$ km$/6,478$ km $= 0.98$, or $\alpha = 79.9°$ or 1.39 radians. The angle β is equal to $\pi/2 - \alpha$, or $\pi/2 - 1.39 = 0.18$ radians (or $10.3°$). The length of the arc $A - A' - A''$ is $L = 2r_E\beta = 2 \times 6,378$ km$\times 0.18 = 2,296$ km, which is the maximum inter-city distance for a low-orbit, 100 km LEO satellite.

(b) $\sin \alpha_{min} = r_E/(r_E + h_{max}) = 6,378$ km$/7,378$ km $= 0.86$, or $\alpha = 59.8°$ or 1.04 radians. The angle β is equal to $\pi/2 - \alpha$, or $\pi/2 - 1.04 = 0.53$ radians (or $30.4°$). The length of the arc $A - A' - A''$ is $L = 2r_E\beta = 2 \times 6,378$ km $\times 0.53 = 6,760$ km, which is the maximum intercity distance for a high-orbit, 1,000-km LEO satellite.

(c) $\sin \alpha'_{min} = r_E/(r_E + h'_{max}) = 6,378$ km$/7,878$ km $= 0.81$, or $\alpha = 54.0°$ or 0.94 radians. The angle β is equal to $\pi/2 - \alpha$, or $\pi/2 - 0.94 = 0.63$ radians (or $36.1°$). The length of the arc $A - A' - A''$ is $L = 2r_E\beta = 2 \times 6,378$ km $\times$

0.63 = 8,036 km, which is the maximum inter-city distance for a high-orbit, 1,500-km LEO satellite.

The Earth circumference is $C = 2\pi r_E = 40{,}074$ km. From the above results, we see that with 10% overlap, it takes $N = 1.1 \times C/L$ satellites (N being the closest higher integer) to cover the circumference, i.e.,

$$N = 1.1 \times 40{,}074/2{,}296 = 20 \quad \text{for low-orbit, 100 km LEO}$$
$$N = 1.1 \times 40{,}074/6{,}760 = 7 \quad \text{for high-orbit, 1,000 km LEO}$$
$$N = 1.1 \times 40{,}074/8{,}036 = 6 \quad \text{for high-orbit, 1,500 km LEO}$$

How many satellites would be required to cover the entire Earth surface? Assuming a 10% overlap ($\times 1.1$) between the different orbit coverages, we have:

- Low orbits (100 km): $1.1 \times 180°/\beta = 198°/10.3° \approx 20$, requiring a total of $20 \times 20 = 400$ satellites
- High orbits (1,000 km): $1.1 \times 180°/\beta = 198°/30.4° \approx 7$, requiring a total of $7 \times 7 = 49$ satellites
- High orbits (1,500 km): $1.1 \times 180°/\beta = 198°/36.1° \approx 6$, requiring a total of $6 \times 6 = 36$ satellites

Thus 6 to 20 orbits with node-longitude differences of $10-36°$ can cover the Earth under the above 10% overlap assumption.

4.3.11 Evaluate the minimal cost of two typical constellation systems made of either $N = 48$ or 66 LEO satellites, assuming satellite (SV) masses of $M = 250$ kg or 700 kg, respectively, assuming for both a unit cost of $U = 75$ k$/kg and a launching cost of $L = 15$ k$/kg.
The cost is simply $C = N \times (U + L) \times M$, or for each system:

$$C_1 = 48 \times (75 + 15k)\, k\$/kg \times 250\, kg = 1{,}080{,}000\, k \approx 1.1B\$ \text{ (billion \$)}$$
$$C_2 = 66 \times (75 + 15k)\, k\$/kg \times 700\, kg = 4{,}158{,}000\, k \approx 4.1B\$ \text{ (billion \$)}$$

4.3.12 Show that for a HAPS at altitude h and seen above the minimum elevation angle γ, the footprint diameter d is approximately given by the formula ($R_E = 6{,}378$ km, $h = 20-25$ km):

$$d = 2R_E \left\{ \cos^{-1}\left(\frac{R_E}{R_E + h} \cos\gamma \right) - \gamma \right\}$$

We first consider the figure below, which shows the geometry of the problem, with the different distances and angles involved. For clarity, the altitude h has been exaggerated with respect to the Earth's radius R_E.

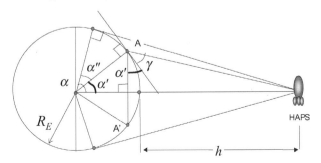

We see from this figure that the footprint diameter is given by the arc length $d = A - A' = 2R_E\alpha'$. By construction, the angle α' is determined by the elevation angle γ. The tangential ray for which $\gamma = 0$ corresponds to the angle $\alpha = \alpha' + \alpha$, which we know from the definition $\cos\alpha = R_E/(R_E + h)$. We thus have only one unknown, which is the angle α'. Going through the geometry leads to different types of untractable equations, so the solution must be much more simple! That is, looking at the result to be demonstrated, something like

$$d = 2R_E\alpha' = 2R_E\{\cos^{-1}[\cos(\alpha' + \gamma)] - \gamma\}$$

reduces the problem to find a good approximation for the unknown $\cos(\alpha' + \gamma)$. We can then develop this cosine according to the following:

$$\cos(\alpha' + \gamma) = \cos\alpha' \cos\gamma - \sin\alpha' \sin\gamma$$
$$= \cos(\alpha - \alpha)\cos\gamma - \sin(\alpha - \alpha)\sin\gamma$$

Then we develop again this result according to

$$\cos(\alpha' + \gamma) = (\cos\alpha\cos\alpha - \sin\alpha\sin\alpha)\cos\gamma$$
$$- (\sin\alpha\cos\alpha - \cos\alpha\sin\alpha)\sin\gamma$$

We can now "get rid" of α considering that it is a small positive angle comprised between zero and a value bounded by $\alpha = \cos^{-1}[R_E/(R_E + h)] = \cos^{-1}[6{,}378/(6{,}378 + 20)] = 4.5° = 0.079$ radians.

Thus we have $\cos\alpha > \cos\alpha = \cos 4.5° = 0.997$, and $\sin\alpha < \sin\alpha = \sin 4.5° = 0.078$, meaning that one can make the approximations $\cos\alpha \approx 1$ and $\sin\alpha \approx 0$, yielding

$$\cos(\alpha' + \gamma) \approx \cos\alpha\cos\gamma - \sin\alpha\sin\gamma$$
$$= \cos\alpha\cos\gamma - \sqrt{1 - \cos^2\alpha}\,\sin\gamma$$
$$\equiv \frac{R_E}{R_E + h}\cos\gamma - \sqrt{1 - \left(\frac{R_E}{R_E + h}\right)^2}\,\sin\gamma$$

which gives the solution angle α' according to

$$\alpha' \approx \cos^{-1}\left\{\frac{R_E}{R_E + h}\cos\gamma - \sqrt{1 - \left(\frac{R_E}{R_E + h}\right)^2}\,\sin\gamma\right\} - \gamma$$

Finally, we can make a further approximation by neglecting the second term, since the square root factor is equal to $\sin 4.5° = 0.078$ (in fact less than this value since $\sin\alpha < \sin 4.5°$). This second approximation is not so coarse when considering small elevation angles (e.g., $\gamma = 15°$) for which the factor $\sin\gamma$ is also relatively small (e.g., $\sin\gamma = 0.25$). Thus

$$\alpha' \approx \cos^{-1}\left(\frac{R_E}{R_E + h}\cos\gamma\right) - \gamma \quad \text{and} \quad d = 2R_E\alpha' \approx 2R_E\left\{\cos^{-1}\left(\frac{R_E}{R_E + h}\cos\gamma\right) - \gamma\right\}$$

which is the result that was to be demonstrated.

Bibliography

General and Introductory

Desurvire, E., *Wiley Survival Guide in Global Telecommunications: Broadband Access, Optical Components and Networks, and Cryptography*, John Wiley & Sons, Hoboken, NJ, 2004 (Companion volume to this book)

Desurvire E. et al., *Erbium Doped Fiber Amplifiers, Device and System Development*, John Wiley & Sons, 2002.

Dodd, A., *The Essential Guide to Telecommunications*, 3rd edition, Prentice-Hall PTR, Upper Saddle River, NJ, 2002

Shepard, S., *Telecom Crash Course*, McGraw-Hill, New York, 2002

Advanced and Technical

Bates, B. and Gregory, D., *Voice & Data Communications Handbook*, McGraw Hill, New York, 1996

Boisseau, M., Demange, M. and Munier, J.-M., *High Speed Networks*, John Wiley & Sons, New York, 1994

Clark, M.P., *Networks and Telecommunications, Design and Operation*, 2nd edition, John Wiley & Sons, New York, 1991

Comer, D.E., *Internetworking with TCP/IP*, Volume I, 3rd edition, Prentice-Hall, Englewood Cliffs, NJ, 1995

Dornan, A., *The essential guide to wireless communications applications*, Prentice-Hall, Upper Saddle River, NJ, 2002

Haykin, S., *Digital Communications*, John Wiley & Sons, New York, 1988

Kaaranen, H., Ahtiainen, A., Laitinen, L., Naghian, S. and Niemi, V., *UMTS networks, architecture, mobility and services*, John Wiley & Sons, New York, 2001

Kartalopoulos, S.V., *Understanding SONET/SDH and ATM*, IEEE Press, 1999

Kraus, J.D., *Antennas*, 2nd edition, McGraw-Hill, New York, 1988

Maral, G., Bousquet, M., and Pares, J., *Les systèmes de télécommunications par satellites*, Masson Ed., Paris 1982 (in French)

Ohmori, S., Wakana, H., and Kawase, S., *Mobile satellite communications*, Artech House, Boston, 1998

Wiley Survival Guide in Global Telecommunications: Signaling Principles, Network Protocols, and Wireless Systems, by E. Desurvire
ISBN 0-471-44608-4 © 2004 John Wiley & Sons, Inc.

Proakis, J.G., *Digital Communications*, 4th edition, McGraw Hill, New York, 2001

Rodriguez, A., Gatrell, J., Karas, J. and Pescke, R., *TCP/IP Tutorial and Technical Overview*, Prentice-Hall, Upper Saddle River, NJ, 2001

Salgues, B., *Les télécoms mobiles*, 2nd edition, Hermes, Paris 1997 (in French)

Schwartz, M., *Information, transmission, modulation and noise*, McGraw-Hill, New York, 1980

Siwiak, K., *Radiowave propagation and antennas for personal communications*, Artech House, Norwood, MA, 1995

Steele, R. and Hanzo, L., *Mobile radio communications, second and third generation of cellular and WATM systems*, 2nd edition, John Wiley & Sons, New York, 1999

Wesolowshi, H., *Mobile communications systems*, John Wiley & Sons, New York, 2002

📖 Recommended Web-Site Links

A list of recommended Web sites in a paper book can only be indicative and never complete or exhaustive, current or even still valid. It is in the spirit of the Internet that people must browse in every direction and generate their own selection bookmark preferences. Here are a few URLs which may prove useful for a fresh start. In each web site, the reader may pay close attention to offered *tutorials*, *white papers*, *FAQs*, *search engines* and *related links*. Direct Internet browsing by keywords is also recommended, but using *Boolean functions* for faster convergence. Note that URL bookmarks/preferences may be renamed for archival and memorizing purposes, using the right click "properties" command. As a final recommendation and advice, one may subscribe to *e-mail newsletters*, provided the "unsubscribe" options be clearly stated. This is the opportunity to acknowledge the content owners and webmasters for all the information freely provided. Some Web sites are real information treasures with document jewels and can make one's day. We apologize if we missed important ones, which we surely did without intent. Again, it takes the reader's patience and exploration to compile an up to date list. The following list is grouped by topics, but topics may overlap each other and URLs appear only once. The thing is to try them out.

Free-Space Optics

www.bakom.ch/en/funk/forschung/laserkommunikation/

www.fsoalliance.com

www.freespaceoptics.com

www.freespaceoptics.org

Global Positioning Systems

www.colorado.edu/geography/gcraft/notes/gps/gps_f.html

www.aero.org/publications/GPSPRIMER/

www.nasm.si.edu/galleries/gps/

www.navtechgps.com/links.asp

www.europa.eu.int/comm/dgs/energy_transport/galileo/programme/services_en.htm

Glossaries of Telecom Terms

www.techweb.com/encyclopedia

www.fcc.gov/glossary

www.glossary.its.bldrdoc.gov/fs-1037

www.atis.org/tg2k

www.alcatel.com/atr

High-Altitude Platforms

www.skytower.com

www.skystation.com

www.aerovironment.com

www.angelhalo.com

www.airship.com

www.bakom.ch/en/funk/forschung/haps

Internet

www.isp-planet.com

www.ibm.com/redbooks

Moore's Law

www.news.com.com (CNET)

www.firstmonday.org/issues/issue7_11/tuomi

Networking

www.techfest.com/networking

www.cis.ohio-state.edu/~jain

Satellite Systems

www.science.nasa.gov/Realtime/JTRACK/3d/JTrack3D.html

www.satobs.org (/iridium)

www.ee.surrey.ac.uk/personal/l.wood/constellations

www.esa.org

www.ico.com

www.iridium.com

www.teledesic.com

www.thuraya.com

www.skybridgesatellite.com

www.spaceflight/nasa.gov/station

www.fourmilab.ch/earthview/satellite.html

www.bakom.ch/en/funk/forschung

www.tbs-satellite.com/tse

Standardization Bodies

www.etsi.org

www.itu.int, www.itu.org

www.fcc.org

Wireless and Mobile Networks

www.lmdswireless.com

www.watmag.com

www.bbwexchange.com

www.wireless-wolrd-research.org

www.weca.com

www.80211b.weblogger.com

www.isp-planet.com/fixed_wireless/technology/

www.wapforum.org

www.3g-generation.com

www.4gmobile.com

www.palowireless.com

Index

Boldface page numbers indicate an emphasis of the subject.

Wiley Survival Guide in Global Telecommunications: Signaling Principles, Network Protocols and Wireless Systems, by E. Desurvire
ISBN 0-471-44608-4   Copyright © 2004 John Wiley & Sons, Inc.